Instant Notes

Ecology

Second Edition

The INSTANT NOTES series

Series editor
B.D. Hames
Department of Biochemistry and Molecular Biology, University of Leeds, Leeds, UK

Animal Biology
Genetics
Microbiology
Chemistry for Biologists
Immunology
Biochemistry 2nd edition
Molecular Biology 2nd edition
Neuroscience
Psychology
Developmental Biology
Plant Biology
Ecology 2nd edition

Forthcoming titles
Bioinformatics

The INSTANT NOTES Chemistry series
Consulting editor: Howard Stanbury

Organic Chemistry
Inorganic Chemistry
Physical Chemistry
Medicinal Chemistry

Forthcoming titles
Analytical Chemistry

Ecology

Second Edition

A. Mackenzie, A.S. Ball
Department of Biological Sciences,
University of Essex, Colchester, UK

and

S.R. Virdee
University of Essex, Colchester, UK

BIOS

© BIOS Scientific Publishers Limited, 2001

First published 1998 (ISBN 1 85996 161 4)
Second edition 2001 (ISBN 1 85996 257 2)

A CIP catalogue record for this book is available from the British Library.

ISBN 1 85996 257 2

BIOS Scientific Publishers Ltd
9 Newtec Place, Magdalen Road, Oxford OX4 1RE, UK
Tel. +44 (0) 1865 726286. Fax +44 (0) 1865 246823
World Wide Web home page: http://www.bios.co.uk/

Distributed exclusively in the United States, its dependent territories, Canada, Mexico, Central and South America, and the Caribbean by Springer-Verlag New York Inc, 175 Fifth Avenue, New York, USA, by arrangement with BIOS Scientific Publishers, Ltd, 9 Newtec Place, Magdalen Road, Oxford OX4 1RE, UK

Cover: A female Anna's Hummingbird (*Calypte anna*) feeding on a cholla cactus (*Opuntia* spp.) flower. Picture courtesy of Joyce and Paul Berquist.

This tiny bird (7.5 cm long, 3 g in weight) is hovering, flapping its wings around 65 times per second, as it feeds on the sugary nectar that the plant provides as an incentive for pollinators to visit it. Anna's Humming bird feeds along the Pacific edge of North America, from northern Mexico to southern Canada. The birds are iridescent bronzy green above and males have a red crown and neck. Females build the nest, mate and lay two jellybean-sized eggs and then rear the young without help from the males. During the breeding season, males and females occupy separate habitats, males establish feeding territories in open chaparral while females build their nests near live Oaks. the red gooseberry (*Ribes speciosum*), that flowers early in spring, is a key food source. It is thought that the bird and plant have co-evolved, allowing this species to have unusually early breeding season.

Why is the bird so small? Why are males colored differently from females? Why do the different sexes use different habitats? What limits its breeding range? How has the relationship between flower and the bird developed? Why does the female not lay one larger egg or more smaller ones? Why does the male not help raise the young? Why are hummingbirds only found in the Americas? These are the kind of questions ecologists are interested in, and which this book considers.

Production Editor: Andrea Bosher
Typeset by Phoenix Photosetting, Chatham, Kent, UK
Printed by Biddles Ltd., Guildford, UK

CONTENTS

ABBREVIATIONS

ADH	alcohol dehydrogenase	IPM	integrated pest management
AE	assimilation efficiency	MEY	maximum economic yield
AIL	aesthetic injury level	MSY	maximum sustainable yield
BOD	biochemical oxygen demand	MVP	minimum viable population
CAM	crassulacean acid metabolism	NPP	net primary productivity
CAT	control action threshold	NVC	National Vegetation Classification
CE	consumption efficiency	PAR	photosynthetically active radiation
DBH	diameter at breast height	PCB	polychlorinated biphenyls
DDD	2,2-bis(p-chlorophenyl)-1,1-dichloroethane	PE	production efficiency
		PVA	population viability analysis
DDT	dichlorodiphenyltrichloroethane	RDZ	resource depletion zone
EIL	economic injury level	SMRS	specific mate recognition system
GPP	gross primary productivity	SOM	soil organic matter
ICTZ	intertropical convergence zone		

PREFACE

The last 40 years have witnessed a vast expansion of both interest and knowledge in ecology. There has been a widening public awareness of the importance of ecological interactions, with particular focus on such issues as the impact of pesticides upon food chains and the loss of global biodiversity owing to habitat destruction. In parallel, a wide suite of advances in ecological understanding have been fuelled by progress in a number of other fields, notably molecular biology.

This great expansion in understanding has lead to a problem for the student of the subject, as leading textbooks have grown greatly in size and complexity to accommodate new theories and new data. Further, ecology courses are now an integral component of a wide range of biological science degrees. This book attempts to distil the key areas of ecology in a way which will help both full-time students of the subject and those studying ecology as a subsidiary subject.

Instant Notes in Ecology is designed to give students rapid, easy access to key ecological material in a format which facilitates learning and revision. The book is divided into 62 topics which cover material at the level we would expect of good first/second year students. Each topic starts with a short list of 'Key Notes' — a revision checklist — before explaining the subject matter. Further reading is included at the end of the book for students who wish to delve more deeply into selected areas of interest.

Section A provides a brief outline of the area of ecology, and includes a list of 'rules' in ecology which derives from some common mistakes and pitfalls that the authors' experiences suggest many students are liable to make. Sections B to G cover the key aspects of the abiotic environment which affect organisms (climate, water, temperature, radiation, nutrients) and how organisms are adapted to their environment. Populations and interactions are covered in Sections H to O, including population ecology, competition, predation, parasitism, mutualism, life histories, behavioral ecology and population genetics. In Sections P to S, ecosystems, biomes and community patterns and processes are dealt with. In the final five Sections, T to X, some key issues in applied ecology are dealt with, including harvesting (with a focus on fisheries), conservation, pollution and global warming and agriculture.

How should you use this book as a student? Your course will almost certainly differ in structure from the layout here. However, the breakdown of the text into small manageable topics, each cross-referenced, will allow you to easily navigate your way to the information you seek.

Aulay Mackenzie, Andy Ball and Sonia Virdee

To Ishbel, Eilidh, Sara, Simon and Katrina

A1 WHAT IS ECOLOGY?

Key Notes

A definition of ecology	Ecology is the study of the interactions between organisms and their environment. The 'environment' is a combination of the physical environment (temperature, water availability, etc.) and any influences on an organism exerted by other organisms – the biotic environment.
Individuals, populations, communities and ecosystems	There are four identifiable subdivisions of scale which ecologists investigate: (i) considering the response of individuals to their environments; (ii) examining the response of populations of a single species to the environment, and considering processes such as abundance and fluctuations; (iii) the composition and structure of communities (the populations occurring in a defined area); (iv) the processes occurring within ecosystems (the combination of a community and the abiotic components of the environment), such as energy flow, food webs and the cycling of nutrients.

A definition of ecology

Ecology is the study of the interactions between organisms and their environment. There are two distinct components to the 'environment': the **physical environment** (comprising such things as temperature, water availablity, wind speed, soil acidity) and **biotic environment**, which comprises any influences on an organism that are exerted by other organisms, including competition, predation, parasitism and cooperation.

Within ecology there are a number of fields, either focusing on specific areas of interest or using particular approaches to address ecological problems. For example, **behavioral ecology** is concerned with explaining the patterns of behavior in animals. **Physiological ecology** explores the physiology of an individual and considers the consequences on function and behavior. A particular emphasis on the impact of evolution on current patterns is the focus of **evolutionary ecology**. A recent development has been the use of molecular biology to directly tackle ecological problems – **molecular ecology**. The domains of **population ecology** and **community ecology** are described below.

Ecological studies are not restricted to 'natural' systems – understanding both the human effects on nature and the ecology of artificial environments (such as crop fields) are important areas of study.

Individuals, populations, communities and ecosystems

Ecology can be considered on a wide scale, moving from an individual molecule to the entire global ecosystem. However, four identifiable subdivisions of scale are of particular interest, (i) **individuals**, (ii) **populations**, (iii) **communities** and (iv) **ecosystems**.

At each scale the subjects of interest to ecologists change. At the individual level the response of **individuals** to their environment (biotic and abiotic) is the key issue, whilst at the level of **populations** of a single species, the determinants of abundance and population fluctuations dominate. **Communities** are the mixture of populations of different species found in a defined area. Ecologists are interested in the processes determining their composition and structure. **Ecosystems** comprise the biotic community in conjunction with the associated complex of physical factors that characterize the physical environment. Issues of interest at this level include energy flow, food webs and the cycling of nutrients.

It should be noted that the terms 'population', 'communities' and 'ecosystems' are often ill-defined. It is often not possible to clearly delineate where one population stops and another starts, and the same problem occurs with communities and ecosystems. To some degree, these terms simply represent convenient simplifications by which we can categorize the natural world.

A2 TEN RULES IN ECOLOGY

Key Notes

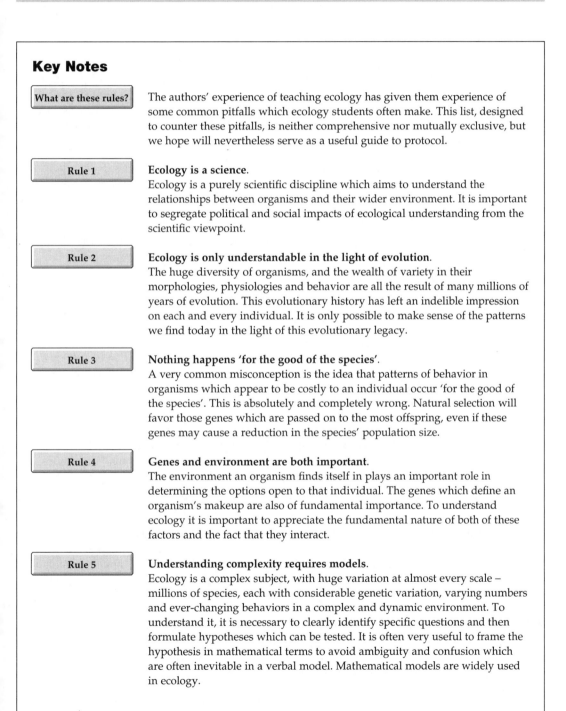

What are these rules?

The authors' experience of teaching ecology has given them experience of some common pitfalls which ecology students often make. This list, designed to counter these pitfalls, is neither comprehensive nor mutually exclusive, but we hope will nevertheless serve as a useful guide to protocol.

Rule 1

Ecology is a science.
Ecology is a purely scientific discipline which aims to understand the relationships between organisms and their wider environment. It is important to segregate political and social impacts of ecological understanding from the scientific viewpoint.

Rule 2

Ecology is only understandable in the light of evolution.
The huge diversity of organisms, and the wealth of variety in their morphologies, physiologies and behavior are all the result of many millions of years of evolution. This evolutionary history has left an indelible impression on each and every individual. It is only possible to make sense of the patterns we find today in the light of this evolutionary legacy.

Rule 3

Nothing happens 'for the good of the species'.
A very common misconception is the idea that patterns of behavior in organisms which appear to be costly to an individual occur 'for the good of the species'. This is absolutely and completely wrong. Natural selection will favor those genes which are passed on to the most offspring, even if these genes may cause a reduction in the species' population size.

Rule 4

Genes and environment are both important.
The environment an organism finds itself in plays an important role in determining the options open to that individual. The genes which define an organism's makeup are also of fundamental importance. To understand ecology it is important to appreciate the fundamental nature of both of these factors and the fact that they interact.

Rule 5

Understanding complexity requires models.
Ecology is a complex subject, with huge variation at almost every scale – millions of species, each with considerable genetic variation, varying numbers and ever-changing behaviors in a complex and dynamic environment. To understand it, it is necessary to clearly identify specific questions and then formulate hypotheses which can be tested. It is often very useful to frame the hypothesis in mathematical terms to avoid ambiguity and confusion which are often inevitable in a verbal model. Mathematical models are widely used in ecology.

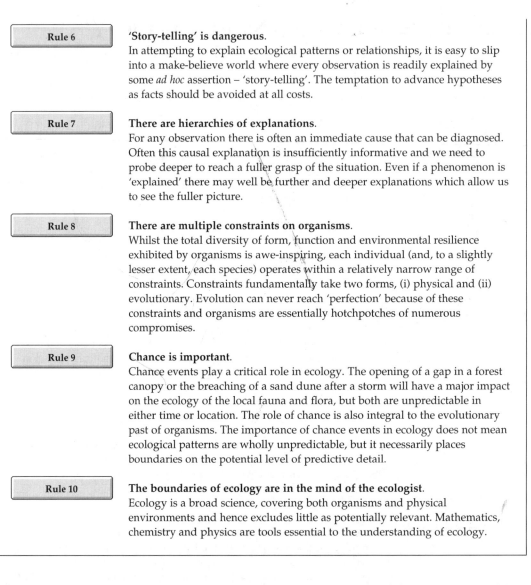

Rule 6	**'Story-telling' is dangerous**. In attempting to explain ecological patterns or relationships, it is easy to slip into a make-believe world where every observation is readily explained by some *ad hoc* assertion – 'story-telling'. The temptation to advance hypotheses as facts should be avoided at all costs.
Rule 7	**There are hierarchies of explanations**. For any observation there is often an immediate cause that can be diagnosed. Often this causal explanation is insufficiently informative and we need to probe deeper to reach a fuller grasp of the situation. Even if a phenomenon is 'explained' there may well be further and deeper explanations which allow us to see the fuller picture.
Rule 8	**There are multiple constraints on organisms**. Whilst the total diversity of form, function and environmental resilience exhibited by organisms is awe-inspiring, each individual (and, to a slightly lesser extent, each species) operates within a relatively narrow range of constraints. Constraints fundamentally take two forms, (i) physical and (ii) evolutionary. Evolution can never reach 'perfection' because of these constraints and organisms are essentially hotchpotches of numerous compromises.
Rule 9	**Chance is important**. Chance events play a critical role in ecology. The opening of a gap in a forest canopy or the breaching of a sand dune after a storm will have a major impact on the ecology of the local fauna and flora, but both are unpredictable in either time or location. The role of chance is also integral to the evolutionary past of organisms. The importance of chance events in ecology does not mean ecological patterns are wholly unpredictable, but it necessarily places boundaries on the potential level of predictive detail.
Rule 10	**The boundaries of ecology are in the mind of the ecologist**. Ecology is a broad science, covering both organisms and physical environments and hence excludes little as potentially relevant. Mathematics, chemistry and physics are tools essential to the understanding of ecology.

What are these rules?

The authors' experience of teaching ecology, and the relayed comments of their colleagues, has given them insight into some common pitfalls which ecology students make. This set of rules is designed to counter these pitfalls and set students on the right course. The list is neither comprehensive nor mutually exclusive, but we hope will nevertheless serve as a useful guide.

Rule 1

Ecology is a science

Ecology is a purely scientific discipline which aims to understand the relation-ships between organisms and their wider environment. Like any science, the outcomes of ecological studies do not dictate ethical or political actions. It is important to make this distinction because the environmental movement has endowed the word 'ecology' with political connotations. It is right that ecology should inform politics, but as a student of ecology it is imperative to consider ecological research from a rigorous scientific viewpoint.

Rule 2

Ecology is only understandable in the light of evolution

The huge diversity of organisms, and the wealth of variety in their morphologies, physiologies and behavior are all the result of many millions of years of evolution. This evolutionary history has left an indelible impression on each and every individual. It is only possible to make sense of the patterns we find today in the light of this evolutionary legacy.

For example if we want to understand why the ostrich, emu, kiwi and rhea are all flightless (an unusual condition in birds), it is critical to know that these birds all share a common ancestor which was flightless, and that the species above became separated on different continents by the break-up of the ancient continent Gondwanaland. Therefore, looking for an independent adaptive reason for flightlessness in each species would be flawed.

At a wider level, the tendency of evolution to optimize organisms' fitness (although see Rule 8) provides ecologists with a valuable tool in hypothesizing organism structure and behavior. Thus, the large size of the peacock's tail suggests that males with larger tails have higher levels of fitness – and indeed data supports this hypothesis.

Some authors have suggested that the environment is the fundamental constraint on organisms and that ecology can more-or-less ignore evolution and genetics. This is a clear misinterpretation of the evidence – there is now ample evidence of short-term evolutionary changes which affect ecological patterns. Some of the best documented examples are the evolution of resistance to pesticides in crop pests and antibiotics in bacteria, and similar patterns have been observed in natural systems. Further, genes control every aspect of an organism, including the way it responds to the environment, so they must be the dominant component. Ecologists considering the behavior of animals should understand that animal behavior is controlled by genes just as much as gut enzymes are. There are now many examples where genes for behavioral traits have been identified.

Rule 3

Nothing happens 'for the good of the species'

A common misconception is the notion that patterns of behavior in organisms which appear to be **costly to an individual** (for example the dying of a female octopus immediately after giving birth or the defensive suicidal attacks of some soldier ants) are **'for the good of the species'**. This argument is absolutely and completely wrong, and only advanced by those who failed to grasp the importance of Rule 2. Natural selection will favor those genes which are passed on to the most offspring. If the genes for suicidal behavior in ants or early death in octopuses were good for the species but bad for the individuals carrying them, evolution would favor their replacement with other genes. Equally fallacious, for the same reason, is the argument that population size is limited via reduced birth rates 'for the good of the species'. Both altruism and population regulation are easily understandable in terms of evolution acting on individuals.

Rule 4

Genes and environment are both important

The environment an organism finds itself in plays an important role in determining the options open to that individual. Environmental conditions will define the birth rate, growth rate and level of mortality of a species. However, the genes which define an organism's makeup are also of fundamental importance. The emergent phenotype of an organism is a joint product of its genetic code and the environmental stimuli that affect it during development:

$$environment + genotype => phenotype$$

To understand ecology it is important to appreciate the fundamental nature of both of these factors, and the fact that they interact.

Rule 5

Understanding complexity requires models
At face value, ecology might appear incomprehensible – millions of species, each with varying numbers and ever-changing behaviors set in the context of a structurally complex and dynamic environment. Clearly, we cannot understand it all at once. The solution is a two-step process, firstly to identify a small and **specific question**, such as 'why do male blackbirds form territories?', and then to test a **specific hypothesis**, for example 'male blackbirds with a territory have a better chance of getting a mate'. What we have done is to construct a verbal model of the world, which we then set about testing.

Sometimes, the model we want to test is more complex, such as 'when collecting food for its nestlings, a starling needs to consider both how far away the nest is and how difficult it is to forage with a beak full of worms'. Starlings become slower and slower at catching prey as their beaks become fuller, but if the nest is a long distance away, it is worth spending more time catching prey. When we have a **complex model** like this, it is best to frame it in terms of simple mathematics otherwise ambiguities and confusions are easily introduced. The predictions of a mathematical model of starling foraging behavior and some data from real starlings are shown in *Fig. 1*. The model appears to provide a good description of starling behavior, quantifying the degree to which starlings increase the load of prey as the distance to the foraging ground increases. Simple (and more complex) **mathematical models** are now widely used in ecology. Even complex models usually have a simple verbal explanation. This book employs a minimum level of mathematics, but remember that such models are integral to ecology.

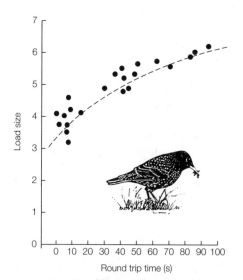

Fig. 1. *The application of a mathematical model in ecology. The dashed line is the pre-dicted relationship between the length of the round trip from nest, to foraging ground, to nest and the number of prey carried. The dots represent observations.*

Rule 6

'Story-telling' is dangerous
In attempting to explain ecological patterns or relationships, it is easy to slip into a make-believe world where every observation is readily explained by some *ad hoc* assertion, such as a classic examination blunder, 'polar bears are white so they are hidden from predators in the snow' (the problem being, obviously, that there are no natural predators of polar bears). Such errors are not restricted to undergraduates – they are commonplace in popular natural history writing and films. Creativity should be encouraged in scientists at the level of constructing hypotheses. Indeed, an imaginative and probing mind is essential. However, the temptation to advance such hypotheses as facts should be avoided.

Rule 7

There are hierarchies of explanations
For any observation there is often an immediate cause that can be diagnosed. Often this causal explanation is insufficiently informative and we need to probe deeper to gain a fuller grasp of the situation. For example, a radio-tagged mole was found to have been stationary for three days. The proximal explanation of this was simple – it was dead. Further investigation found that it had a high density of gut parasites, which were the likely cause of death. It may have been possible to further investigate whether the mole was genetically predisposed to parasite infection or whether the environmental conditions had favored parasite survival, or some further explanation. The level of explanation required depends on the question being asked. For example, we might ask 'Why do males of many duck species adopt a bright or high-contrast breeding plumage?' The immediate causal explanation is that testosterone levels rise in spring and cause the changes. For an ecologist, this explanation is insufficient, but a fuller explanation is rooted in the fact that drakes with dull plumage generally have low mating success. Clearly there are further explanations beyond this ('Why do dull drakes fail?', 'Why do drakes show seasonal plumage variation when many birds do not?'). The point is that even if a phenomenon is 'explained' there may well be further explanations which shed a different light on the observation, without the original explanation being in any way wrong.

Rule 8

There are multiple constraints on organisms
Whilst the total diversity of form, function and environmental resilience exhibited by organisms is awe-inspiring, each individual (and to a slightly lesser extent, each species) operates within a relatively narrow range of constraints. Constraints fundamentally take two forms, (i) those imposed by the laws of physics – **physical constraints** – and (ii) those caused by the vagaries of evolutionary history and the limitations on genetical flexibility – **evolutionary constraints**. Physical laws dictate what is and is not possible for organisms to achieve. Thus, it is not possible for an elephant to have the limb proportions of a gazelle, because, whilst an elephant is only roughly four times as long as a gazelle, it is about 64 times the weight (and hence needs limbs 64 times stronger), because volume (and hence weight) increases cubically as length increases linearly ($4^3 = 64$). There is thus a **trade-off** between nimbleness and large size. Similarly, there is an upper size limit on single-celled organisms, such as bacteria, which rely on diffusion over their outer surface to uptake assimilates, because as the size of an organism increases, the volume increases more rapidly than the surface area. Big bacteria would not be able to transfer assimilates to their centers. Physical constraints thus impose ubiquitous trade-offs.

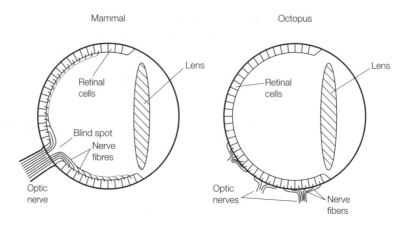

Fig. 2. A diagrammatic comparison between the eye of an octopus and that of a mammal, indicating how different evolutionary histories have produced a blind spot in mammals but not in octopuses. The size of the retinal cells is greatly exaggerated.

Evolutionary constraints are equally ubiquitous, but much less predictable. For example, the **blind spot** in the vertebrate eye occurs because of a basic design fault – the nerve connections of the photoreceptive cells are on the inner edge of the retina (the side which the light hits) and nerve fibers merge, forming the optic nerve, and exit the retina together. At this spot no light can be detected, resulting in a zone of missing vision – a blind spot. There is nothing inevitable about this arrangement, and indeed in the octopus (a mollusc, which evolved advanced eyes independently of vertebrates) the retinal cells are the 'right' way up, with nerve connections on the outer edge (*Fig. 2*). Where the nerve fibers merge to form the optic nerve, they are already on the outer edge of the retina and so there is no blind spot. Another example is that of the **laryngeal nerve**, which in fish passes directly from the brain to behind the sixth gill arch. During evolution, the gill arches have been transformed and the sixth gill arch in mammals is now represented by a blood vessel near the heart. The result is that in mammals the laryngeal nerve takes a detour, going from the brain to the heart and returning to the larynx. In long-necked mammals, such as giraffes, this detour is considerable. A much better design would be a direct connection from brain to larynx, but evolution has not offered this opportunity.

Evolution can never achieve 'perfection' because of these constraints and organisms are essentially hotchpotches of numerous compromises, despite the many examples of amazing design.

Rule 9

Chance is important
Chance events play a critical role in ecology. The opening of a gap in a forest canopy after a storm will have a major impact on the ecology of the fauna and flora of the forest floor, but it is unpredictable in either time or location. Similarly, the dynamics of sand dune and rocky shore species is dominated by the random destruction and creation of new, bare colonizing surfaces. The population decline of the passenger pigeon (*Ectopistes migratorius*), which was so numerous in North America in the eighteenth century that flocks passing overhead blocked out the sun for minutes at a time, was largely caused by over-hunting and subsequently, diminished breeding success at low population densities. However, extinction was finally caused by the chance combination of

disease and a harsh winter. The role of chance is also integral to the evolutionary past of organisms, as the blind spot of the vertebrate eye (see Rule 8 and *Fig. 2*) illustrates. The importance of chance events in ecology does not mean ecological patterns are wholly unpredictable, but it necessarily places boundaries on the potential level of predictive detail.

Rule 10

The boundaries of ecology are in the mind of the ecologist
A science which covers both organisms and physical environments necessarily excludes little as potentially relevant. Of course, few ecologists have direct need for an understanding of astrophysics or the behavior of quarks, but, depending on the question studied, it may be essential to understand, for example, the chemistry of clay when investigating the availability of nutrients to plant roots or the physics of flight when studying the energetics of hummingbirds. A good ecologist regards mathematics, chemistry, physics and other disciplines as tools essential to the understanding of ecology.

B1 ADAPTATION

Key Notes

Fitness	Fitness is a measure of the ability of an individual to produce viable offspring and contribute to future generations. Individuals vary in their relative fitness, and this variation is due partly to genetic differences among individuals and partly to environmental influences.
Natural selection	The individuals in a species which have the highest fitness will contribute disproportionately to the subsequent generations. If fitness differences have a genetic component, then the genetic make-up of the subsequent generations will be altered. This process is known as natural selection or 'survival of the fittest'.
Adaptation	Any *heritable* trait possessed by an organism which aids survival or reproduction is an adaptation. Such traits may be physiological, morphological or behavioral. Adaptation is the result of natural selection.
Genotype and phenotype	The genotype is the genetic composition of an individual. The phenotype is the individual organism, a product of the interaction between its genotype and its environment. The ability of the phenotype to vary due to environmental influences on its genotype is known as *phenotypic plasticity* (e.g. human suntan, wind-shaped plants, locust morph (solitary or migratory)).
Related topics	Ten rules in ecology (A2) Speciation (O2) Genetic variation (O1)

Fitness

The fitness of an individual will be high if that individual gives rise to **many offspring** which themselves are reproductively successful. Fitness will not necessarily be maximized by producing the most offspring – it may be better to produce fewer, larger offspring which will have better survivorship. Individuals may have higher fitness due to their possession of **genes** which give advantage. An example of this occurs in insecticide resistance. In the Australian sheep blowfly, *Lucilia cuprina,* resistance to the organophosphate insecticide malathion was found to be conferred by a gene called *Rmal*. Flies homozygous for this gene tolerate a high level of malathion, which will kill flies not possessing the gene. Clearly, if malathion was used as a control agent, we would anticipate the *Rmal* gene to increase in frequency in subsequent generations.

An individual's contribution to future generations may also be influenced by **nongenetic factors**. Environmental differences between individuals (such as the quality of food an animal receives during development) will influence fitness. However, as these differences are **not inherited** by their offspring, no adaptation will result. Evolution can only occur when differences are inherited.

Natural selection   Those individuals within a population which survive and reproduce most successfully (i.e. those with the highest fitness) will contribute more offspring to subsequent generations than individuals with lower fitness. If fitness differences have a genetic component, then the genes which the fittest individuals possess will become commoner, whilst the genes which the least fit individuals possess will become rarer. Thus, differential fitness amongst individuals results in genetic change in the population. Populations of the fly *Drosophila melanogaster* are often found in association with wine production, where they are exposed to unusually high levels of ethanol. Such populations have an elevated ethanol detoxification ability, as natural selection has lead to possession of higher activity levels of the alcohol dehydrogenase (ADH) enzyme.

Adaptation   Any **heritable** trait (one which is capable of being transmitted to the next generation), be it behavioral, morphological or physiological, which **aids survival or reproduction** in a particular environment is an **adaptation** to that environment. Adaptation is the result of natural selection acting on heritable differences in fitness. It may be noted that nonheritable traits, such as the age of an individual, may also influence survival and reproduction, but such phenomena have no effect on evolution.

In a North American fish, the desert sucker (*Catostomus clarki*), an esterase enzyme occurs in two forms which differ in their optimal temperatures. Northerly populations possess the alleles coding for the low temperature enzyme, whilst in southerly populations the alleles coding for the high temperature enzyme predominate (*Fig. 1*). In the peppered moth, *Biston betularia*, there are genetically based color differences (*Fig. 2*). Pale individuals predominate in unpolluted areas of Britain where they are camouflaged from avian predators against lichen-covered trees, whilst melanic (dark-colored) individuals are favored in polluted areas where lichens are absent and tree trunks are darkened by soot.

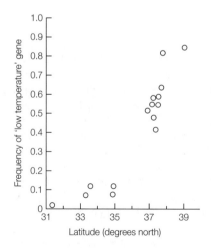

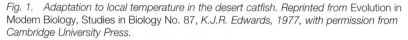

Fig. 1. Adaptation to local temperature in the desert catfish. Reprinted from Evolution in Modern Biology, Studies in Biology No. 87, *K.J.R. Edwards, 1977, with permission from Cambridge University Press.*

Fig. 2.   Adaptation to different colored backgrounds in the peppered moth, Biston betularia.
Each form is cryptically colored in the habitat it is found in.

Genotype and phenotype

The genotype is the genetic composition of an individual. In sexual outcrossing species, usually most individuals have differing genotypes. The phenotype is the individual organism, a product of the interaction between its genotype and its environment. In asexual species (such as aphids), a group of individuals may share a genotype, but, due to differing environmental influences, exhibit different phenotypes. For example, if plant quality falls, wingless aphids will produce winged offspring, with the same genotype as the mother.

This ability of the phenotype to vary due to environmental influences on its genotype is known as **phenotypic plasticity**. Other examples are suntan in humans, wind-shaped plants, and different morphs in locusts (solitary or migratory), which are cued by temperature and humidity.

B2 COPING WITH ENVIRONMENTAL VARIATION

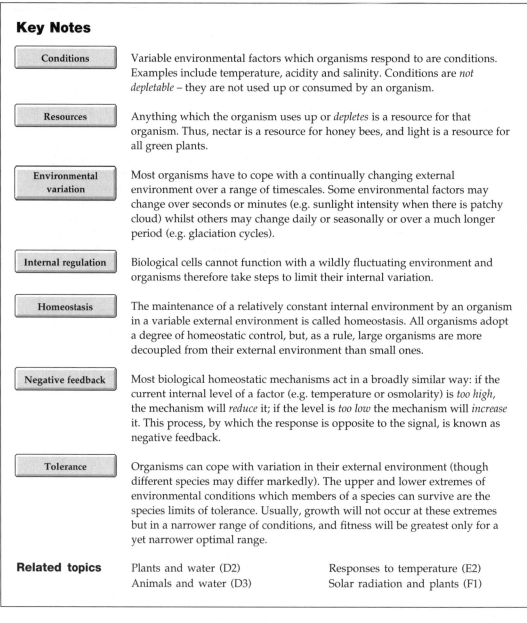

Key Notes

Conditions	Variable environmental factors which organisms respond to are conditions. Examples include temperature, acidity and salinity. Conditions are *not depletable* – they are not used up or consumed by an organism.
Resources	Anything which the organism uses up or *depletes* is a resource for that organism. Thus, nectar is a resource for honey bees, and light is a resource for all green plants.
Environmental variation	Most organisms have to cope with a continually changing external environment over a range of timescales. Some environmental factors may change over seconds or minutes (e.g. sunlight intensity when there is patchy cloud) whilst others may change daily or seasonally or over a much longer period (e.g. glaciation cycles).
Internal regulation	Biological cells cannot function with a wildly fluctuating environment and organisms therefore take steps to limit their internal variation.
Homeostasis	The maintenance of a relatively constant internal environment by an organism in a variable external environment is called homeostasis. All organisms adopt a degree of homeostatic control, but, as a rule, large organisms are more decoupled from their external environment than small ones.
Negative feedback	Most biological homeostatic mechanisms act in a broadly similar way: if the current internal level of a factor (e.g. temperature or osmolarity) is *too high*, the mechanism will *reduce* it; if the level is *too low* the mechanism will *increase* it. This process, by which the response is opposite to the signal, is known as negative feedback.
Tolerance	Organisms can cope with variation in their external environment (though different species may differ markedly). The upper and lower extremes of environmental conditions which members of a species can survive are the species limits of tolerance. Usually, growth will not occur at these extremes but in a narrower range of conditions, and fitness will be greatest only for a yet narrower optimal range.
Related topics	Plants and water (D2) Responses to temperature (E2) Animals and water (D3) Solar radiation and plants (F1)

Conditions

Except for a very few organisms living in extremely stable environments (e.g. the deep ocean), organisms are exposed to variable environmental factors which have direct physiological and behavioral impacts. The most obvious and general of these is **temperature** variation, although acidity, salinity, osmolarity and radiation are also important to various organisms, as is disturbance caused by, for example, falling raindrops or spating rivers. **Conditions** are not used up or consumed by an organism and are hence **not depletable**, in contrast to **resources.**

Resources

A resource is anything which an organism depletes. This is not simply food, but light and inorganic nutrients (for plants) and, importantly, space (e.g. a great tit's nesting hole is, once used, unavailable to other great tits). The single most important resource is solar radiation, which is the sole source of energy for green plants.

It should be noted that the dichotomy between conditions and resources only exists for a particular organism. Thus, solar radiation is a condition for an insect, but a resource for a plant. Further, occasionally the same factor can be both a resource and a condition. For example, a plant uses water as a resource, but heavy rainfall, which might lodge the plant or wash it away, represents a condition.

Environmental variation

Environmental conditions are constantly changing. The Earth's diurnal rotation and the annual rotation of the Earth around the sun drive changes in temperature and radiation and these, coupled with the centrifugal motion of the planet and convection currents, cause climatic changes on timescales from seconds to years. Glaciation cycles occur over yet larger timescales; the Pleistocene which began 1.7 million years ago has experienced about 18 glaciations, each lasting about 90 000 years and separated by a warm interglacial period (like the present one) lasting about 10 000 years.

Conditions may also be altered by the impact of other organisms. For example, in the absence of oxygen, heterotrophic bacteria will reduce the pH of soil (i.e. increase its acidity).

Internal regulation

Biological processes at the cellular level are sensitive to environmental conditions and can only operate within relatively narrow ranges of temperature, pH and osmolarity. For example, the human body must maintain internal temperatures very close to 37°C. Deviations of only a few degrees above or below this will be fatal if sustained. Maintenance of a constant internal environment (**homeostasis**) requires continuous monitoring and exchange of energy and materials with the external environment.

Homeostasis

To maintain a relatively constant internal environment in the presence of very variable external conditions requires homeostasis. Even the simplest organisms take steps to limit their internal variation; for example, bacteria may regulate internal ion concentrations. Larger organisms, because they have a lower surface area:volume ratio, are more decoupled from the environment. As they also tend to possess more sophisticated homeostatic machinery than the simplest prokaryotes, they can maintain homeostatic balance in a wider range of conditions. Homeostatic regulation can occur by physiological mechanisms (as in the example below) or by behavioral strategies – for example, lizards sun themselves in the morning to elevate their body temperatures.

Negative feedback

Most homeostatic mechanisms rely on a negative feedback process. Any homeostatic mechanism has three basic components: a receptor, a control center and an effector. The receptor detects a change in a key internal condition, for example the blood osmolarity of a mammal (*Fig. 1*). The control center processes this information and compares it to a set point, or optimal value. If the variable is not tolerably close to this set point, the control center will direct a response by an effector. In the example in *Fig. 1*, the hypothalamus serves as both a receptor and a control center for monitoring the blood osmolarity. If the blood osmolarity is too high (i.e. the blood is too concentrated), the hypothalamus will direct a thirst response, which, when satiated, will cause the blood osmolarity to fall. On the other hand, if the blood osmolarity is too low (i.e. the blood is too dilute), then a response of water loss will be directed in the kidneys, which will cause blood osmolarity to rise again.

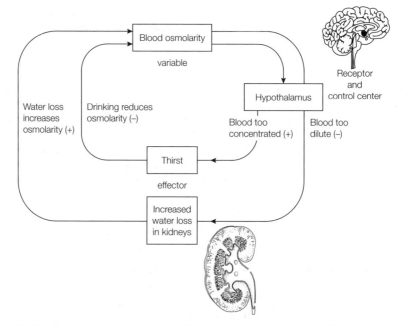

Fig. 1.   A negative feedback system: the maintenance of mammalian blood osmolarity.

Tolerance

An organism adapted to its environment can cope with variation around some optima for a range of environmental conditions. Some species are more particular than others in their requirements. For example, the sticky catchfly (*Silene viscosa*) germinates over a range of 23°C, whilst the closely related ragged robin (*Lychnis flos-cuculi*) only germinates over a range of 13°C.

The degree of environmental variation that a species can cope with is that species' range of **tolerance**. Organisms will perform best in an optimal environment and will tend to reproduce only in conditions close to that (*Fig. 2*). As conditions deviate further from the optimal, individuals will become too stressed to be able to reproduce, though growth may occur. As conditions approach the upper and lower extremes of environmental conditions which individuals can survive (the species limits of tolerance), no growth occurs. The growth and survival temperature response of the spider beetle *Ptinus tectus* is shown in *Fig. 3*.

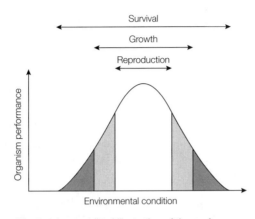

Fig. 2. *A generalized illustration of the performance of a species in respect of an environmental condition.*

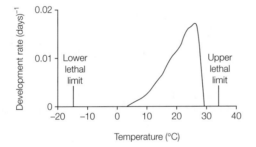

Fig. 3. *Development rate and survival in response to temperature in the Tasmanian spider beetle,* Ptinus tectus. *Survival is possible over a wider range of temperatures compared to growth.*

B3 THE NICHE

Key Notes

Niche	The ecological niche of an organism is the position it fills in its environment, comprising the *conditions* under which it is found, the *resources* it utilizes and the time it occurs there.
Habitat	The habitat of an organism is the physical environment it is found in, for example, a temperate broad-leaved woodland. Each habitat provides numerous niches.
Multidimensional niche space	Each condition or resource which defines the niche of an organism contributes one dimension to the space in which the organism can occur. Considering all dimensions together defines fully the organism's niche, and is the multidimensional niche space, or '*n*-dimensional hypervolume'.
Fundamental niche	The niche space an organism can fill in the absence of competition or predation is known as the fundamental niche.
Realized niche	The niche space occupied by an organism when competition and predation occur is the realized niche, which is always a subset of the fundamental niche.
Related topics	The nature of competition (I1) Resource partitioning (I3) Intraspecific competition (I2)

Niche

Organisms of a given species can maintain viable populations only under a certain range of conditions, and will utilize only particular resources. They also may only occur in a given environment during particular times (e.g. insectivorous bats are nocturnal, when few insectivorous birds are feeding). The intersection of these factors describes the niche, which is the position the organism fills in its environment. Organisms may change niches as they develop, for example common toads *Bufo bufo* occupy an aquatic environment (and are grazers of algae and detritus) prior to metamorphosing into adults, whence they become terrestrial (and are insectivorous).

Habitat

In contrast to the niche of an organism, a habitat is the physical environment in which an organism is found. Habitats typically contain many niches and support many different species. Thus, temperate broad-leaved woodland provides a vast number of niches for a range of birds (e.g. nuthatches, great tits, woodcocks), mammals (e.g. wood mice, foxes, common shrews); insects (e.g. butterflies, moths, beetles, spiders, aphids) and plants (e.g. wood anemones, bluebells, mosses, lichens).

Multidimensional niche space

Each condition that affects an organism or resource that it utilizes can be regarded as a single axis or **dimension**, within which there can be defined a range that the organism will occur in. By considering a number of such dimensions at once, an increasingly refined picture of the organism's niche can be reached. For example, the temperature range which a chaffinch will tolerate will overlap with that of many other species. However, if we consider the prey size and the foraging height as further dimensions we will differentiate the chaffinch niche from that of many other species (*Fig. 1*). In *Fig. 2*, the niche of an American warbler, the blue-gray gnat catcher, is shown in two dimensions. It is theoretically possible (though difficult to measure or represent on a page) to add every dimension of resource or condition which affects the organism, resulting in a fully defined ecological niche – the '*n*-dimensional hypervolume' (where *n* is the number of axes). Simple theory suggests that this fully defined niche is expected to be unique to that one species (or even one lifestage of that species), although recent work shows this need not be so in a dynamic or patchy environment (see Topic I3). A practical weakness in the *n*-dimensional hypervolume theory is that it is impossible to be sure that all dimensions have been considered, but, nevertheless, this is a useful concept.

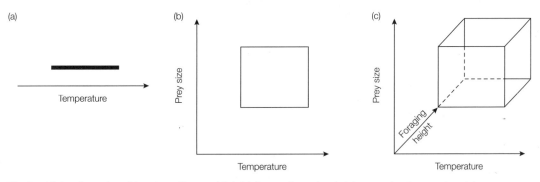

Fig. 1.   Niche dimensions for an insectivorous bird. (a) A one-dimensional niche covering temperature tolerance; (b) a two-dimensional niche incorporating temperature and prey size; (c) a three-dimensional niche incorporating temperature, prey size and the height at which foraging occurs.

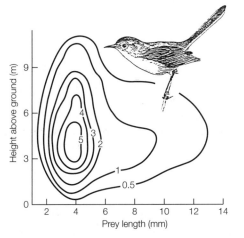

Fig. 2.   The feeding niche of the blue-gray gnat catcher based on prey size and feeding height in Californian oak woodlands. Contours represent feeding frequencies. Redrawn from Ecology and Field Biology, 5th edn, R.L. Smith, 1996, Addison Wesley Longman.

Fundamental niche

The niche space a species can occupy is affected by the degree of competition and predation that impact upon it. Commonly, in the absence of these stresses, a species can thrive in a wider range of resource levels and conditions. This potential niche space is the fundamental niche.

Realized niche

Normally, species are exposed to competitors and predators and are hence limited to a more restricted niche space, known as the realized niche. The impact of competition on the fundamental niche is illustrated by a classical experiment on two species of *Galium* spp. by the plant ecologist Tansley. Heath bedstraw (*Galium saxatile*) is found on acidic soils, whilst slender bedstraw (*G. pumilium*) is found on calcareous soil. When grown alone, both species thrive in both soil types. However, when the species were grown together, slender bedstraw was excluded on the acidic soil, whilst heath bedstraw was excluded from the calcareous soil. Clearly, competition affects the observed niche. Experiments with guppies (*Poecilia reticulata*) and sticklebacks (*Gasterosteus aculeatus*) suggest that the proximity of a predator constrains the fish to feed in sheltered areas where predation risk is lowered. In another example, the pattern of feeding times of the three American bat species shown in *Fig. 3* suggests that the realized temporal niche of these animals is also shaped by interspecific interactions.

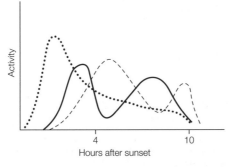

Fig. 3.   Foraging activities of bats in Iowa woodlands – dotted line, brown bat; solid line, silver-haired bat; dashed line, hoary bat.

It may be noted that mutualists also affect the realized niche of an organism, but in contrast to predators and competitors, the presence of a mutualist will tend to expand the realized niche rather than contract it. In the more extreme cases of obligate mutualism, such as that between many orchid species and their fungal mycorrhizal root mutualists, the niche of the single species may be nonexistent – orchids without their mycorrhizae simply do not grow.

Niche theory is explored in more depth in Section I.

C1 SOLAR RADIATION AND CLIMATE

Key Notes

Solar radiation	Solar energy drives climatic processes. Energy from the sun strikes the Earth where high-energy wavelengths of light are absorbed and re-emitted in the form of radiant heat. Warmed air rises, and expands as it does so. This expansion takes energy from the air, resulting in temperature reduction, a process known as adiabatic cooling.
Global wind patterns	The major wind systems of the Earth result from the upward movement of warm air around the equator, which is replaced by cooler air coming from the north and south, forming the trade winds. The Coriolis effect (caused by the Earth's rotation) deflects moving air to the right in the Northern Hemisphere and to the left in the Southern Hemisphere. The trade winds meet just to the north of the equator in the intertropical convergence zone (ITCZ). At 40° north and southwesterly, upper atmosphere 'jet stream' winds occur. Ice packs at the poles increase the surface reflectance (or 'albedo') which reduces the heating of the air, resulting in zones of dense cold air.
The circulation of oceans	The world's oceans are stirred by winds. Trade winds pile up water against continents, causing an imbalance in sea levels. For example, in North America sea levels are 1 or 2 m higher on the Atlantic side than the Pacific side. This difference drives oceanic currents. Thus, for instance, warm water piled up in the Caribbean moves northwards along the American coast as the Gulf Stream and then veers towards northern Europe, where it provides a strong warming influence.
Rain	Rain falls when moist air cools. Warm air can hold more water than cool air, so cooling causes water droplets to condense and fall as rain. If, for example, air travels over sea and then rises over a mountain, the air will cool at the adiabatic lapse rate, which is 6–10°C km^{-1}, depending on water content, and rainfall will result. After crossing the mountains, the air will descend and warm as it is compressed, leading to a rain shadow on the lee side of a mountain.
Havoc	Static conditions in tropical oceans can lead to storms. Minimal movement of air over warm water over a period of days can be disrupted by columns of warm air suddenly rising, resulting in surface winds being sucked into the rising column. The rising air is saturated with water and as it rises the air expands and cools. The water vapor turns into droplets, which release their heat of condensation and provide energy to further fuel the process, which may develop into a hurricane.
Related topics	Microclimate (C2) Ecosystem patterns (S1)

Solar radiation   The Earth is a rotating sphere which orbits the sun. The equator lies just a few
                        degrees away from being parallel with the plane of the Earth's orbit (*Fig. 1*).
                        This means that the **irradiance** per unit area is always greater at the equator
                        than it is at high latitudes. High latitudes therefore receive less heat than low
                        latitudes. It is this inequality in energy supply that drives the Earth's climate
                        and therefore determines the distribution of organisms over the Earth. The plane
                        of the Earth's rotation is 23°27′ from being parallel with the plane of the orbit.
                        It is this angle that is responsible for seasonality in the Northern and Southern
                        Hemispheres. The variations in solar radiation from summer solstice to winter
                        solstice at three locations, a temperate region, a tropical region and a high lati-
                        tude region are shown in *Fig. 2a*.

                        A second inequality is the distribution of night and day. All areas of the Earth
                        have exactly the same amount of night and day each year. However, while at
                        the equator night and day come in 12-hour cycles, at the poles day and night

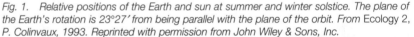

Fig. 1.   *Relative positions of the Earth and sun at summer and winter solstice. The plane of
the Earth's rotation is 23°27′ from being parallel with the plane of the orbit. From* Ecology 2,
P. Colinvaux, 1993. Reprinted with permission from John Wiley & Sons, Inc.

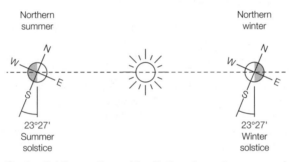

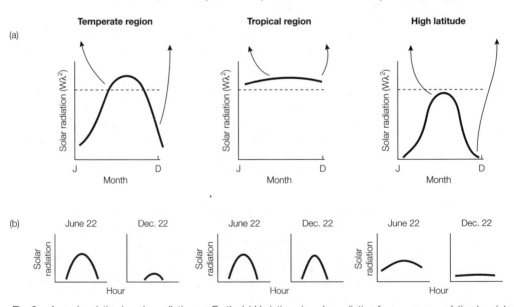

Fig. 2.   *Annual variation in solar radiation on Earth. (a) Variations in solar radiation from summer solstice to winter
solstice at three locations: a temperature region, a tropical region, and a high latitude region; (b) diurnal variations
in solar radiation on 2 days in the year: the summer solstice and the winter solstice. Redrawn from* Elements of
Ecology, *4th edn, R.L. Smith and T.L. Smith, 1998, Benjamin-Cummings Publishing.*

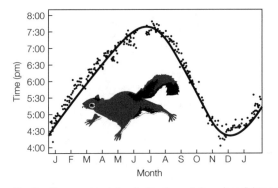

Fig. 3.   Seasonal variation in the time of day when flying squirrels become active. Redrawn from Elements of Ecology, 4th edn, R.L. Smith and T.L. Smith, 1998, Benjamin-Cummings Publishing.

come in 6-month cycles. These diurnal variations can be seen in *Fig. 2b*. These changes in seasonal rhythms of night and day have a major effect on animal behavior. The flying squirrel, for example, starts its daily activity with night-fall, regardless of the season. As the short days of winter turn to the longer days of spring, the squirrel begins its activity a little later each day (*Fig. 3*).

The result of these variations is a more constant heat at low latitudes. Low latitudes are therefore always warmer than polar regions. This results in the mass transfer of heat from the equator to the poles in the flow of air and water. A major part of this heat transfer is by ocean currents, due to the high specific heat capacity of water. However, movements in air are more rapid, giving rise to the properties of climate.

Energy from sunlight strikes the Earth where the high-energy wavelengths of light are absorbed and re-emitted in the form of radiant heat. The primary heating of the atmosphere comes, therefore, from the ground and ocean surfaces. The warm air rises and as it does so the air expands and cools. The cooling of the rising air as it expands results from the expenditure of energy by the warm air during expansion. This process is called **adiabatic** cooling. The cooling is a function of altitude and the adiabatic lapse rate in dry air is approximately 10°C per 1000 m of elevation. In moist air the lapse rate is less because heat is gained as water condenses, falling to approximately 6°C per 1000 m.

Global wind patterns

The major wind systems of the Earth result from the fact that masses of air around the equator are forced to rise by ground heating, causing cooler air to move in from higher latitudes, which in turn is replaced by descending air to fill the void. However, this simple system is modified by the Coriolus effect (see below) and the properties of scale, resulting in the patterns shown in *Fig. 4*.

The revolution of the Earth causes winds to be dragged by the surface, thus deflecting their courses. An object at the equator revolves at a speed of approx-imately 1500 km h^{-1}. At higher latitudes, the object would be moving more slowly as the distance travelled for one revolution of the earth is less. Any mass of air moving away from the equator has an initial velocity of 1500 km h^{-1}, and as the air travels into areas where the surface moves more slowly it veers in the direction of its superior momentum. This diversion of winds through their own angular momentum is called the **Coriolis effect**, named after

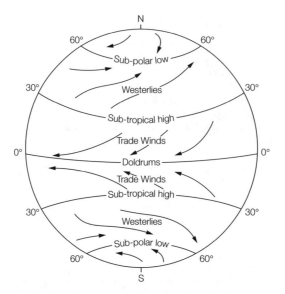

Fig. 4.   *Winds at sea level. The doldrums at the intertropical convergence zone (ITCZ) are shown at the geographical equator as they would be on an ideal Earth. From* Ecology 2, *P. Colinvaux, 1993. Reprinted with permission from John Wiley & Sons, Inc.*

a seventeenth century French mathematician. Southward-traveling air in the Northern Hemisphere will be deflected to the right, just as northward-traveling air. In the Southern Hemisphere the directions are reversed, with deflection to the left. At tropical latitudes, the surface air that fills the equatorial void from the north is deflected to the right and becomes the northeast trade wind. Close to the equator it meets air coming from the south that was deflected left, the southeast trade wind. Because there is less land to slow the air in the Southern Hemisphere, the trade winds do not meet at the geographical equator, but instead converge at the **intertropical convergence zone (ITCZ)**, north of the equator. Heated air rises rapidly along the ITCZ leading to heavy rains typical of Caribbean islands.

At latitudes higher than 30°, the sinking of upper air forces winds from 30° to 40° latitude to set off poleward. However, they are directed to the right in the Northern Hemisphere and to the left in the Southern Hemisphere by the Coriolus effect. This results in westerly winds in both hemispheres at about 40° latitude. In the Southern Hemisphere, latitude 40° has virtually no land to slow the winds, resulting in the 'roaring forties', winds of high velocity which track the 40th parallel. Upper atmosphere winds along both 40th parallels are also westerlies, the **jet streams**. Because atmospheric pressure is higher at low latitudes, upper air flows poleward to where pressure is lower. Coriolis effects divert the wind eastward in both hemispheres leading to westerly jet streams.

At the poles the lack of irradiance has severe consequences for climate. One pole is covered by floating ice packs while the other is covered by glacial ice. The ice acts as a reflecting mirror, increasing the **albedo** (reflectivity) and reducing the heating of the air. Dense, cold air therefore sits over the poles, flowing out into subpolar regions as the cold east winds of northern Europe.

The circulation of oceans

The world's oceans are also stirred by the winds. The drive provided by the trade winds pushes masses of water up against continental dikes downwind.

For example, the north Atlantic Ocean begins as water piled up in the shallow Caribbean Sea, being heated by the tropical sun. This piling up of water by the northeast trades means that the sea level is one or two meters higher on the Atlantic side than the Pacific side. The water piles up in the Caribbean and escapes northward along the American coast as the **Gulf Stream**. However, moving water is also affected by the Coriolis effect, and the Gulf Stream veers right towards northern Europe and warms the British Isles, which accordingly has a much warmer climate than places in Canada well to the south of it. The Gulf Stream continues to veer right until the trade winds once again take control. In the Southern Hemisphere, ocean currents rotate in the opposite direction, counterclockwise. This is most clearly demonstrated by the way ocean currents follow the roaring forties in a perpetual cycling of the globe. As currents turn back towards the equator, they move along the lee of a continent. For example, in the north Pacific the warm current flowing north and veering right is the Japan Current which bathes southern Alaska and British Columbia with water which may be warmer than the adjacent land. These coasts therefore have fog. But then the ocean currents veer right again and travel down the coast of California, where the Coriolus effect turns them away from the coast and out to sea. What happens then is that colder water must be taken from below the surface to replace this water. Cold bottom water surfacing in this way is called an **upwelling** and these are areas of high biomass productivity due to the presence of nutrients brought to the surface.

Rain

Rain falls when moist air cools. This will happen when high irradiance of the sea along the ITCZ raises moist air to great heights, resulting in condensation, then rain. Rain also falls when winds are deflected by mountain ranges. *Figure 5* shows the progress of events. Air arriving from the sea rises up a mountain,

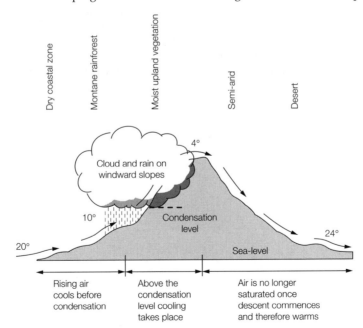

Fig. 5.   Precipitation along mountain ranges. The rain shadow downwind of a mountain results because the descending air picks up moisture. From Ecology 2, *P. Colinvaux, 1993. Reprinted with permission from John Wiley & Sons, Inc.*

cooling as it goes with the **adiabatic lapse rate** (6–10°C km^{-1}, depending on water content) until water condenses. Rain forms high against the mountain-side. The air continues to rise, but now cools only at the adiabatic lapse rate of 6°C km^{-1}. The air crosses the mountains and begins to descend. The descending air warms as it is compressed (10°C km^{-1}) while also taking up moisture, giving rise to a rain shadow on the lee side of a mountain. Descending air will always absorb moisture. Therefore at 30° latitude, where the air is sinking, the area will always be dry. Major deserts such as the Sahara, the Grand Gobi and the American Southwest lie in this belt.

Rising air is required for rain. However, air can be prevented from rising over large geographic regions by a process called **inversion**. An inversion is warm air floating over cold, forming a stable relationship. The commonest form of inversion results when winds blow across cold ocean water, cooling the air at the bottom. The cold winds blow on and, because of the low temperature and surface wind, the winds are dry. The Galapagos Islands are desert due to inversion even though they are located at the equator. The hot air rising from the heated Galapagos Islands causes a thin layer of stratus to form at the inversion. The most complete deserts are on the western coasts of continents, such as northern Chile, where cold upwellings along the shore cause inversions.

Havoc

Water piled up in the tropical seas (such as the Caribbean) by the trade winds can be heated throughout by the sun. If there is minimal lateral movement of air and water for a critical number of days then columns of air can rise suddenly over the warm sea, causing surface winds to be sucked in to feed the rising column. However, the rising air is saturated with water and as it rises the air expands and cools; the water vapor turns into droplets, releasing their **heat of condensation** high above the surface of the sea. Now the rising column is heated from near the top as well as from the bottom. To supply the rising warm air, yet more air is sucked up at the surface. The surface winds angle in due to Coriolis effects, and the whole structure begins to spin. This vortex is now called a **hurricane**. Once over land, the hurricane loses its source of warm, moist air and will eventually collapse. However the air in the hurricane continues to rise, expand and cool and rain falls. Surface air also strikes at speeds in excess of 100 miles h^{-1}, causing devastation.

C2 MICROCLIMATE

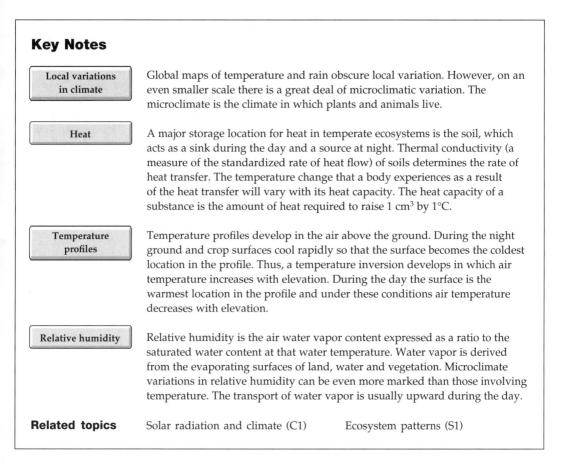

Key Notes

Local variations in climate	Global maps of temperature and rain obscure local variation. However, on an even smaller scale there is a great deal of microclimatic variation. The microclimate is the climate in which plants and animals live.
Heat	A major storage location for heat in temperate ecosystems is the soil, which acts as a sink during the day and a source at night. Thermal conductivity (a measure of the standardized rate of heat flow) of soils determines the rate of heat transfer. The temperature change that a body experiences as a result of the heat transfer will vary with its heat capacity. The heat capacity of a substance is the amount of heat required to raise 1 cm^3 by 1°C.
Temperature profiles	Temperature profiles develop in the air above the ground. During the night ground and crop surfaces cool rapidly so that the surface becomes the coldest location in the profile. Thus, a temperature inversion develops in which air temperature increases with elevation. During the day the surface is the warmest location in the profile and under these conditions air temperature decreases with elevation.
Relative humidity	Relative humidity is the air water vapor content expressed as a ratio to the saturated water content at that water temperature. Water vapor is derived from the evaporating surfaces of land, water and vegetation. Microclimate variations in relative humidity can be even more marked than those involving temperature. The transport of water vapor is usually upward during the day.
Related topics	Solar radiation and climate (C1) Ecosystem patterns (S1)

Local variations in climate

Global maps of temperature, winds and rain hide a great deal of variation at the local level. However, even on a smaller scale there is a great deal of **microclimatic variation**. For example, the sinking of dense, cold air into the bottom of a valley at night can reduce the air temperature by some 30°C below the air temperature 100 m above the valley. On a yet smaller scale, air in the immediate vicinity of a low alpine plant may be 20°C higher than air only 0.3 m above the plant. Hence, we must not merely confine our attention to global or geographic patterns when examining the influence of temperatures on the distribution and abundance of organisms. The **microclimate** is the climate in which plants and animals live, and is scaled to the organism in that the microclimate of a tree is different in scale to the microclimate of beetle larvae in soil. Microclimate differs from the climate which prevails above the first few meters over the ground, primarily in the intensity of the changes with elevation and the changes with time that occur there.

Heat

Soil constitutes a major storage location for heat, acting as a sink for energy during the day and a source to the surface at night. The flux of heat in and out of the soil is a process of **thermal conductivity**. Thermal conductivity in soil depends upon porosity, moisture content and organic matter content. It is defined as the quantity of heat flowing in unit time through a 1 cm² cross section of soil in response to a temperature gradient of 1°C cm⁻¹ of depth. The thermal conductivity of soils determines the rate of heat transfer. The temperature change that a body experiences as a result of the heat transfer will vary with its **heat capacity**. The heat capacity of a substance is the amount of heat required to raise 1 cm³ by 1°C. Heat is continually moving into or out of the soil, and the thermal energy is being continually redistributed in the soil. In addition, the pattern of soil temperature profiles changes rapidly during a normal day (*Fig. 1*). The soil surface is coldest in the early morning and warmest in the early afternoon. The amplitude of the daily soil temperature variation decreases with depth into the soil. At midday, heat is directed downward through the upper 100 cm of soil. Heat exit from the soil begins after sunset, but some heat flow continues downward throughout the night. During the summer there is a net daily gain of heat in the soil. *Figure 2* shows the temperature variation of an arctic plant *Novosieversia glacialis* on a sunny day.

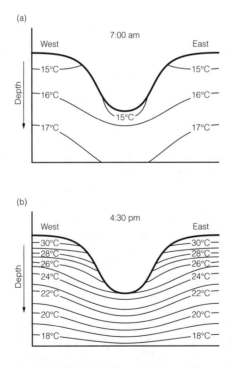

Fig. 1.   *Soil temperature profiles in the top 15 cm at 42°N in late spring, (a) at dawn and (b) in the mid-afternoon. Note that the temperature gradient reverses over the day.*

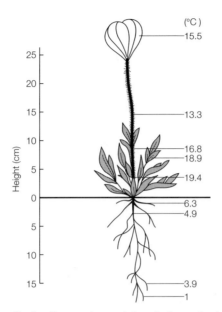

Fig. 2. *Temperature variations in the arctic plant* Novosieversia glacialis *on a sunny day with an air temperature of 11.7°C. Redrawn from* Elements of Ecology, *4th edn, R.L. Smith and T.L. Smith, 1998, Benjamin-Cummings Publishing.*

Temperature profiles

Temperature profiles also develop in the air above the ground as well as in the soil. During the night the surfaces of both the ground and crops cool quickly so that the surface becomes the coldest location in the profile (*Fig. 3*). Air in contact with the surface loses energy to the surface, becoming chilled and heavy. Thus a temperature inversion develops in which air temperature increases with elevation. During the day the surface is the warmest location in the profile and under these conditions air temperature decreases with elevation. Temperature profiles extended down into plant canopies differ from those in free air. Differences in shading and air movement occur, resulting in changed air temperature profiles. The same variations in temperature are also caused by depth in water or soil. Depth has two effects on fluctuations of temperature; firstly the fluctuations are diminished or 'damped' and secondly they lag behind the fluctuations at the surface. The strength of these effects increases with depth and decreases with thermal conductivity of the medium (low in soil, higher in water).

Relative humidity

Relative humidity is the ratio of actual to saturation water vapor. Water vapor is derived from the evaporating surfaces of land and water. The transport of water vapor into and through the layer of air adjacent to the ground is analogous to heat transport. Microclimate variations in relative humidity can be even more marked than those involving temperature. The transport of water vapor is usually upward during the day. It is not unusual, for example, for the relative humidity to be almost 100% (i.e. saturated) at ground level amongst dense vegetation, while the air immediately above the vegetation less than 0.5 m away has a relative humidity of only 50%. However, the shape of the relative humidity of the air profile may vary throughout the day. At dawn the

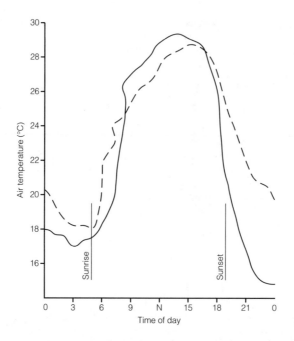

*Fig. 3.   The pattern of air temperature at 1 m (solid line) and 16 m (dotted line) above a
soybean crop. From* Microclimate: The Biological Environment, *Rosenberg, 1974. This
material is used by permission of John Wiley and Sons Inc.*

relative humidity may be constant with elevation. As the day progresses evaporation at the surface increases the relative humidity of air at the surface, but
by early evening evaporation decreases and the relative humidity of the air is
lowered at the surface.

D1 THE PROPERTIES OF WATER

Key Notes

Chemical properties of water

Water is the universal internal medium of all organisms, comprising more than 90% of living matter. The shape of the water molecule, with an HOH angle of 105° results in the side with the hydrogen being positively charged (electropositive) whereas the other side is negatively charged (electronegative). This explains many of water's properties relative to physical and chemical reactions. It also explains why water is attracted to charged ions.

Penetration of light through water

The penetration of visible light through water bodies follows a negative exponential relationship called 'Beer's law'. Since water strongly absorbs infrared radiation, that portion of the solar spectrum will be sharply depleted as water depth increases. Furthermore, turbidity due to suspended solids (silt) will strongly deplete visible light.

Water and temperature

Two features of water make it particularly suitable as a medium for life-forms to inhabit. Water has (i) a high heat capacity and (ii) maximal density at 4°C. A high heat capacity means that it can absorb heat energy with only a small increase in temperature. As a result, aquatic life forms are buffered from temperature fluctuations. Water becomes increasingly dense and therefore heavier at lower temperatures, with the maximum density occurring at 4°C. Thus, ice floats on water and a body of water freezes from the top down. This phenomenon protects aquatic organisms because ice acts as an insulator to prevent further decreases in temperature in the underlying water.

Energy transfer and water phases

At terrestrial temperatures, water passes easily from vapor to liquid and solid phases with a large release or absorption of heat. For example, the evaporation of 1 g of water requires about 2430 J of heat, the latent heat of vaporization. The vapor is carried in air until the water condenses, releasing $2430 J g^{-1}$. When water freezes, about $335 J g^{-1}$ is released as the heat of fusion and the same amount of energy is required to melt the snow. Thus, these energy-consuming and releasing processes in water phase changes provide the mechanisms for the transportation of large quantities of heat to and from the surface of the earth.

Related topics   Plants and water (D2)   Animals and water (D3)

Chemical properties of water

Water is the only inorganic liquid that occurs naturally on earth. It is also the only chemical compound that occurs naturally in all three physical states: solid, liquid and vapor. Water existed on the planet long before any form of life evolved but since life developed in water, it is the universal internal medium of all organisms. Living matter is made up of more than 90% water. Water is also the external medium of all aquatic lifeforms and can function as a resource, condition and a habitat. The unique relationship between water and living organisms stems from the fact that water is a universal solvent: almost anything will dissolve in water to some degree. The shape of the water molecule, with an HOH angle of 105°, results in the side with the hydrogen being positively charged, **electropositive**, whereas the other side is negatively charged, **electronegative**. This explains many of water's properties relative to physical and chemical reactions. It also explains why water is attracted to charged ions. Cations such as Na^+, K^+ and Ca^{2+} become hydrated because of their attraction to the negatively charged oxygen end of the water molecule. The polar nature of water also explains hydrogen bond formation by water. The bonding of each water molecule to other water molecules and to other biological components explains the solution properties.

Penetration of light through water

The penetration of visible light through water bodies follows a negative exponential **Beer's law** relationship:

$$R_s = R_{sc}e^{-ax}$$

Where R_{sc} is the solar constant and R_s is the solar radiation after passage through a depth x of water of extinction coefficient a. Since water strongly absorbs infrared radiation, that portion of the solar spectrum will be sharply depleted on penetration through water bodies. Furthermore, turbidity due to suspended solids (silt) will strongly deplete visible light. Although the prediction of visible light attenuation in pure water is relatively simple, in natural waters it is much more complex due to the variability in the nature and content of suspended materials, and therefore the simple Beer's law relationship may not be easily applied. This means that photosynthesis can occur in water, at least to certain depths. Consequently, primary producers such as plants and cyanobacteria, the basis of any food chain, can grow.

Water and temperature

Water surfaces are poor reflectors of solar radiation and therefore serve as a good sink for solar energy. However, water has a high **heat capacity**. The heat capacity of a substance is the amount of heat required to raise 1 cm^3 by 1°C. A high heat capacity means that it can absorb heat energy with only a small increase in temperature when compared with other substances. Because of this high heat capacity and the ability of water to mix and therefore to dissipate heat quickly, aquatic life forms do not have to be adaptable to a wide range of temperatures or to rapidly changing temperatures. Water becomes increasingly dense and therefore heavier at lower temperatures down to 4°C. This is why deeper water feels cooler in a lake. Because the **maximum density** of water occurs at 4°C, water becomes increasingly lighter at 3°C, 2°C, 1°C and 0°C (freezing). The density of liquid water at 0°C is greater than the density of frozen water at the same temperature. Thus, water is heavier as a liquid than as a solid, so that solid water and warm water can float over cold water (*Fig. 1*). Another important consequence of these properties is that water freezes from the top down. This protects aquatic organisms because ice acts as an insulator to prevent further decreases in temperature in the remaining water and decreases the chances that ponds and lakes will freeze solid.

Energy transfer and water phases

To the meteorologist, water vapor is the single most important constituent of the atmosphere. This is because at terrestrial temperatures, water passes easily from vapor to liquid and solid phases with a large release or absorption of heat. For example, the evaporation of 1 g of water requires about 2430 J of heat, the

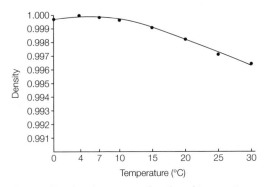

Fig. 1. Density of water as a function of temperature.

latent heat of vaporization. The vapor is carried in air which rises until a level is attained at which condensation occurs with its concomitant release of the 2430 J g^{-1}. When water freezes, about 335 g^{-1} is released as the **heat of fusion** and the same amount of energy is required to melt the snow. **Sublimation**, the direct transition from solid to vapor phase (and *vice versa*) without passing through an intermediate liquid phase, involves the consumption or release of 335 J g^{-1}. Thus these energy-consuming and releasing processes in water phase changes provide the mechanisms for the transportation of large quantities of heat to and from the surface of the Earth. These processes play an important role in energy transfer and availability of energy to biological systems.

D2 PLANTS AND WATER

Key Notes

Soil water	For terrestrial plants the main source of water is the soil, which serves as a reservoir. Water enters the reservoir as rain or melting snow and passes into the soil pores. The upper limit of the water-holding capacity of a soil is called the *field capacity*. This is the amount of water which can be held by soil pores against the force of gravity. Plants cannot extract all the water held in the soil, as they cannot exert sufficient suction force to extract water from the narrower soil pores. The lower limit of water availability is thus determined by the physiology of the plant species and is known as the *permanent wilting point* – the soil water content at which plants wilt and are unable to recover.
The uptake of water by roots	Roots can capture water from the soil in two ways: either water may move through the soil towards a root or the root may grow through the soil towards the water. As a root withdraws water from the soil capillary pores at its surface, it creates water depletion zones around it. If a root draws water from the soil very rapidly, the resource depletion zone (RDZ) will receive water from the surrounding soil at a slow rate, restricting water availability, so plants may wilt even in soil containing abundant water.
Aquatic plants and water	Water is apparently freely available in aquatic environments. However, the osmotic regulation of internal fluids can be energetically expensive, especially in saline environments. The salinity of an aquatic environment and of terrestrial habitats bordering the sea has an important influence on plant distribution and abundance. Plants which grow in high salinity, halophytes, accumulate electrolytes in their vacuoles, but the concentration in the cytoplasm and organelles is kept low.
Water availability and plant productivity	Precipitation is a key determinant of plant productivity in forests, whilst in arid regions there is an approximately linear increase in primary productivity with increasing precipitation. The amount of water that would be transpired from a site, assuming no soil water limitation and complete vegetation cover, is the *potential evapotranspiration rate*. The difference between this index and the precipitation rate defines whether the environment is moist or arid.
Related topics	The properties of water (D1) Responses to temperature (E2) Animals and water (D3) Temperature and species distribution (E3)

Soil water

The volume of water that becomes incorporated into the body of a plant is infinitesimal compared with what passes through the plant in transpiration. Hydration is a necessary condition for metabolic reactions to proceed within the organism; water is the medium in which the reactions occur. For terrestrial plants the main source of water is the soil, which serves as a reservoir. Water enters the reservoir as rain or melting snow and passes into the soil pores. Soil water is not always available to the plants. This is dependent on the size of the pores which may hold water by capillary action against gravity. If the pores are wide, as in a sandy soil, much of the water will drain away passing down through the soil profile until it reaches an impermeable rock where it accumulates as a rising water table or drains away and eventually enters streams or rivers. The water held by soil pores against the force of gravity is called the **field capacity** of the soil. This is the upper limit of the capacity of a soil to hold water as a useable resource for plant growth. The lower limit is determined by the ability of plants to exert sufficient suction force to extract water from the narrower soil pores, and is known as the **permanent wilting point** – the soil water content at which plants wilt and are unable to recover. *Figure 1* shows the relationship between soil water status and pore size in soil. The soil water content at the permanent wilting point does not differ significantly among plant species. Solutes in the soil solution add osmotic forces to the capillary forces that the plant must match when it absorbs water from the soil. These osmotic forces become particularly important in saline solutions in arid environments. Here, most water movement is upwards from the soil to the atmosphere and salts rise to the surface, creating osmotically lethal salt pans.

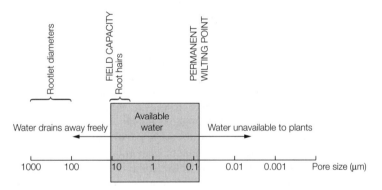

Fig. 1.   The status of water in the soil as measured by the diameter of soil pores that remain water-filled.

The uptake of water by roots

There are two ways in which roots can capture water in the soil. Water may move through the soil towards a root or the root may grow through the soil towards the water. As a root withdraws water from the soil pores at its surface, it creates water depletion zones around it. These determine gradients of water potential between the interconnected soil pores. Water flows in these **capillary pores**, along the gradient into the depleted zones, supplying further water to the root. Water then enters the plant through the epidermis of roots and then moves across the root cortex, passing into the stele and then flows up xylem vessels to the shoot system. This transport of water from roots to stems and leaves is driven by pressure. This simple process is made more complex because the more the soil around the roots is depleted of water, the more resistance there is to water flow. As the root starts to withdraw water from the soil, the first water that it obtains is from the wider pores because they hold the water with weaker capillary forces. This leaves the narrower capillaries through which water can flow, and so increases the resistance to water flow. Thus, when the root draws water from the soil very rapidly, the **resource depletion zone** (RDZ) becomes sharply defined and water can move across it only slowly. Because of this, rapidly transpiring plants may wilt in soil containing abundant water. The fineness and degree of ramification of the root system through the soil are important in determining the access of the plant to the water in the soil reservoir. This means that different parts of the same root system further down in the soil may encounter water held at different forces. In arid regions where rainfall occurs as occasional showers, the surface layer may be brought to field capacity while the remainder of the soil stays at or below wilting point. This is potentially hazardous to seedlings which may germinate after rain in the wet surface but the soil mass will be unable to support its further growth. A variety of specialized dormancy-breaking mechanisms are found in species living in such habitats which protect against too quick a response to insufficient rain.

Most roots elongate before producing laterals and this ensures that exploration precedes exploitation. Branch roots usually develop so that they emerge on radii of the parent root; secondary roots radiate from these primaries and tertiaries from the secondaries. These rules of growth maximize the exploration of soil while preventing the chance that two branches will enter each others depletion zones. The root system that a plant develops early in its development can determine its responsiveness to future events. Plants which are waterlogged early in their development have a superficial root system which is inhibited from growth into anaerobic, water-filled parts of the soil. Later in the season when water is in short supply, these same plants may suffer from drought because their root system has not developed into the deeper soil layers. In an environment in which the main supply of water comes from occasional showers falling on a dry surface, a seedling which develops an early taproot system will gain little from subsequent showers. In contrast, in an environment in which heavy showers occur, the early development of a taproot will ensure continued access to water in times of drought. The effectiveness of a root as a resource forager is in part due to the ability of roots to adapt during development. This is in marked contrast to shoot development. *Figure 2* illustrates this point and shows how the root of a wheat plant developed when passing through a layer of water-logged clay in a sandy soil.

Aquatic plants and water

In aquatic environments, the availability of water as such is not a problem. In freshwater or brackish habitats, however, there is a tendency for water to enter the plants from the environment by osmosis. In marine environments, the majority of plants are isotonic to their environment so that there is no net flow of water. However, there are plants in this environment which are hypotonic so that water flows out from the plant into the environment, putting them in a similar position to terrestrial plants. Thus, for many aquatic plants the regulation of internal fluids is a vital and sometimes energetically expensive process. The salinity of an aquatic environment can have an important influence on plant distribution and abundance, especially in places such as estuaries where there is a sharp gradient between marine and freshwater habitats.

Salinity can also have an important effect on plant distribution in terrestrial habitats bordering the sea. There are great differences in the sensitivity of plant species to salinity. Avocados are sensitive to low salt concentrations (20–50 mM) while some mangroves can tolerate salt concentrations 10–20 times greater.

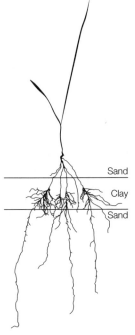

Fig. 2. The root system of a wheat plant grown in a sandy soil containing a layer of clay. Reprinted from Ecology, *2nd edn, Begon* et al., *1990, with permission from Blackwell Science.*

These species meet high osmotic pressures in their soil solution and so face problems with water uptake. Many such **halophytes** accumulate electrolytes in their vacuoles, but the concentration in the cytoplasm and organelles is kept low. In this way such plants maintain high osmotic pressures yet escape damage.

Water availability and plant productivity

A general relationship exists between plant productivity and precipitation for many of the world's forests. Water is essential as a cell constituent as well as a requirement for photosynthesis. Large quantities of water are lost in transpiration as the stomata must be left open for CO_2 uptake. In arid regions, there is an approximately linear increase in primary productivity with increasing precipitation. However, in more humid forest climates there is a plateau beyond which productivity does not continue to rise. Not all water falling as precipitation will be available to plants; water in excess of field capacity will tend to drain away.

The **potential evapotranspiration rate** is the amount of water that would be transpired from a site assuming no soil water limitation and complete vegetation cover. It may be calculated from the factors which will cause a plant to lose water. The key variable is temperature, but radiation, air humidity, and windspeed may also be important. Potential evapotranspiration minus precipitation provides a crude index of how far the water available for plant growth falls below what might be transpired by actively growing plants. Moist sites will have higher precipitation rates than potential evapotranspiration, whilst in arid areas, the opposite will apply. It may be noted that if two sites receive the same amount of rainfall, one may be moist and the other arid by this plant-orientated perspective.

Table 1. Productivity for selected vegetation types

Site	Productivity per unit weight (g/ leaf biomass year^{-1})	Productivity per unit area (g/ biomass m^{-2} year^{-1})
Desert	2.33	90
Deciduous forest	2.22	1200
Coniferous forest	1.64	1300
Grassland	1.21	600

A characteristic feature of many plant communities exhibiting low productivity is that potential evapotranspiration far exceeds precipitation; in other words drought is a key cause of poor productivity.

Water shortage also leads to the development of less dense vegetation. Less dense vegetation intercepts less light, causing low productivity rather than any reduced photosynthetic rate of droughted plants. This can clearly be illustrated by comparing the productivity per unit weight of leaf biomass instead of per unit area of ground (*Table 1*). Deserts are unproductive per unit area, but desert plants are slightly more productive than others on a weight-for-weight basis.

D3 ANIMALS AND WATER

Key Notes

Water balance in fish

Maintaining water balance is problematic in an aquatic environment, which is countered by osmoregulatory mechanisms. Freshwater fish have to continually excrete excess water because the fish is hypertonic relative to its surroundings (the concentration of solutes in body fluids is higher than the solute concentration of the water), and they produce a large volume of very dilute urine. Bony fishes living in seawater have the opposite problem, being hypotonic to their surroundings. The kidneys of marine fish secrete very little urine, and instead function mainly as a means of removal of divalent ions such as Ca^{2+}, Mg^{2+} and SO_4^{2-}.

Water balance in amphibians

Amphibian kidneys function much like those of freshwater fishes. However, on land, where dehydration is the most important problem in terms of osmoregulation, frogs conserve body fluid by reabsorbing water across the epithelium of the urinary bladder.

Water conservation by terrestrial animals

A major problem faced by terrestrial organisms is the loss of a continuous supply of water necessary to keep tissue surfaces moist. When air is inhaled, it passes along the respiratory tract into the lungs, where it is in contact with the moist respiratory tissues. If the moist air was exhaled, water would be lost. The recovery of respiratory moisture by most terrestrial animals involves *countercurrent exchange*. Exhaled air from the lungs encounters a countercurrent-like gradient on the way out. This interaction between the departing air and the respiratory surfaces results in an efficient return of moisture to the tissues.

Water conservation by mammalian kidneys

The water-conserving ability of the mammalian kidney represents a key terrestrial adaptation. Water recovery from the urine before it leaves the kidney takes place in the loop of Henle. Mammals adapted to the desert (such as kangaroo rats) that excrete highly concentrated hypertonic urine, have exceptionally long loops of Henle. In contrast, beavers, which spend much of their time in fresh water, have nephrons with very short loops, resulting in dilute urine. The kidneys of reptiles are less sophisticated and produce urine that is, at best, isotonic to body fluids. This means that the solute concentration of urine is equal to the solute concentration of the body fluids.

Related topics

Solar radiation and climate (C1)
Microclimate (C2)
The properties of water (D1)

Plants and water (D2)
Ecosystem patterns (S1)

Water balance in fish

In aquatic organisms, water will either tend to enter the organism from the environment (if they have a higher solute concentration than the water) or leave the organism (if the water has higher solute concentrations than the organism).

These processes are countered by **osmoregulation**. Osmoregulation is an adaptation to control the water balance in organisms living in hypertonic, hypotonic or terrestrial environments. A freshwater fish has the problem of excreting excess water because the animal is **hypertonic** to its surroundings. This means that the concentration of solutes in body fluids from these fish is greater than the solute concentration of the surrounding water. Instead of conserving water, the nephrons in the kidney use cilia to sweep a large volume of very dilute urine from the body.

Bony fishes living in seawater are **hypotonic** to their surroundings and have the opposite problem of freshwater fishes. This means that the concentration of solutes in body fluids from these fish is lower than the solute concentration of the seawater. The kidneys of marine fish secrete very little urine, and instead function mainly as a means of removal of divalent ions such as Ca^{2+}, Mg^{2+} and SO_4^{2-}, which the fish takes in by its incessant drinking of seawater. The gills of these fish excrete monovalent ions such as Na^+ and Cl^- and the bulk of its nitrogenous waste in the form of ammonium (NH_4^+).

Water balance in amphibians

Amphibian kidneys function in much the same way as those of freshwater fishes. When in fresh water, the skin of the frog accumulates salts from the water while the kidneys excrete dilute urine. Conversely, on land, in terms of **osmoregulation,** dehydration is the most important problem facing the frog. Under these conditions, frogs conserve body fluid by reabsorbing water across the epithelium of the urinary bladder.

Water conservation by terrestrial animals

Because no animal is completely watertight, its water content needs continual replenishment. One of the most serious problems organisms faced in evolving to a terrestrial existence was the loss of a continuous supply of water necessary to keep tissue surfaces moist. The production of the amniotic egg by vertebrates represented a mechanism by which vertebrates could prevent water loss during development and therefore allow vertebrates to colonize land.

Because gas exchange depends on a large, moist surface, measures for minimizing respiratory water loss are critical if an organism is to survive. When plants run short of moisture, they can block most water loss by closing the stomata. For animals, they must breathe and so water loss from exchange surfaces is a problem for animals. When air is inhaled, it passes along the respiratory tract into the lungs where it is exposed to the large surface area of the alveoli. The air is in contact with the moist respiratory tissues. If the moist air was exhaled, water would be lost. The recovery of respiratory moisture by most terrestrial animals involves **countercurrent exchange**. The strategy is most obvious in animals from arid habitats, where selection of water recovery has been most intense. Inhaled air is exposed to an increasing moisture gradient as it moves to the lungs. Exhaled air from the lungs encounters a countercurrent-like gradient on the way out. This interaction between the departing air and the respiratory surfaces results in an efficient return of moisture to the tissues. In camels, for example, totally dry desert air is fully humidified en route to the lungs, but the countercurrent removes 95% of all the moisture from the exhaled air.

Other animals may adjust their behavior to allow water conservation. For example, some animals show nocturnal activity patterns to reduce water loss during hunting and feeding.

Water conservation by mammalian kidneys

Most terrestrial animals drink free water and/or obtain water from their food. Some water is generated from the metabolism of food and body materials. The availability of water may place strict limits on the distribution and abundance of animals. The water-conserving ability of the mammalian kidney, therefore, represents another key terrestrial adaptation. The solute concentration (expressed as molarity), or **osmolarity**, of human blood is 300 mosm l^{-1}, but the kidney can excrete urine up to four times as concentrated. The cooperative action of the loop of Henle and the collecting duct maintain the gradient of osmolarity in the interstitial tissue of the kidney, making the concentration of urine possible (*Fig. 1*). Mammals that excrete the most hypertonic urine (i.e. urine with a greater solute concentration than other body fluids), such as kangaroo rats and other mammals adapted to the desert, have exceptionally long loops of Henle. Also, marine mammals tend to have long loops of Henle. Long loops maintain steep osmotic gradients in the kidney, resulting in urine becoming very concentrated. In contrast, beavers, which spend much of their time in fresh water and rarely face problems of dehydration, have nephrons with very short loops, resulting in diluted urine.

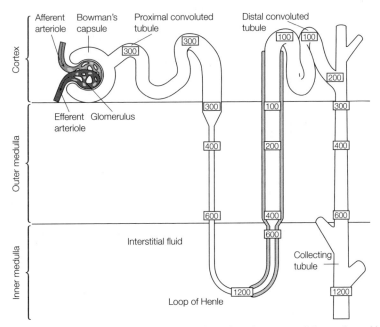

Fig. 1. The movement of water and ions through various parts of the nephron. Numbers indicate the concentration of salt in the filtrate in milliosmols per liter.

Birds, like mammals, have kidneys with nephrons that specialize in conserving water. However, bird nephrons have much shorter loops of Henle than mammalian nephrons. Bird kidneys, therefore, cannot concentrate urine to the osmolarities achieved by mammalian kidneys.

The kidneys of reptiles produce urine that is, at best, **isotonic** to body fluids. This means that the solute concentration of urine is equal to the solute concentration of the body fluids. However, most reptiles excrete nitrogenous waste in an insoluble form known as uric acid, which helps conserve water because it does not contribute to the osmolarity of the urine.

E1 TEMPERATURE AND METABOLISM

Key Notes

Homeotherms and poikilotherms	It is possible to categorize organisms according to their temperature regulation processes. One possible division is between homeotherms and poikilotherms. As environmental temperature rises, homeotherms maintain an approximately constant body temperature, while the body temperature of poikilotherms varies with environmental temperature. One problem with this classification is that even classic homeotherms experience periods of reduced temperature. An alternative distinction between organisms is described below.
Ectotherms and endotherms	Ectotherms are organisms such as plants, reptiles and protista which are largely reliant on external sources of heat to raise their body temperature. Endotherms are organisms capable of generating heat internally in order to raise their body temperature. Birds and mammals make up this group. The *thermoneutral zone* is the range of environmental temperatures in which an endotherm has only to exert a minimum metabolic effort in order to maintain a constant body temperature. The further away from the thermoneutral zone that the environmental temperature moves, the more energy the endotherm has to expend to maintain body temperature.
Heat exchange	All organisms gain heat from and lose heat to their environment as well as producing heat. A variety of physiological and behavioral mechanisms are used to regulate heat. Despite these mechanisms, the body temperature of an ectotherm varies significantly with environmental conditions.
Temperature thresholds	High temperatures may lead to enzyme inactivation or the unbalancing of components of metabolism; for example, in plants, respiration may proceed faster than photosynthesis, leading to death. However, the most frequent effect of high temperature on ectotherms is dehydration. All terrestrial ectotherms must conserve water but at high temperatures rates of water loss can be lethal. There are large differences between the low temperature tolerances of differing species, associated with the processes of freezing, chilling and hardening. Many species are killed by temperatures below –1°C due to the damaging effects of ice-crystal formation within cells; those that live through freezing winters often do so at a resistant, dormant stage of their life cycle.
Related topics	Solar radiation and climate (C1) Responses to temperature (E2) Plants and water (D2) Temperature and species Animals and water (D3) distribution (E3) Ecosystem patterns (S1)

Homeotherms and poikilotherms

When examining the relationships between organisms and environmental temperature, it is usual to subdivide organisms. One possible division is between the 'warm-blooded' and the 'cold-blooded'. However, these terms are subjective; a more satisfactory classification divides organisms into **homeotherms** and **poikilotherms**. As environmental temperature rises, homeotherms maintain an approximately constant body temperature, while the body temperature of poikilotherms varies with environmental temperature. One problem with this classification is that even classic homeotherms such as mammals and birds experience periods of reduced temperature (e.g. during hibernation) while some poikilotherms (e.g. Antarctic fish) experience only small body temperature variations because their environmental temperature remains constant. Also, many poikilotherms are capable of some body-temperature regulation. A better distinction between organisms is described below.

Ectotherms and endotherms

A clearer classification of the relationships between organisms and environmental temperature is to subdivide organisms into **ectotherms** and **endotherms**. Endotherms regulate their body temperature by producing heat within their own bodies; ectotherms rely on an external heat source. This represents a distinction between birds and mammals (endotherms) and all other organisms. Although there are numerous exceptions to this (for example, some reptiles and insects can elevate their body temperatures to facilitate activity) the distinction is nevertheless valuable. Ectotherms and endotherms differ in the extent to which they are able to maintain a constant body temperature (*Fig. 1*). Over a certain temperature range (**the thermoneutral zone**) an endotherm consumes energy at a basal rate. However, at environmental temperatures further and further away from this zone, the endotherm consumes more and more energy in maintaining a constant body temperature. Moreover, even in the thermoneutral zone endotherms typically consume energy much more rapidly than ectotherms. Endotherms produce heat at a rate controlled by the brain. They usually maintain a constant body temperature between 35°C and 40°C, and therefore they tend to lose heat to the environment. However, this loss is moderated by insulator material (fur, fat or

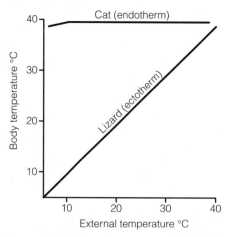

Fig. 1. *Relationship between the internal (body) and external temperatures for cat and lizard. The cat maintains a more or less constant body temperature irrespective of changes in the external temperature. The lizard's internal temperature depends on the temperature of its immediate surroundings. From* Biology: A Functional Approach, *4th edn, Roberts, 1986. With permission from Thomas Nelson and Sons.*

feathers) and by controlling blood flow near the skin surface. To increase heat loss, mechanisms such as panting are used. These mechanisms enable a high degree of control of body temperature to be maintained, enabling a consistency of peak performance. The price paid is high energy expenditure.

Heat exchange

All organisms either gain or lose heat to their environment as well as producing heat through metabolic processes. Almost all ectotherms modify heat exchange using the avenues of heat exchange shown in *Fig. 2*. Among the mechanisms used, some are fixed properties of particular species and some are behavioral responses, while others are more sophisticated behavioral patterns; other mechanisms are aspects of their physiology. Despite these mechanisms, the body temperature of an ectotherm varies significantly with environmental temperature for three main reasons:

- the ability of many ectotherms to regulate temperature is very low;
- ectotherms are to some extent always dependent on the external source of heat;
- energy must be expended to modify the heat budget. The extent to which an organism regulates its temperature will therefore be a compromise between cost and benefit.

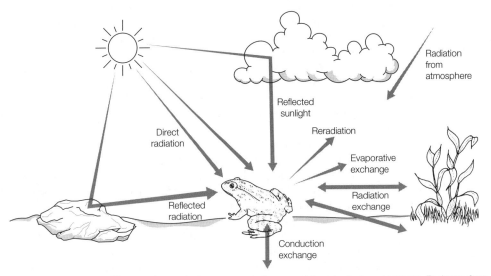

Fig. 2.   *The avenues of heat exchange between an ectotherm and the physical environment. Redrawn from Hainsworth, F.R., Animal Physiology, 1981, Addison Wesley Longman.*

Temperature thresholds

There are three main temperature ranges of interest: very low, very high and the temperatures in between. The most dangerous thing about high temperatures is that they lie only a few degrees above the animal's metabolic optimum, a result of the physio-chemical properties of their enzymes. High temperatures may therefore lead to enzyme inactivation or the unbalancing of components of metabolism (e.g. respiration proceeding faster than photosynthesis and leading to starvation). However, the most frequent effect of high temperature on organisms is **dehydration** (*Fig. 3*). One of the problems with dehydration in animals is the loss of the ability of the animal to cool itself due to a reduction in the volume of blood reaching the body surfaces. All terrestrial ectotherms must conserve water, but at high

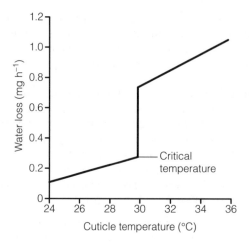

Fig. 3.   The rate of water loss from a cockroach with increasing temperature. From Biology: A Functional Approach, *4th edn, Roberts, 1986. With permission from Thomas Nelson and Sons.*

temperatures rates of water loss can be lethal. Plants living in hot environments may suffer severe water shortage and as a consequence they cannot use the latent heat of evaporation of water to keep leaf temperature down. In such plants (e.g. desert succulents) the risk of overheating can be minimized by a low surface-to-volume ratio. Even if atmospheric humidity is high, the risk of overheating exists through reduced evaporative cooling. Nevertheless, most problems with over-heating are found in hot, dry environments. There may be stages in an organism's life cycle that are particularly resistant to the effects of high temperatures. This is true of dormant structures such as bacterial endospores and seeds of plants, mainly as a result of their naturally dehydrated state.

At low temperatures, there are large differences between the tolerances of dif-fering species associated with the processes of freezing, chilling and hardening. Many species are killed by temperatures below –1°C due to the damaging effects of ice-crystal formation within cells. Those species that survive lower tempera-tures do so because they have evolved mechanisms that prevent ice-crystal for-mation within cells. The crystals may either damage cell integrity or absorb water leaving a concentrated solute suspension which may be lethal. Some ectothermic animals which are exposed to low temperatures accumulate solutes which act as an antifreeze, preventing crystal formation. Low temperature tolerance in plants is almost always associated with a period of acclimatization or hardening. Also, resistance to freezing injury changes according to the plant's stage of develop-ment. Most seeds are resistant to low temperatures. Similar examples can be found among ectothermic animals; those that live through freezing winters do so at a resistant, dormant stage of their life cycle. Even at temperatures above freezing, metabolic reactions slow down and may almost cease. Ectothermic ani-mals become moribund and cease to carry out basic maintenance functions. This may weaken an organism making it more susceptible to sources of mortality. Specifically, plants are liable to injury by chilling at temperatures around 10°C, brought about probably through disruption of membrane structure. There are also many animals which are susceptible to chilling damage, especially those that are not usually exposed to low temperatures.

E2 RESPONSES TO TEMPERATURE

Key Notes

Temperature and rates of enzyme reaction	The rate of an enzyme catalyzed reaction increases with temperature. In ectotherms this means that metabolic activity will be faster at higher environmental temperatures. The temperature coefficient (Q_{10}) is an index of the effect of a 10°C temperature rise on metabolic rate, and is often near 2.0.
Rates of development and growth	Within the nonlethal temperature range the most important effect on ectotherms of temperature is likely to be its effect on rate of development and growth. When rate of development is plotted against body temperature there exists an extended range of temperatures over which the relationship is linear. 'Physiological time' is a measurement combining temperature and time and applied to ectothermic organisms, which reflects that growth and development of these organisms is dependent on environmental temperature as well as time.
Acclimation and acclimatization	Temperature may also act as a stimulus, determining whether the organisms will begin development. Vernalization is the induction of flowering by low temperatures. Exposure of an organism to higher (or lower) temperatures in the laboratory can alter the organisms temperature response. The habituation of an organism's response to changes in laboratory environmental conditions is termed *acclimation*. *Acclimatization* is the habituation of an organism's physiological response to changes in natural environmental conditions.
Related topics	Solar radiation and climate (C1) Temperature and species Temperature and metabolism (E1) distribution (E3) Ecosystem patterns (S1)

Temperature and rates of enzyme reaction

In ectotherms the metabolic rate is relatively slow at low temperatures and more rapid as the environment becomes warmer. Endotherms buffer their internal organs from fluctuations in the environmental temperature, and hence do not exhibit such effects. The increase in metabolic rate with temperature can be described by a **temperature coefficient (Q_{10})**, which is given by:

$$Q_{10} = \frac{\text{metabolic rate at body temperature T°C}}{\text{metabolic rate at body temperature (T–10)°C}}$$

The value of the coefficient indicates the increase in reaction rate caused by a 10°C rise in temperature, and is commonly about 2.0, although Q_{10} is not constant across all temperatures, showing deviations towards the upper and lower thermal limits of an organism.

Rates of development and growth

Within the nonlethal temperature range for an ectotherm, the most important effect of temperature is likely to be its effect on the rate of development and growth of the organism. When rate of development is plotted against body temperature there exists an extended range of temperatures over which the relationship is linear. *Figure 1* shows the development of the cabbage white butterfly, *Pieris rapae*, from egg hatch to pupa. As most organisms spend all of their time below their nonlinear high temperature zone, it is assumed that the rate of development rises linearly with temperature above a **developmental threshold temperature**. The consequence of this relationship is that, unlike ourselves and endotherms, ectotherms cannot be said to require a certain length of time for development. What they require is a combination of time and temperature, **physiological time**. The importance of this concept lies in the ability to understand the timing of events, and thus population dynamics. In practice, however, there are difficulties in determining an organism's physiological time-scale, mainly due to the effects of fluctuating temperatures. The linear relationship itself is never more than an estimate, and there are problems with the monitoring of body temperature in the field.

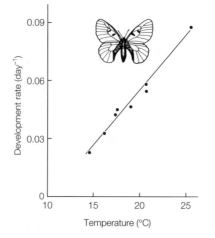

Fig. 1.   *Development of the cabbage white butterfly,* Pieris rapae, *from egg hatch to pupa requires 174 day-degrees above a threshold temperature of 10.5°C. Reprinted from* Ecology, *2nd edn, Begon et al. 1990, with permission from Blackwell Science.*

Acclimation and acclimatization

Temperature may also act as a stimulus, determining whether the organisms will begin development. For many plants a period of chilling or freezing is necessary before germination. For example, seeds of winter wheat only develop and flower if they have been pre-chilled. This phenomenon – the induction of flowering by low temperatures – is called **vernalization**. Temperature may also interact with other stimuli (e.g. photoperiod) to break dormancy. The responses of an ectotherm to temperature are based on the temperatures the organism has experienced in the past. Exposure of an ectotherm to a relatively high temperature for several days can shift its entire temperature response upwards; conversely, several days' exposure to low temperatures can shift its response downwards. This process is known as **acclimation** if these changes are laboratory induced or **acclimatization** if they occur naturally. This process takes time. The variations that occur within species are relatively slight compared with interspecific differences. For example, some mosses and lichens can withstand temperatures of 70°C, while there are bacteria which can live and reproduce at temperatures above 100°C. However, within species there are differences in temperature responses between populations from different locations, usually as a result of genetic differences rather than just acclimatization. In practice, geographic differentiation is usually limited enabling species to be described as having one temperature response.

E3 TEMPERATURE AND SPECIES DISTRIBUTION

Key Notes

Species distribution and temperature	The distribution of the major biomes over the Earth reflects the major temperature zones. However, it is more difficult to attribute a role to temperature when considering species distribution. In certain cases the distribution limits of a species can be attributed to a lethal temperature which precludes the species' existence. However, a more widespread type of relationship is one which has a correspondence between the distribution limits of a species and an isotherm. An isotherm is a line on a map that joins locations having the same mean temperatures. Overall, an organism's limit to distribution is determined not by lethal temperatures, but by conditions that make it a poor competitor.
Evolved response to temperature	The effects of temperature on individuals may be moderated by evolved differences. Allen's rule states that endothermic animals from cold climates tend to have shorter extremities (ears and legs) compared with animals from warmer climates, thus reducing their surface area:volume ratio. This rule has widespread applicability. Bergmann's rule states that mammals tend to be larger in colder areas than warm climates, again to reduce their surface area:volume ratio. This rule does not have widespread interspecific applicability due to other important factors involved in determining body weight, but is often true intraspecifically.
Related topics	Temperature and metabolism (E1) The nature of competition (I1)
	Responses to temperature (E2) Ecosystem patterns (S1)

Species distribution and temperature

It is appropriate to begin this section with a definition of the potential role of temperature in determining the distribution and abundance of organisms. Variations can be defined under seven main headings (*Table 1*).

The distribution of the major biomes over the Earth represent a reflection of the major temperature zones. Similarly, changes in species with increasing altitude are again a reflection of changing temperatures. However, it is more difficult to attribute a role to temperature when considering single species. In certain cases, the distribution limits of a single species can be attributed to a lethal temperature which precludes the species' existence. For example, frost damage is probably the single most important factor in determining plant distribution. However, a more widespread type of relationship is one which has a close, but not ideal correlation between the distribution limits of a species and an isotherm. An **isotherm** is a line on a map that joins places having the same mean temperature at a particular time of the year. Whilst some correlation can usually be found, only limited significance can be attached to the isotherm in terms of a complete explanation. Even very mobile animals such as birds may

Table 1. *Variations in the temperature profile of regions*

Variation	Description
Latitudinal/seasonal	These two variations cannot be separated. The tilting of the Earth to the sun results in generalized temperature zones, with the hottest temperatures occurring in the middle latitudes (38°C) rather than the equator (35°C).
Altitudinal	Superimposed on these broad geographical trends is the influence of altitude. There is a drop of 1°C for every 100 m increase in altitude in dry air, and a drop of 0.6°C in moist air as a result of the adiabatic expansion of air.
Continentality	The effects of continentality are largely attributable to different rates of heating and cooling of land and sea. Land surfaces reflect less heat than water so quickly warm up, whilst also cooling quicker. The sea therefore has a moderating effect on coastal regions.
Microclimate	On a small scale, local variations can give rise to microclimate variations. For example the air temperature in a patch of vegetation can vary by 10°C over a vertical distance of 2.6 m from the soil surface to the top of the canopy.
Depth	Depth, either in soils or water can have two effects on temperature fluctuations. Firstly, the fluctuations are dampened, and secondly, they lag behind surface fluctuations. The strength of these effects increases with depth and decreases with the thermal conductivity of the medium (low in soil, higher in water). A metre below the soil, daily temperature fluctuations are damped out completely.
Diurnal	The daily rhythm of solar radiation causes changes in temperature.

have their distributions closely linked to temperature, as in the case of the eastern phoebe (*Sayornis phoebe*), a migratory bird of eastern and central North America. The wintering population of the eastern phoebe is confined to that part of the US in which the mean minimum January temperature exceeds –4°C. The close correlation of the bird's winter range to this isotherm probably relates to its energy balance. An organism's limit to distribution is determined not by lethal temperatures, but by conditions that make it a poor competitor. Competition is not the only biological interaction that combines with temperature to limit distribution. Many animals have a distribution which correlates with temperature and the occurrence or quality of their food.

Variations in temperature may also be intimately associated with another environmental condition or resource such that the two are inseparable. The most widely used example of this is the relationship between relative humidity and temperature. For aquatic organisms the relationship between temperature and dissolved oxygen concentration is important. The solubility of oxygen decreases with increasing temperature. There exists a close correlation between these environmental factors and species distribution patterns, in which it is impossible to separate the effects of temperature and oxygen concentration.

Evolved response to temperature

As already discussed, the effects of temperature on individuals may be moderated by acclimatization or by evolved differences. These factors also influence distribution and abundance. For example, endothermic animals from cold climates tend to have shorter extremities (ears and limbs) compared with animals from warmer climates (**Allen's rule**). Also, there is a tendency for birds and mammals to be larger in colder areas (**Bergmann's rule**). The explanation in both cases is that endothermic organisms in colder climates should have a smaller surface area relative to volume across which they lose heat. While Allen's rule has widespread applicability, Bergmann's rule is much less universal, probably due to the number of other factors which affect body size, but it is a valuable predictor at the intraspecific level.

F1 SOLAR RADIATION AND PLANTS

Key Notes

Radiant energy and photosynthesis

Radiant energy is the sole energy source that can be used by green plants. When a leaf intercepts radiant energy it may be absorbed, reflected or transmitted. Part of the fraction absorbed reaches the chloroplast, fuelling photosynthesis, the process where radiant energy is used to convert water and CO_2 into sugars. Solar radiation contains a spectrum of different wavelengths. However, only a restricted band of this spectrum is effective for photosynthesis. This is the band of photosynthetically active radiation (PAR) and for green plants lies between 380 and 710 nm.

Efficiency of radiant energy conversion

It is possible to calculate the efficiency of photosynthesis. As a biochemical process photosynthesis is efficient; 35% of usable radiant energy entering a reaction site is converted to potential energy. The actual efficiency at the plant level varies between 0.5% and 3.0%, depending on the plant and the environment.

C3 and C4 plants

A major difference in the photosynthetic capacity of plants is that between C3 and C4 plants. C4 plants are able to capture CO_2 with greater water use efficiency than C3 plants, but this advantage comes at an energy cost. In C4 plants the rate of photosynthesis increases with light intensity, whilst photosynthesis tails off with increasing light intensity in C3 plants.

Changes in the intensity of radiation

Plants rarely achieve their full photosynthetic potential, due to water shortage and to variation in the intensity of radiation. The systematic variations in light intensity are the diurnal and annual rhythms of solar radiation. Less systematic variations in light intensity are caused by the positioning of leaves in relation to each other.

Strategic and tactical response of plants to radiation

A major strategic difference between plant species in their response to the intensity of radiation is exhibited by 'sun species' and 'shade species', which possess a range of adaptations to high and low light levels, respectively. Also, plants may grow leaves which develop differently under different light conditions as part of a tactical response to the light environment. This is most clearly seen in the formation of sun leaves and shade leaves within a leaf canopy of a single plant.

Control of photosynthesis

The leaf stomata are the route for the uptake of CO_2 for photosynthesis. However, if stomata are left open to allow CO_2 to enter the leaf, water will leave the leaf via transpiration. As water is in short supply in most terrestrial ecosystems for at least some of the time, some form of photosynthetic control must be operated. Plants have a number of strategic responses to this dilemma.

Radiant energy and photosynthesis

Solar radiation is the sole energy source that can be used in metabolic activity by green plants. The efficiency with which plants harvest energy is the efficiency with which energy is transformed into the 6-carbon sugar, glucose, by photosynthesis. Because the fitness of each plant is dependent on the efficiency with which it can harvest glucose, plants need to be as efficient as possible at this process. **Radiant energy** from the sun reaches the plant either directly or after it has been diffused by the atmosphere or reflected from other objects. The relative amounts of diffused and direct radiation reaching a plant depends upon a number of factors including the amount of dust present and the thickness of the scattering air layer between the sun and the plant; the most direct radiation is available to plants at low latitudes. When a leaf intercepts radiant energy, it may be absorbed, reflected unchanged or it may be transmitted after modification. Part of the fraction absorbed may reach the chloroplasts and fuel the photosynthetic process. During **photosynthesis** radiant energy is converted into energy-rich carbon compounds. However, if the radiant energy is not trapped by the leaf then this energy is lost. In contrast to important elements such as carbon or nitrogen which can be cycled continuously, radiant energy can only be used once.

Solar radiation contains a spectrum of different wavelengths. *Figure 1* shows the solar spectrum reaching the atmosphere. Around 20% of the incoming radiant energy is degraded in the atmosphere via selective absorption by atmospheric gases (e.g. CO_2 and H_2O). However, the photosynthetic system is only

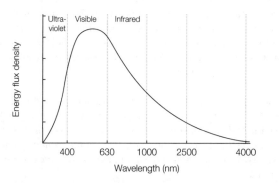

Fig. 1.   The solar spectrum striking the Earth's atmosphere.

able to gain access to a restricted band of this spectrum. Green plants depend on chlorophyll pigments for carbon fixation and these pigments fix radiation between wavelengths 380 nm and 710 nm. This band corresponds to the greatest band of energy flux. This is the band of **photosynthetically active radiation** (PAR). The absorption spectrum of chlorophyll *a* is shown in *Fig. 2*. Only 44% of the incident radiation at the Earth's surface lies within this PAR region. However, pigments produced by other organisms can operate outside the PAR region of green plants. Photosynthetic bacteria produce a pigment, bactochlorophyll, which has its peak absorption between 800 nm and 890 nm.

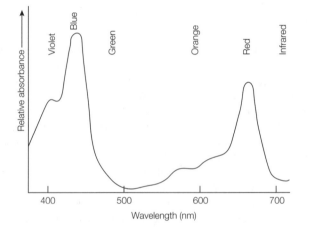

Fig. 2.   Absorption spectrum of chlorophyll a.

Efficiency of radiant energy conversion

It is possible to calculate the **efficiency of photosynthesis**. The total free energy supplied by photosynthesis is given by the chemical equation:

$$6CO_2 + 12H_2O => C_6H_{12}O_6 + 6H_2O + 6O_2$$

(carbon dioxide + water => glucose (ENERGY) + water + oxygen)

This can be compared with the energy put into the system by estimating the numbers of electrons that pass through the pathway during the synthesis of glucose. The number is 24, but each electron is energized twice, firstly in photosystem II and then photosystem I; so a total of 48 photons of light are required per molecule of glucose. The longest wavelength these photons are likely to have is 700 nm (*Fig. 1*), giving a minimum of 172 kJ mol^{-1} per photon, yielding an energy input to all 24 electrons of 8240 kJ. The biochemical efficiency can then be calculated by dividing the increase in free energy of the reaction (2872 kJ) by 8240 kJ, giving an efficiency of 35%. Therefore, as a biochemical process photosynthesis is efficient; 35% of usable sunlight entering a reaction site is converted to potential energy.

C3 and C4 plants

Numerous measurements have been made to measure the actual efficiency with which plants convert solar radiation into potential energy (*Table 1*). **Photosynthetic capacity** can be defined as the rate of photosynthesis when incident radiation is saturated, temperature is optimal, relative humidity is high and atmospheric CO_2 and O_2 concentrations are normal. When leaves of different species are compared under these ideal conditions, those exhibiting

Table 1. A comparison of energy conversion efficiencies of various plants

Plant or ecosystem	Radiant energy converted (kJ m^{-2} d^{-1})	Conversion efficiency (%)
Tropical rain forest	550	3.5
Microscopic algae	300	3.0
Coral reefs	160–630	2.4
Sugar cane	310	1.8
Cornfield	140	1.6
Tropical forest plantation	120	0.7
Open sea	12	0.09
Arctic tundra	8	0.08
Hot deserts	2	0.05

the greatest photosynthetic capacities are from environments where nutrients, light and water are not usually limiting. These include many agricultural crops. Species from environments with limiting conditions (e.g. desert perennials, shade plants) have low photosynthetic capacity even under optimal conditions. A major difference in the photosynthetic capacity of plants is that between **C3** and **C4** plants. C4 plants are able to capture CO_2 with greater water use efficiency than C3 plants, but this advantage comes at an energy cost. C4 plants are more common in tropical and subtropical floras. In C4 plants (e.g. *Zea mays* (corn), *Sorghum vulgare* (sorghum)), the rate of photosynthesis increases with the intensity of PAR (*Fig. 3*). In the more common C3 plants (e.g. *Triticum vulgare* (wheat), *Fagus grandifolia* (beech)), the rate of photosynthesis levels off (*Fig. 3*). Of course environmental conditions rarely reach optimum levels and so plants do not reach their maximum photosynthetic capacity in natural environments. Plants that do approach their photosynthetic capacity are annuals and grasses in deserts after rain, when high intensity of radiation and plentiful supplies of water coincide. The potential photosynthesis of plants may also fall short because periods of radiation are short and temperatures are low.

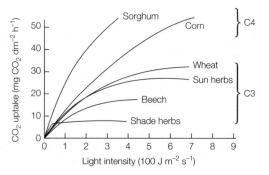

Fig. 3. The response of photosynthesis to light intensity in a range of plants at optimal temperature and natural CO_2 concentrations for both C3 and C4 plants.

Changes in the intensity of radiation

One of the main reasons why plants rarely achieve their photosynthetic capacity is the variation in the intensity of radiation. Plants exist in a world where radiation varies throughout the day and the year, and where the radiation received may be modified by external factors (e.g. shading). The ways in which an organism reacts to a systematic supply of a resource is a reflection of its past evolution and its present physiology. The systematic elements in the variations in light intensity are the **diurnal and annual rhythms of solar radiation**. A plant passes through periods of famine and glut of its light resource every 24 hours (except near the poles) and seasons of famine and glut every year (except in the tropics). This annual cycling of light intensity is in part reflected in the shedding of leaves by deciduous trees in temperate regions. The daily movement of leaves from many species also reflects the changing light intensity and direction of light. When the environmental variation is systematic and repeated, a response pattern is generated by the plant; it is an evolved fixed pattern.

Less systematic variations in light received by a leaf are caused by the positioning of the leaf in relation to others. Each leaf through light interception creates a **resource depletion zone** (RDZ) – a moving area of shadow in which other leaves may lie. These light depletion zones become less well defined deep in a canopy because light loses its original direction through diffusion and reflection. Also, the composition of radiation that has passed through leaves in a canopy is also altered and it becomes less useful because the PAR component will have been reduced. The typical reaction of plants to nonsystematic variations is a degree of plasticity among individual responses.

Strategic and tactical response of plants to radiation

The major **strategic difference** between plant species in their reaction to the intensity of radiation is that which has evolved between plant species that are characteristic of sunny, open habitats (sun species) and those that are characteristic of shaded habitats (shade species). Shaded species of plants use radiation at low intensities more efficiently than sun species but the rate of photosynthesis reaches a plateau at lower intensities (*Fig. 3*). This difference between sun and shaded plants comes about due to differences in the leaf physiology and plant morphology. The leaves of sun plants are usually exposed at an acute angle to the midday sun; this results in an incident beam of radiation spreading over a larger leaf area, thereby reducing radiation intensity. Furthermore, the leaves of sun plants are often arranged as a multi-layer canopy so that in bright sunshine even shaded leaves can make a contribution to plant assimilation rates. In contrast, shaded plants commonly have leaves arranged at a horizontal angle and in single-layered canopies. This variation in strategic responses between plants partly explains the variation in assimilation efficiency discussed earlier.

Along with these strategic responses, during plant growth leaves may develop differently as part of a **tactical response** to the light environment in which the leaf develops. This can most clearly be seen in the formation of 'sun leaves' and 'shade leaves' within the leaf canopy of a single plant. Typically, sun leaves are smaller, thicker and contain more cells, chloroplasts and denser veins leading to an increased dry weight per unit area of leaf. Shade leaves are generally large and more translucent with a low dry weight. These leaves may only have one-fifth of the photosynthetic activity of sun leaves. The production of sun and shade leaves by plants is a response to the environment in which the plant finds itself. A delay is required while the plant senses the leaf environment and then responds.

Control of photosynthesis

The rate at which a leaf photosynthesizes is also dependent on the demands required by other parts of the plant. If there is only a little demand for energy then the plant may reduce the rate of photosynthesis even though conditions are ideal. It must also be remembered that photosynthesis can only be carried out in the presence of incident radiation when water and CO_2 are available.

The prime route for the uptake of CO_2 is through the stomata; but if the stomata are open, water will evaporate through them via **transpiration** and part of the incident radiation will provide the latent heat for this evaporation. If water is lost faster than it can be taken up, the plant will wilt and die. As water is in short supply in most terrestrial ecosystems during some part of the year, this has important implications for plants trying to optimize photosynthetic activity. Plants have a number of strategic responses to this dilemma:

- plants with a short life have high photosynthetic activity during periods of water availability but remain dormant as seeds during the rest of the year (e.g. many desert annuals);
- plants with a long life produce leaves during periods when water is available but shed them during drought (e.g. *Acacia* spp.);
- plants may produce leaves that are long-lived, but transpire slowly, tolerating a water deficit, but only being able to photosynthesize at low rates even when water is abundant;
- C4 plants can reduce the concentration of CO_2 in their intracellular spaces far below that of C3 plants. This results in greater photosynthetic rates with less water loss. However the downside to this response is that C4 plants are inefficient at low light intensities. Most C4 species are therefore found in hot, dry regions or in the tropics;
- amongst the CAM (crassulacean acid metabolism) plants the stomata are open at night and CO_2 is fixed by the conversion of C3 acids into C4 acids. The stomata close during the day and the C4 acids release CO_2, which then enters the normal C3 photosynthetic pathway. This results in water conservation at the expense of photosynthetic potential. Hence CAM plants are found mainly in arid open environments.

The major tactical response to the control of photosynthesis and water loss is through the changes in stomatal conductance that can occur rapidly during the day. The conductance of water by the stomata and the rate of photosynthesis are very highly correlated.

Measurement of photosynthesis
The rate of photosynthesis is a gross measurement of the rate at which a plant is able firstly to capture radiant energy and then secondly to fix it into carbon compounds. For ecologists it is more interesting (and easier) to measure the net gain. **Net assimilation** is the increase in dry matter that results from the difference between photosynthetic assimilation and losses due to respiration and plant death. Net assimilation is negative when plants are in darkness, and increases with the increasing PAR. The **compensation point** is the intensity of PAR at which the gain from photosynthesis exactly balances the losses.

G1 SOURCES AND CYCLES

Key Notes

Nutrient sources	Nutrients are gained and lost by communities in a variety of ways. Mechanical weathering, chemical weathering, particularly carbonation, are important processes. Simple dissolution of minerals in water also makes nutrients available from rock and soil. Water may carry nutrients in watercourses providing an important source in downstream areas. Atmospheric gases provide nutrient sources too, particularly carbon dioxide and nitrogen. Other nutrients from the atmosphere become available to communities as wetfall (rain, snow and fog) and dryfall (settling of particles during dry periods).
Nutrient budgets in terrestrial communities	A particular nutrient atom taken up by an organism may cycle continuously until eventually the nutrient will be lost through any one of a number of processes that remove nutrients from the system. Release may be direct to the atmosphere or via bacterial action (as in the case of methane). For many elements, the most substantial pathway of loss is in streamflow. Other pathways of nutrient loss include fire, the harvesting of crops and deforestation.
Nutrient budgets in aquatic communities	Aquatic systems obtain the bulk of their supply of nutrients from streamflow. In streams, rivers and lakes with a stream outflow, export in outgoing stream water is a major factor. Commonly, phases of fast inorganic nutrient displacement alternate with periods when the nutrient is locked in biomass. In lakes, plankton play a key role in nutrient cycling. Oceans contain warm suface waters, where most plant life is found and cold deep waters (which make up 90% of the total water volume). Nutrients in the surface waters come from two sources: (i) upwellings from deep water (which comprise over 95% of the nutrient budget), and (ii) river input.
Geochemistry	The pools of chemical elements on earth exist in various compartments: in rocks (the lithosphere), and soil water, streams, lakes or oceans (which, combined, constitute the hydrosphere). In all these cases, the chemical elements exist in the inorganic form. In contrast, living organisms and dead and decaying organic matter are compartments which contain elements in the organic form. Studies of the chemical processes occurring within these compartments and the flux of elements between them (which are fundamentally altered and affected by biotic processes) is termed biogeochemistry.
Global biogeochemical cycles	Terrestrial plants utilize atmospheric CO_2 as their carbon source for photosynthesis, whereas aquatic plants use dissolved carbonates (hydrosphere carbon). Respiration releases the carbon locked in photosynthetic products back to the atmospheric and hydrospheric carbon compartments. The atmospheric phase is predominant in the global nitrogen

cycle, in which nitrogen fixation and denitrification by microorganisms are of particular importance. The main stocks of phosphorus occur in soil water, rivers, lakes, rocks and and ocean sediments, whilst sulfur has both atmospheric and lithospheric components.

Related topics	Plants and consumers (G2)	Primary and secondary production
	Soil formation, properties and	(P2)
	classification (G3)	

Nutrient sources   Nutrients are gained and lost by all communities in a variety of different ways. Nutrient budgets can be produced if we can calculate gains and losses for nutrients. In some communities these budgets may be more or less in balance, while in others inputs exceed outputs leading to the accumulation in the compartments of living biomass and dead organic matter. This is obvious during community succession. Also, outputs may exceed inputs if the biota is disturbed by an event such as fire, deforestation or crop harvest. Weathering of parent bedrock and soil is the dominant source of nutrients such as calcium, phosphorus and potassium, which may then be taken up by plants. **Mechanical weathering** is caused by processes such as freezing of water and growth of roots. **Chemical weathering**, however, is the most important process, particularly carbonation, in which carbonic acid (H_2CO_3) reacts with minerals to release ions such as potassium and calcium. Simple dissolution of minerals in water also makes nutrients available from rock and soil.

Atmospheric carbon dioxide is the source of carbon of terrestrial communities. Similarly, gaseous nitrogen from the atmosphere provides most of the nitrogen content of communities. Several types of bacteria and blue-green algae possess the enzyme nitrogenase and convert N_2 to NH_4^+, which can then be taken up by plant roots. Communities containing plants such as legumes and alder trees (*Alnus* spp.), with their root nodules containing symbiotic nitrogen-fixing bacteria, may receive a substantial proportion of their nitrogen in this way (100–300 kg ha^{-1} year^{-1}). Other nutrients from the atmosphere become available to communities as **wetfall** (rain, snow and fog) and **dryfall** (settling of particles during dry periods). Rain contains a number of chemicals from a variety of sources:

● solution of trace gases such as sulfur and nitrogen oxides;
● solution of aerosols containing particles rich in sodium, sulfate, magnesium and chloride;
● dust particles from fires, storms rich in calcium, sulfate and phosphate.

The constituents of rainfall that serve as nuclei for raindrop formation make up the **rainout component**, whereas other constituents, both particulate and gaseous, are collected from the atmosphere as rain falls (the **washout component**). Nutrients dissolved in precipitation become available to plants when the water enters the soil and reaches plant roots, although some absorption by leaves is also involved. **Dryfall** is an important process in communities with long dry seasons, and can account for up to 50% of the atmospheric input of sulfate, nitrate, calcium and potassium. Finally, in some cases **streamflow** can provide a significant input of nutrients when material is deposited in floodplains.

Nutrient budgets in terrestrial communities
A particular nutrient atom taken up by a plant may then be eaten by a herbivore which dies and decomposes, releasing the nutrient atom back into the soil from where it can be taken up once again by plant roots. This cycle may continue until eventually the nutrient will be lost through any one of a number of processes that remove nutrients from the system.

Release to the atmosphere is one pathway of nutrient loss. In many communities there is an approximate balance between the carbon released to the atmosphere from respiration (CO_2) and carbon fixed by photosynthesis. Other gases, such as methane and hydrogen sulfide, are released by the action of anaerobic bacteria. The production of these gases is greatest in water-logged soils and floodplain forests. Also, in these locations some bacteria are capable of reducing nitrates and nitrites to gaseous nitrogen during the process of **denitrification**. Plants may also be sources of gaseous release. Tropical forest trees appear to emit aerosols containing phosphorus, potassium and sulfur. Ammonia gas is released during the breakdown of vertebrate excreta and has been shown to be an important loss of nitrogen.

For many elements, the most substantial pathway of loss is in **streamflow**. The water that drains from the soil into a stream carries many nutrients. With the exception of iron and phosphorus which are not mobile in soils, the loss of nutrients is predominantly in solution. Particulate matter in streamflow occurs as dead organic matter and as inorganic particulates. Total loss of nutrients is greatest in years with high rainfall. If bedrock is permeable, losses occur not only in streamflow but also in the water that drains into groundwater. These values are difficult to quantify as groundwater may discharge into a water body some distance away from the community.

Other pathways of nutrient loss include fire, which can turn a large proportion (approximately 40%) of a community's organic carbon and nitrogen pool into gaseous carbon dioxide and nitrogen which are then lost to the atmosphere. Similar losses of nutrients occur during harvesting of crops or through deforestation.

Nutrient budgets in aquatic commmunities
Aquatic systems obtain most of their nutrients through streamflow. In streams, river communities and lakes with a stream outflow, export in outgoing stream water is an important factor. In lakes without an outflow and in oceans, nutrient accumulation in permanent sediments is often the major export pathway.

In streams and rivers only a small fraction of available nutrients take part in biological interactions. Bacteria, fungi and algae growing on the substratum of the stream bed are responsible for the uptake of inorganic nutrients. Nutrients are then passed through the food web via invertebrates that graze on the microbes. Ultimately, decomposition of the biota releases inorganic nutrient molecules.

In lakes, plankton plays a key role in nutrient cycling. In a large temperate lake on a warm sunny day the major nutrient fluxes will be:

- uptake of dissolved nutrients by phytoplankton;
- the loss of grazing by zooplankton;
- recycling to the water column through excretion by plankton and by the decomposition of dead micro-organisms.

Under these conditions there is relatively little loss to the sediment and most nutrients in the lake cycle between the dissolved nutrient compartment and the biota. In contrast, in a small lake with a relatively large throughput of water, nutrient cycling will be small compared to the throughput of nutrients to the outflowing stream. The return of nutrients from the sediment can be a major contributor to nutrients in the water column in shallow lakes in the summer when high rates of decomposition at the sediment surface release nutrients into the overlying water. Many lakes in arid regions lacking a stream outflow lose water only through evaporation. The nutrients present in waters of these **endorheic** lakes (internal flow) are thus more concentrated than their freshwater counterparts, being particularly rich in sodium and phosphorus. They are usually fertile, having dense populations of blue-green algae (e.g. *Spirulina platensis*). The largest of all the endorheic lakes is the ocean. Because of its size, oceans tend to have a constant chemical composition. Geochemists who study the oceans divide it into two compartments – the warm suface waters, where most plant life is found and the cold deep waters (making up 90% of the total water). Nutrients in the surface waters come from two sources – river input and upwellings from deep water. About 30 times more water arrives at the surface via upwellings than from rivers. Nutrients arriving at the surface are taken up by phytoplankton or bacteria and are passed along the food chain. Detritus particles are constantly sinking to deep waters where most are decomposed, releasing nutrients which eventually find their way back to the surface by an upwelling. This process of nutrient cycling occurs over vast distances – many of the nutrients arriving as upwelling water in the Antarctic are probably derived from oceans in the Northern Hemisphere.

Geochemistry

The pools of chemical elements available exist as pools in various compartments. Some compartments occur in the atmosphere (e.g. nitrogen in gaseous form, carbon as carbon dioxide). Others occur in rocks of the lithosphere (e.g. potassium in feldspar, calcium as calcium carbonate), while some are in soil water, streams, lakes or oceans – the hydrosphere (e.g. carbon in carbonic acid, phosphorus as phosphates, nitrogen as nitrate). In all these cases, the chemical elements exist in the inorganic form. In contrast, living organisms and dead and decaying organic matter are compartments which contain elements in the organic form (e.g. nitrogen in protein, phosphorus in ATP, carbon in starch). Studies of the chemical processes occurring within these compartments and the flux of elements between them is termed **biogeochemistry**.

Many geochemical fluxes can occur in the absence of life because all geological formations above sea-level are eroding and degrading. Volcanoes release sulfur into the atmosphere irrespective of the presence of organisms. Alternatively, organisms alter the rate of flux and the differential flux of the elements by extracting and recycling some chemicals from the underlying geochemical flow.

The flux of matter can be investigated at a variety of levels. Studies can be carried out at the community level, examining a local pool of chemicals; other investigators are interested in the global scale, considering atmospheric and oceanic fluxes of chemicals.

Global biogeochemical cycles

Nutrients can be moved over vast areas by the atmosphere or by water currents. There are no boundaries to this movement. It is therefore appropriate to examine **global biogeochemical cycles**. *Figure 1* illustrates the major global pathways of nutrients between inorganic and organic materials and between biotic and abiotic reservoirs. The biota of a habitat obtains some of its nutrient elements

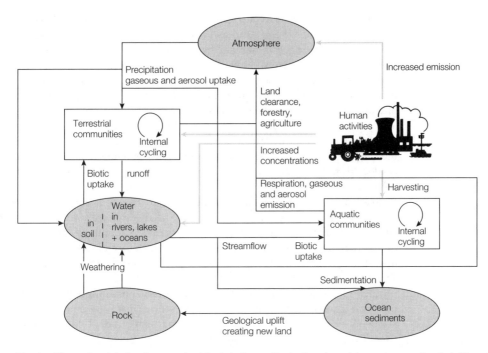

Fig. 1.   The major global pathways of nutrients between the hydrosphere (atmosphere and water), lithosphere (rocks and sediments) and the biotic reservoirs (terrestrial and aquatic communities). Reprinted from Ecology, 2nd edn, Begon et al., 1990, with permission from Blackwell Science.

(e.g. phosphorus) from the weathering of rock. Other nutrients such as carbon and nitrogen are derived mainly from the atmosphere. Sulfur is derived from both atmospheric and lithospheric sources. Let us look briefly at the cycling of four important elements, carbon, nitrogen, sulfur and phosphorus.

Photosynthesis and respiration are the two opposing processes that drive the global **carbon cycle**, with CO_2 as the main vehicle of flux between atmosphere, hydrosphere and biota. Historically, the lithosphere plays only a minor role; fossil fuels lay as dormant reservoirs of carbon until man's intervention.

Terrestrial plants utilize atmospheric CO_2 as their carbon source for photosynthesis, whereas aquatic plants use dissolved carbonates (**hydrosphere carbon**). The two subcycles are linked by CO_2 exchanges between the atmosphere and oceans as follows:

$$\text{atmospheric } CO_2 \Leftrightarrow \text{dissolved } CO_2$$

$$CO_2 + H_2O \Leftrightarrow H_2CO_3 \text{ (carbonic acid)}$$

In addition, carbon is found in inland waters and oceans as bicarbonate resulting from the **weathering** (carbonation) of calcium-rich rocks such as limestone and chalk:

$$CO_2 + H_2O + CaCO_3 \Leftrightarrow CaH_2(CO_3)_2$$

Respiration by plants, animals and microorganisms releases the carbon locked in photosynthetic products back into the atmospheric and hydrospheric carbon compartments.

The atmospheric phase is predominant in the global **nitrogen cycle**, in which nitrogen fixation and denitrification by microorganisms are of particular importance. Atmospheric nitrogen is also fixed by lightning discharges during storms and reaches the ground as nitric acid dissolved in rain water, but only 3–4% of fixed nitrogen is derived from this pathway. The amount of nitrogen fluxed in streamflow from terrestrial to aquatic environments is relatively small, but it is important for aquatic systems. This is because nitrogen is one of the two elements (along with phosphorus) that most often limit plant growth. Finally, there is a small annual loss of nitrogen to ocean sediments.

The main stocks of **phosphorus** occur in soil water, rivers, lakes and rocks along with ocean sediments. The phosphorus cycle can be described as a **sedimentary** (or open) cycle because of the tendency for mineral phosphorus to be carried from the land to the oceans where it becomes incorporated into the sediment. Sedimentation removes about 1.3×10^7 tons of phosphorus per year from ocean water. A typical phosphorus atom released from rock by weathering may enter and cycle within the terrestrial community for years before it enters a stream via groundwater. It is then carried, spiralling to the ocean where, on average, it makes 100 trips between the surface and deep water (through upwellings), each lasting up to 1000 years before it settles on the sediment surface in particulate form.

In the global phosphorus cycle the **lithospheric** phase predominates, whereas the nitrogen cycle has a dominant **atmospheric** phase. By contrast, **sulfur** has atmospheric and lithospheric phases of similar magnitude. Three natural biogeochemical processes release sulfur to the atmosphere:

● the formation of sea spray aerosols (4.4×10^6 tons year^{-1});
● volcanic activity (relatively minor);
● anaerobic respiration by sulfate-reducing bacteria (33–230 tons year^{-1}).

Sulfate-reducing bacteria release reduced compounds such as H_2S from waterlogged communities such as marshes, and from marine communities associated with tidal flats. A reverse flow from the atmosphere involves oxidation of sulfur compounds to sulfate, which returns to earth as both wetfall and dryfall. About 2.1×10^7 tons return per year to land and 1.9×10^7 tons to the ocean. The weathering of rocks provides about half the sulfur draining off land into rivers and lakes, the remainder being derived from atmospheric sources. On its way to the ocean, a proportion of the available sulfur is taken up by plants, passed along by food chains and via decomposition processes, and becomes available to plants again. However, in comparison to phosphorus and nitrogen, a much smaller fraction of the flux of sulfur is involved in internal recycling in terrestrial and aquatic communities. Finally, there is a continuous loss of sulfur to ocean sediments, mainly through abiotic sediments, such as the conversion of H_2S by reaction with iron to ferrous sulfide (giving marine and freshwater sediments their black color).

G2 PLANTS AND CONSUMERS

Key Notes

The fate of matter in the community
The main elemental component of living matter is carbon. Carbon enters the trophic structure of a community when CO_2 is fixed through photosynthesis, the utilization of the energy of sunlight to combine CO_2 and water into sugars. It then becomes incorporated in net primary production (NPP). This is the total energy accumulated per unit time by plants during photosynthesis. When the high-energy molecule in which the carbon was stored is finally used to provide energy for work, the carbon is released into the atmosphere as CO_2.

Producers
Autotrophs, the main producers in most terrestrial ecosystems, use energy from the sun or from the oxidation of inorganic substrates to produce organic molecules from inorganic ones. Plants are the main producers in terrestrial systems, while phytoplankton are the most important in open oceans. In these deep waters, photosynthetic production occurs in the upper levels, the photic zone, where light can penetrate. In deeper waters, light does not penetrate and no photosynthesis occurs in this aphotic zone.

Consumers
Plants are eaten by primary consumers, such as grazing mammals and insects. In oceans, phytoplankton are consumed by zooplankton. These primary producers are in turn eaten by secondary consumers, carnivores such as mammals and spiders. In oceans, many fish are secondary consumers as they feed on zooplankton.

Decomposers
The organic material that composes living organisms in an ecosystem is eventually recycled, broken down and returned to the abiotic environment in forms that can be used by plants. Decomposers, which feed on nonliving organic material are the organisms in this recycling. The most important decomposers are bacteria and fungi.

Related topics
Sources and cycles (G1)
Primary and secondary production (P2)

Food chains (P3)

The fate of matter in the community

The major elemental component of all living matter is carbon, contributing more than 95% to the composition of matter. Carbon compounds are involved in the accumulation and storage of energy. This energy is eventually dissipated when the carbon compound is oxidized to carbon dioxide through metabolism. Carbon enters the trophic structure of a community when CO_2 is fixed through **photosynthesis**. Photosynthesis is the utilization of the energy of sunlight to combine CO_2 and water into sugars. It then becomes incorporated in **net primary production** (NPP). This is the rate of total energy accumulated per unit time by plants during photosynthesis. The carbon is then available for consumption as part of a carbohydrate, a protein or a fat. It then undergoes a succession of cycles, being consumed, assimilated and defecated. When the high-energy molecule in which the carbon was stored is finally used to provide energy for work, although the energy is lost from the system as heat the carbon is released into the atmosphere as CO_2. This molecule may then once again be used in photosynthesis. Carbon and all other nutrients (e.g. potassium, phosphorus, nitrogen) are available to plants as simple organic molecules or ions in the atmosphere, or as dissolved ions (potassium, phosphate, nitrates). Each can be incorporated in complex carbon compounds in biomass. When the C compounds are metabolized to CO_2 the mineral nutrients are released again in a simple inorganic form. Again, these may pass repeatedly through these cycles. *Figure 1* summarizes this cycle and highlights the interdependence of producers, consumers and decomposers.

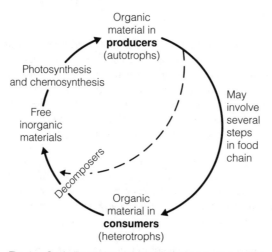

Fig. 1.   Cycle illustrating the interdependence of producers and consumers in an ecosystem. The photosynthetic and chemosynthetic activities of the producers (green plants, etc.) produce organic materials which are fed on by the consumers. The saprotrophic activities of the decomposers (bacteria, etc.) free inorganic materials from the dead bodies of the producers and the consumers, thereby ensuring a continual supply of raw materials for the producers. From Biology: A Functional Approach, 4th edn, Roberts, 1986. With permission from Thomas Nelson and Sons.

Producers

The main producers in most terrestrial ecosystems are plants. Plants are **autotrophs**; autotrophs use energy from the sun or from the oxidation of inorganic substrates to make organic molecules from inorganic ones. In streams, much of the organic matter used by consumers is also supplied by terrestrial plants, entering the ecosystem as debris that falls into the water or is washed in by runoff. In open oceans, phytoplankton are the most important autotrophs, whereas algae and aquatic plants are more important producers in shallow and near-shore waters of both freshwater and marine ecosystems. In the deep waters of oceans light will not penetrate and no photosynthetic production can occur, in this **aphotic zone**. However, life here depends on photosynthetic production in the **photic zone**, where photosynthesis can occur and nutrients and energy from this zone fall into the aphotic zone in the form of detritus.

Consumers

The **primary consumers**, or herbivores, are organisms of an ecosystem that eat plants or algae. On land this group are mostly insects, snails and certain vertebrates including grazing mammals and birds. In aquatic systems, phytoplankton are consumed mainly by zooplankton, various small invertebrates and some fish. Examples of **secondary consumers**, carnivores feeding on herbivores, in terrestrial ecosystems include spiders, frogs, insect-eating birds, carnivorous mammals and animal parasites. In aquatic habitats, many fish feed on zooplankton and then are in turn preyed upon by other fish.

Decomposers

The organic material that composes living organisms in an ecosystem is eventually broken down, recycled and then returned to the abiotic environment in forms that can be used by plants. **Decomposers**, which feed on nonliving organic material, are the organisms in this recycling. The most important decomposers are bacteria and fungi, which first secrete enzymes that digest organic material and then absorb the breakdown products. Earthworms, crayfish and cockroaches are also decomposers, but these animals digest organic material internally after ingesting it. Decomposers often form a major link between primary producers and secondary consumers in an ecosystem. For example, an earthworm may feed on plant detritus in soil and then be eaten by a bird.

G3 SOIL FORMATION, PROPERTIES AND CLASSIFICATION

Key Notes

| Soil formation | Underlying terrestrial ecosystems is the soil; a thin layer of the Earth's crust that has been remade by life and weather. Soil provides habitats with a nutrient delivery system, a recycling system and a waste-disposal system. The study of soils is called pedology. Soil organisms participate in the formation of the habitat and constitute one of the five interactive factors in soil formation; the other four are climate, topography, parent material and time. The initial colonizers of soil parent material are usually the cyanobacteria, capable of photosynthesis and nitrogen fixation. After higher vegetation has become established, a variety of soil processes produces the dynamic mixture of living and dead cells, soil organic matter (SOM). |

| The soil profile | At the surface of soil is a litter of dead or rotting plant parts, with one or more distinctly different layers underneath. These layers have come to be called 'horizons'. Soil horizons form as rotting plant parts mix with the upper layers of the mineral soil. The subsoil underneath is the earth from which the soil was made and is called the parent material. This layered appearance is called the soil profile. Soil scientists recognize three kinds of soil horizon above the parent material; the A, B and C horizons. Under these three groups of horizons is the unaltered parent material. These are known as R horizons. The soil profile is an instant indicator of important ecosystem processes. |

| Primary classification: the great soil groups | On a world scale, the great soil groups are the most easily mapped of any class in a soil classification. Color and banding of the soil profile distinguish the great soil groups. World soil maps represent a rough plot of the great soil groups and are very similar to maps of climate or vegetation. Soil surveyors map soils on the scale of farm fields or parts of counties. Usually only one of the great soil groups will be present. The units used, 'soil series', are a group of soils developed from the same kind of parent material, by the same generic combination of processes, and whose horizons are quite similar in their arrangement and general characteristics. |

Related topics   Succession (R1)   Ecosystem patterns (S1)

Soil formation   Underlying all terrestrial ecosystems is a thin layer of the Earth's crust that has been remade by life and weather – the soil. Soil consists of mineral particles of various sizes, shapes, and chemical characteristics, together with plant roots, the living soil population, and an organic matter component in various stages of decomposition. Soil provides habitats with a nutrient delivery system, a recycling system and a waste-disposal system. For plants, soil is a site of germination, support and decay as well as being a site for the storage of water and nutrients. For animals and decomposers, soil is a refuge, and for animals also a sewer. The study of soil is called **pedology** (from the latin word meaning 'foot'). Soil microorganisms participate in the formation of the habitat wherein they live. They, together with other soil biota, especially the higher vegetation, constitute one of the five interactive factors in soil formation; the other four are climate, topography, parent material and time. The physical and chemical breakdown of rocks to fine particles with large surface areas and the accompanying release of plant nutrients initiate the soil-forming process. Two major nutrients that are deficient in the early stages of the process are carbon and nitrogen; therefore, the initial colonizers of soil parent material are usually the cyanobacteria, capable of photosynthesis and nitrogen fixation. After higher vegetation has become established, a variety of soil processes produce the dynamic mixture of living and dead cells, soil organic matter (SOM), and mineral particles in sufficiently small sizes to permit the colloidal interactions characteristic of soil. Soil gases, water, and dissolved minerals complete the soil habitat. Soils can vary considerably in different areas. Some soils drain, some are well supplied with nutrients. Typical soil in the temperate belt, where western agriculture originated, is grayish-brown. Where the natural vegetation is deciduous forest, such as in northern France, the soil can be brown when wet, turning grey when dry. But in large parts of the tropics the soil is red, wet or dry. Gray-brown soils tend to be associated with temperate forests and red soils with tropical rain forests or savannas. In soil found under much of the boreal forest, the soil is ash white. Russian peasants call such land 'ash soil' (podzol).

The soil profile

The side of a trench in any well-vegetated soil reveals a succession of layers in the soil. At the surface is a litter of dead or rotting plant parts, with one or more distinctly different layers underneath separating the surface litter from the subsoil a few feet down. These layers have been formed by weathering processes working down from the surface. It is convenient to describe them, and they have come to be called 'horizons'. Soil horizons form as rotting plant parts mix with the upper layers of the mineral soil and drainage water percolates down through the litter slowly washing the lower horizons. The thickness of earth affected by these processes constitutes the soil. The subsoil underneath is the earth from which the soil was made and is therefore called the **parent material**. This layered appearance is called the soil profile which is the set of **soil horizons** between the undifferentiated parent material and the surface litter. Soil scientists recognize three kinds of soil horizon above the parent material: the A, B and C horizons. It is usually easier to separate a soil into these three horizons just by looking at it. The principles of this simple classification are that:

- A horizons have lost material from leaching, although they have gained organic matter as deposits;
- B horizons have gained material from leaching and by synthesis *in situ*, particularly of clay minerals;
- C horizons are parent material that has been weathered, usually being oxidized or, in dry climates, having deposits of evaporites.

At the top of the soil profile (*Fig. 1*), underneath the litter of leaves, the mineral soil is colored and structured by the organic particles mixed into it by soil animals or roots and by the presence of various organic materials produced by decomposition, called '**humic acids**'. The percolating water dissolves anything

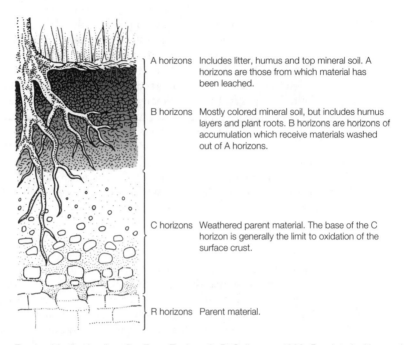

A horizons Includes litter, humus and top mineral soil. A horizons are those from which material has been leached.

B horizons Mostly colored mineral soil, but includes humus layers and plant roots. B horizons are horizons of accumulation which receive materials washed out of A horizons.

C horizons Weathered parent material. The base of the C horizon is generally the limit to oxidation of the surface crust.

R horizons Parent material.

Fig. 1. Idealized soil profile. From Ecology 2, *P. Colinvaux, 1993. Reprinted with permission from John Wiley & Sons, Inc.*

soluble in the surface of the mineral soil and carries it down to deeper horizons from the A horizon. Below the A horizons are a second set of horizons, the B horizons, which contain material washed down from the top layers that has been caught and redeposited. C horizons contain weathered parent material. Under these three groups of horizons is the unaltered plant material. These are known as R horizons. The parent material of an R horizon can be any geological formation: bedrock, gravel, sand, clay etc. The thickness and complexity of a soil profile is a function of time and weather, as well as the plants that grow there. In the desert the soil profile may be just a thin surface layer. The soil profile is therefore an instant indicator of important ecosystem processes.

Primary classification: the great soil groups

On a world scale the great soil groups are the most readily mapped of any type of soil classification system. Color and banding of the various soil profiles distinguish the **great soil groups**. The Russians were the first to classify soils and, as a consequence, the standard names for soil groups were Russian. However, in the 1960s the US Soil Conservation staff developed an alternative system in which the great soil groups were replaced by a different unit, the soil order, with the Russian names replaced with new Greek- and Latin-derived names. *Table 1* lists the soil orders recognized by the US Soil Survey Manual and gives the corresponding great soil group. In the following section we will examine briefly the characteristics of the most important major soil groups.

Under boreal forests, the soil profile of podzols (spodzols) are richly banded. At the top is a carpet of brown needles (the A0 horizon). Beneath them the needles are black and slimy, being transformed into humus. Then comes a horizon of mineral soil stained dark with humus (the A1 horizon). Beneath this layer is a pale or even white layer, the A2 horizon. The B horizons are as richly

Table 1. Soil orders recognized by the US Soil Survey Manual

Soil order	Derivation	Classic great soil group
Entisol		Azonal (simple profile)
Inceptisol	Beginning	Brown forest soils
Aridisol	Dry	Desert soils
Mollisol	Soft	Chernozems and prairie soils
Spodizol	Wood ash	Podzols
Alfisol		Gray-brown podzolic soils
Ultisol	Last	Red-yellow podzolic soils
Oxisol	Oxide	Latosols and laterites
Histosol	Tissue	Bog soils

banded, with red layers of ferric oxide, and another stained blue-black with organic matter. No other soils are as strikingly banded as podzols.

The profile of a tundra entisol soil is permanently frozen (permafrost), and a thick layer of dead organic matter insulates the frozen mineral soil from the sun. Soil under tundra vegetation will not thaw down to more than a few centimeters within the mineral parent soil, preventing any deep development of soil profiles.

The gray-brown podzolic soils (alfisols) found in deciduous forest biomes are the soils that underlie the brown fields of classical agriculture in western Europe. At the top is mulchy leaf litter, rotting underneath into humus that is mixed with the surface mineral soil. This mixture may span 10–20 cm of dark, fertile-looking earth. The absence of any bleached A2 horizon is certainly related to the higher pH of soil under broad leaves than coniferous needles, as well as to the presence of earthworms. The horizons grade into each other with subtle changes of hues: brown litter, the deep brown B horizon of mineral accumulation and then to parent material.

Laterites (oxisols) are the thick, red soils characteristic of tropical forest biomes. These soils have been enriched with oxides and hydrates of iron and aluminium, giving the soil the characteristic red color. A laterite formation can be tens of meters in thickness. The profile consists of a few centimeters of leaf litter (A0), an A horizon identified by a browning of the humus. It is difficult to separate A and B horizons, these soils have very little organic matter. They are almost ubiquitous in the tropics.

Under grasslands such as the Russian steppe lie mollisols and desert soils. Here, rainfall is so limited that water rarely percolates all the way down through the soil. Dead grass decomposes slowly so that a thick organic layer builds up at the top of the soil, forming the famous black earth of the wheat lands. This peaty layer grades into mineral soil. A white mineral band often marks the level where the evaporating water has left its load of dissolved matter. Below this is the parent material. A prairie soil is the result of this process in the wetter grassland climates, where higher production leads to greater amounts of organic matter. The variant called a chernozem (Russian for 'black earth') is found in slightly drier climates. Where rainfall is only very slight (hot deserts), a B horizon can scarcely be recognized and a simple desert soil (aridisol) has only A and C horizons.

World soil maps represent a rough plot of the great soil groups and are very similar to maps of climate or vegetation. This is reasonable as both soil and vegetation are influenced by climate. However, a true map of the great soil groups would not match boundaries with plants very closely. Practical soil surveyors map soils on the scale of farm fields or parts of counties, where usually only one of the great soil groups will be present. These surveys may include a **soil series**, which is a group of soils developed from the same kind of parent material, by the same generic combination of processes, and whose horizons are quite similar in their arrangement and general characteristics.

H1 POPULATIONS AND POPULATION STRUCTURE

Key Notes

Population	A population is a group of organisms of the same species which occupies a given area. The boundaries between populations can be arbitrary. Populations may be categorized as consisting of either unitary or modular organisms. In unitary populations, each zygote gives rise to a single individual. In modular organisms, the zygote develops into a unit of construction which gives rise to further modules and a branching structure. The structure may then fragment producing many individual ramets.
Population size	The population size for unitary organisms, such as mammals, is simply the number of individuals in a given area. For modular organisms, such as plants and corals, the situation is more complex. In this case the number of 'pieces' (ramets) or the number of shoots (modules) may give a more meaningful indication of abundance than the number of different individuals.
Age and stage structure	The age structure of a population is the number of individuals in each age class expressed as a ratio, and is usually displayed as an age pyramid diagram. A population which is neither expanding nor contracting will have a stationary age distribution. A growing population will have more young, while a declining population will be dominated by older age classes. Where organisms pass through discrete growth stages (e.g. insect larval instars), the number of individuals at each stage (the 'stage structure') may provide a useful description of the population. In species where growth rates are indeterminant (such as plants), size classes may be more informative.
Related topics	Natality, mortality and population growth (H2) The nature of competition (I1) Density and density dependence (H3) The nature of predation (J1) Population dynamics – fluctuations, cycles and chaos (H4)

Population

A population is a group of organisms of the same species occupying a given space at the same time. Sometimes the population is clearly defined, as in sticklebacks in a pond or mice on an island. More often population boundaries are drawn by the ecologists and depend on the question under investigation. Populations may consist of either **unitary** or **modular organisms**. In populations of **unitary** organisms, each individual is produced directly from a zygote and the form and development of individuals is highly predictable. Mammals, birds, amphibians and insects are all examples of unitary organisms. In contrast, in populations consisting of **modular** organisms the zygote develops into a unit of construction, or module, which then gives rise to further modules to form a branching structure. The form and timing of development are not predictable. Most plants, sponges, hydroids and corals are modular.

Higher plants grow by accumulating modules that usually consist of one leaf, a bud and a piece of stem. Flowers are a type of module. Some modular organisms such as trees and sea fans concentrate on vertical growth, while rhizomatous grasses and encrusting sponges spread laterally over a substrate.

Connections between the parts of spreading modular organisms may die and rot away leaving individuals that are separated but genetically identical and derived from a single zygote; these are known as **ramets** (*Fig. 1*). Floating aquatic plants such as duckweed literally fall apart into ramets as they grow. In some cases a single **clone** (or **genet**) may extend over large areas, as has been recorded in bracken fern (*Pteridium aquilinum*) in Finland where one clone, comprising many thousands of ramets, covers nearly 14 hectares and is estimated to be 1400 years old.

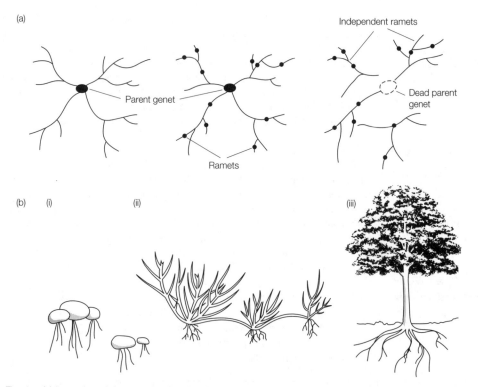

Fig. 1.   (a) Lateral modular growth, showing the original genet and its ramets. Because all ramets are genetically identical, the ramets collectively make up the genet. (b) Examples of modular organisms: (i) duckweed, which separates into individual ramets as it grows; (ii) grasses, ramets attached by stolons or rhizomes; and (iii) trees, which grow by accumulating persistent modules.

Population size   It is easy to define the population size when considering unitary organisms – it is simply the number of individuals present in a given area. However, it is rather more difficult when we consider modular organisms. If we are interested in the number of **evolutionary individuals**, then we should consider the number of clones. On the other hand the **immediate ecological impact** of a population might be better assessed by the number of ramets.

Age and stage structure

The age structure describes the number of individuals in each age class as a ratio of one class to another. Age classes can be specific categories, such as years or months, or life history stages, such as eggs, larvae, pupae and instars.

A **stable age distribution** results where the ratio of one age group to the next remains the same and the shape of the **age pyramid** does not change over time. The shape stays constant because the birth and death rates for each age class are constant.

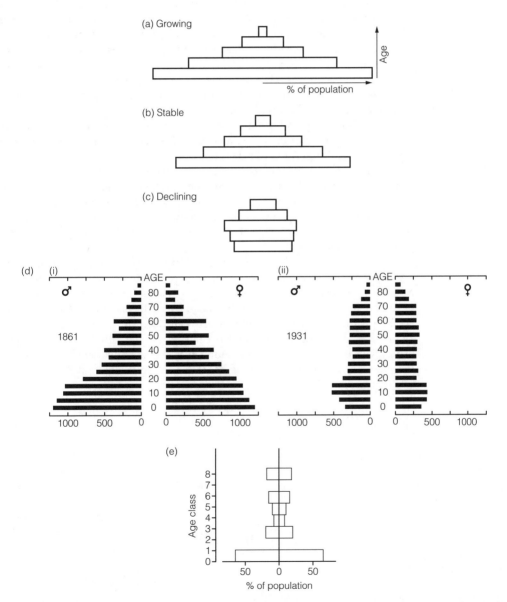

Fig. 2.   Age pyramids. (a), (b) and (c) Typical age pyramids of populations in different stages of growth (see text). (d) Age pyramids from the human population on the Isle of Skye, Scotland over a 70-year period when the population changed from stable to declining (from Darling, West Highland Survey, 1955, by permission of Oxford University Press). (e) An age pyramid for the cactus ground finch (Geospiza scandens) in 1984 on Isla Daphne major in the Galapagos, illustrating annual variation in recruitment (see text).

A population that is neither increasing nor decreasing has a **stationary age distribution**, the age structure is stable and the population size constant (*Fig. 2b*). A population can be increasing or decreasing and still have a stable age structure. Growing populations are characterized by a large number of young, giving the pyramid a broader base (*Fig. 2a*). Populations with a high proportion of individuals in the older age groups are aging and usually declining in numbers (*Fig. 2c*).

If the stable age distribution is disrupted by any temporary event, such as a harsh winter, disease or starvation, the age composition will gradually restore itself when conditions and the birth and death rates return to normal.

Age distributions provide a valuable insight into the history of a population. For example, the age structure of the cactus ground finch, *Geospiza scandens*, found in the Galapagos, clearly shows the years where rainfall was low and reproduction failed. In contrast, heavy rainfall in 1983 is apparent from a very large year class 1 in 1984 (*Fig. 2d*).

Many organisms pass through discrete growth stages, such as larval instars in insects. The number of individuals at each stage, the **stage structure**, may provide a useful description of the population.

Among many plant species, age-class structure provides only a limited picture. As growth rates are **indeterminant**, not tightly linked to age (unlike in most mammals, for example) some plants will achieve a greater size than others of the same age and species. **Size classes**, such as mass, area covered or (in trees) diameter at breast height (DBH) may be more useful ecological measures than age in these cases.

The body of a plant, as a modular organism, also has an age structure and is composed of young and developing, actively functioning and senescent parts. Moreover, leaves, shoots and roots change in activity as they age. For an ecologist to treat all leaves as equal would be to ignore the fact that other organisms such as herbivores will discriminate between them. In the sand sedge, *Carex arenaria*, the age structure of shoots is changed dramatically by the application of fertilizer even though the total number of shoots is unaltered. The population becomes dominated by young shoots and older shoots die.

H2 NATALITY, MORTALITY AND POPULATION GROWTH

Key Notes

Natality	Natality is the birth of new individuals. The realized natality is the actual successful reproduction per female over a period of time. The age-specific birthrate is the number of offspring produced per unit time by females in specific age classes.
Mortality	The death rate, or mortality rate, is the number of individuals dying during a given time interval divided by the average population size over that time interval. This is an instantaneous rate and can be estimated for the population as a whole or for specific age classes to give the age specific mortality rate. The probability of dying is the number dying per individual present at the start of the time period.
Survivorship	Survivorship is the converse of mortality. Survivorship data are often shown as a survivorship curve for a particular population; a graph showing the proportion of survivors on a logarithmic scale through each phase of life. There are three generalized patterns of age-specific survivorship depending on whether the probability of dying is highest later in life (Type I), constant through life (Type II) or highest for young stages (Type III).
Life tables	Life tables summarize the fate of a group of individuals born at approximately the same time from birth to the end of the life cycle. Such a group is known as a cohort and investigation of this kind is termed cohort analysis. Life tables show the number of individuals present at different life stages or ages together with age-specific survival rates and age-specific mortality rates calculated for each stage. Mortality at each stage is expressed by k-values which are derived from logarithms and can be summed to give total mortality.
k-Factor analysis	This technique allows the identification of key factors contributing to mortality. Stage-specific k-values obtained over successive years are compared to the values for total mortality (k_{total}). k-Factor analysis highlights those stages suffering the greatest mortality which are responsible for fluctuations in loss rate and hence population size.
The fecundity schedule	Fecundity is the number of eggs, seeds, or offspring in the first stage of the life cycle produced by an individual. The fecundity schedule allows the calculation of the basic reproductive rate R_0. This is the number of offspring produced per original individual by the end of the cohort. In an annual population, it indicates the overall extent to which the population has increased or decreased over that time.

| **Population growth** | The changes in population size over time can be calculated by adding birth (B) and the number of immigrants (I) to the original population at time t, (N_t), and subtracting the number of deaths (D) and emigrants (E) to give a new population size at the time $t + 1$ (N_{t+1}). This is represented by the equation: |

$$N_{t+1} = N_t + B + I - D - E$$

For a particular set of conditions, an individual has a maximum potential for reproduction which is its intrinsic natural rate of increase, r. This is the theoretical maximum that may be reached in a given environment if the population is not resource-limited.

| **Density-independent population growth** | Unlimited growth of this kind is described by a continuous population model and expressed in terms of the rate of change in population numbers at time t: |

Rate of change of population   Intrinsic rate of increase
size at time t =   × Population size

$$\frac{dN}{dt} = rN$$

| **Density-dependent growth – the logistic equation** | The logistic equation describes the growth of a simple population in a confined space, where resources are not unlimited. In the early stages resources are abundant, the death rate is minimal and reproduction can take place as fast as possible allowing the individuals to attain their intrinsic rate of increase. The population increases geometrically until the maximum number of individuals the environment can sustainably support is approached. This maximum number is called the carrying capacity (K). The population growth rate declines to zero as the population becomes more crowded and the population size stabilizes. This can be described as the logistic equation |

Rate of change of   Intrinsic rate   Population   Density dependent
population size at time t =   of increase   ×   size   ×   factor

$$\frac{dN}{dt} = rN\left(1 - \frac{N}{K}\right)$$

where the density-dependent factor, $\left(1 - \frac{N}{K}\right)$ approaches zero as the population approaches the carrying capacity and intraspecific competition becomes more intense. This equation predicts growth of a population over time to be sigmoidal, as is commonly observed in real populations.

Related topics   Populations and population   The nature of competition (I1)
                          structure (H1)   The nature of predation (J1)
                          Density and density dependence (H3)
                          Population dynamics – fluctuations,
                          cycles and chaos (H4)

Natality   **Natality** is the birth of new individuals. The realized natality is the actual successful reproduction per female over a period of time and reflects the seasonality of breeding, the number of broods per year, the length of gestation etc. It is influenced by the condition of the individual and is often density dependent (see Topic H4). The age-specific birth rate is the number of offspring produced per unit time by females in different age classes.

Mortality

The death rate, or **mortality** rate, is the number of individuals dying during a given time interval divided by the average population size over that time interval. This is an instantaneous rate and can be estimated for the population as a whole or for specific age classes to give the age-specific mortality rate. For example, a population with 1000 individuals at the beginning of the time interval and 600 by the end of the period has an average size of 800. The mortality rate is given by 400/800, which is 0.5. The probability of dying is the number dying per individual present at the start of the time period, in this case 400/1000, which is 0.4.

Survivorship

Survivorship is the converse of mortality. The number of survivors is usually of more interest than the number dying, therefore mortality is often expressed as **life expectancy** – the average number of years to be lived in the future by population members of a given age. Survivorship data are often shown as a **survivorship curve** for a particular population, a graph showing the proportion of survivors on a log scale through each phase of life. *Figure 1a* shows such a plot for a cohort of the Dall mountain sheep. Survivorship falls away abruptly between 0 and 1 year and again after 9 years, corresponding with phases of high mortality early and late in life. Survival data are shown on the same axes for the great tit which has a constant probability of surviving with age.

There are three idealized patterns of survivorship, expressed as a semi-log plot in *Fig. 1b*. Type I shows good survival of young with high death rates only in old age (stereotyping the kind of pattern found in large mammals), Type II shows a steady mortality throughout life (reflecting the pattern found in some bird species), and Type III represents very high mortality in the young (as occurs, for example, in spawning fish and fungi). Whilst these conventions are useful to demonstrate major differences in survival patterns, it should be recognized that most species exhibit higher mortality in the youngest age classes, and that increased mortality of the old (due to senescence) can only be avoided if the death rate is so high that no individuals reach old age. Whilst birds in the field often show Type II mortality, old captive birds exhibit senescence.

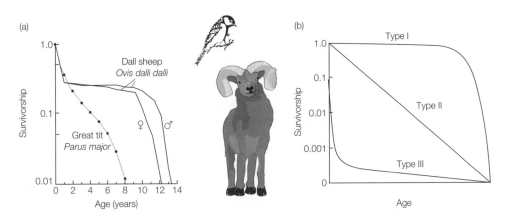

Fig. 1.   (a) Survivorship curves for a bird (the great tit) and a mammal (Dall sheep). (b) Generalized survivorship curves.

Life tables

Information on mortality is obtained through a population census which investigates the numbers dying at different life stages. These data are presented and analyzed by means of a **life table**. Life tables summarize the fate of a group of individuals born at approximately the same time from birth to the end of the life cycle. Such a group is known as a **cohort** and investigation of this kind is termed **cohort analysis**.

Table 1 shows the life table for the grasshopper *Chorthippus brunneus*. Eggs hatch into first instar nymphs and progress through a number of instars, shedding the larval skin and replacing it with a new, larger one each time. By mid-summer the fourth instars will have moulted into adults and by mid-November all the adults will have died, leaving their eggs in the soil. n_x is the number present at the start of each of the six identified stages.

The proportion surviving to the start of the next stage (l_x) is the **age-specific survival rate**. This has been standardized as a proportion, such that l_x for the first stage is 1 (no mortality has occurred yet) and subsequent values give the proportion of the original number of eggs surviving to that stage. This allows studies conducted at different times with different numbers of individuals to be compared.

The next column is the proportion of the original eggs dying *during* that stage (d_x). This is the difference between l_x at one stage and the next. These values do not allow comparison of the importance of mortality at different stages because d_x depends on how many individuals there are at each stage. The more individuals there are, the more there are available to die and the d_x value will be larger.

The proportion dying at each stage gives the **age-specific mortality rate** (q_x). This is d_x as a fraction of l_x. This is also the chance of an individual dying. The q_x values give an excellent indication of the intensity of mortality at each stage, but because they act mutliplicatively, they cannot be summed down the table to give total nymphal mortality. k-Values get around this problem. k-Values are simply the logarithm of the numbers in one stage minus the logarithm of the number in the next stage. Because adding two logarithms is the same as multiplying the unlogged numbers together, by converting the number of survivors to logs and calculating k-values we are able to add these together to find the total effect of mortality (k_{total}) and how it was distributed among life stages. The k-value of a life stage is referred to as its **killing power**. Thus, the killing power of the nymphal stages is 0.15 + 0.12 + 0.12 + 0.05 = 0.44, compared with 1.09 for the egg stage (i.e. the egg stage is the major contributor to total mortality).

k-factor analysis

With a series of life tables for cohorts born in successive years, we can see at which stage mortality has the greatest influence on population size. That is, we can see which **key factor** makes the largest contribution to k_{total}. This technique is known as k-factor analysis.

Figure 2 is derived from the first three years of life for brown trout, *Salmo trutta*, in the Lake District, England. Life tables were compiled each year for 17 years with mortality identified at each of six stages. Eggs hatch and emerge as alevins which survive for several weeks on their yolk sacs before developing into immature fish or parr. The figure shows a close relationship between the killing factor of the alevin stage (k_{alevin}) and total mortality (k_{total}). Therefore, it can be concluded that variations in alevin mortality are behind fluctuations in the total loss rate and hence population size.

Table 1. A cohort life table for the common field grasshopper, Chorthippus brunneus (after Richards & Waloff, 1954)

Stage (x)	Number observed at start of each stage (n_x)	Proportion of original cohort surviving to start of each stage (l_x)	Proportion of original cohort dying during each stage (d_x)	Mortality rate (q_x)	$\log_{10} n_x$	$\log_{10} l_x$	$\log_{10} n_x - \log_{10} n_{x+1}$ $= k_x$	Eggs produced in each stage (F_x)	Eggs produced per surviving individual in each stage (m_x)	Eggs produced per original individual in each stage ($l_x m_x$)
Eggs (0)	44 000	1.000	0.920	0.92	4.64	0.00	1.09	–	–	–
Instar I (1)	3513	0.080	0.022	0.28	3.55	−1.09	0.15	–	–	–
Instar II (2)	2529	0.058	0.014	0.24	3.40	−1.24	0.12	–	–	–
Instar III (3)	1922	0.044	0.011	0.25	3.28	−1.36	0.12	–	–	–
Instar IV (4)	1461	0.033	0.003	0.11	3.16	−1.48	0.05	–	–	–
Adults (5)	1300	0.030	–	–	3.11	−1.53	–	22 617	17	0.51

$$R_0 = \Sigma\, l_x m_x = \frac{\Sigma F_x}{a_0} = 0.51$$

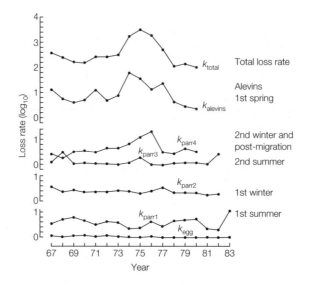

Fig. 2.   k-values for the brown trout life cycle.

The fecundity schedule

An example of a fecundity schedule is given in the last three columns of *Table 1*. The first column gives the number of offspring (eggs) produced during each life stage. Only adults produce eggs so there is only one entry, F_x. The next column gives the individual **fecundity**, or birth rate, which is the mean number of eggs produced per surviving individual. The fecundity schedule allows an important term, the **basic reproductive rate R_0** to be calculated. This is the number of offspring (eggs) produced per original individual by the end of the cohort and can be calculated by dividing the number of eggs produced by the number of individuals present at the start:

$$R_0 = \frac{\Sigma f_x}{n_0}$$

This is the same as summing the number of eggs produced per original individual for each of the life stages, or $R_0 = \Sigma \, l_x m_x$. R_0 is the average number of offspring produced by an individual in its lifetime. In an annual species (without overlapping generations) R_0 indicates the extent to which the population has increased or decreased over the duration of the life table. The value of 0.51 for the common field grasshopper (*Table 1*) indicates that the population has declined. If this were to continue the grasshopper population would rapidly shrink. However, values of R_0 alter from year to year and data for one year only do not allow long-term patterns to be predicted.

Population growth

A population that is increasing at its intrinsic rate will undergo a geometric increase in population number and will follow the characteristic geometric curve (*Fig. 3a*). This can only happen in reality if the population does not run out of resources. Although population numbers increase rapidly, the *per capita* rate of increase (r) remains constant. The geometric increase occurs because as more individuals are added to the population, more are increasing at that rate. At each time interval, the number of individuals added to the population will be greater than in the previous time interval.

Individuals are added to a population through birth (B) and immigration (I) and are removed as a result of death (D) and emigration (E). If the gains exceed the losses (i.e. if $B + I > D + E$), the population will grow in numbers. The changes in population size over time can be calculated by adding birth and immigration to the original population at time t (N_t) and subtracting the number of deaths and emigrants to give a new population size at the time $t+1$ (N_{t+1}). This is represented by the equation:

$$N_{t+1} = N_t + B + I - D - E$$

In closed populations, experiencing no exchange of individuals, population growth will depend on birth and death rates. Populations can be **regulated** by density. That is their growth rate depends on the size of the population and how close it is to the maximum that the habitat can support. Unregulated and regulated growth are described in more detail below.

For a particular set of conditions, an individual has a maximum potential for reproduction which is its **intrinsic natural rate of increase, r**. Its name is slightly misleading because its value will be different in different environments as death and birth rates differ. The intrinsic rate of increase is the theoretical maximum that may be reached in a given environment if the population is not resource limited. Populations with finite resources may have positive, negative or zero values of r, where the population is, respectively, increasing, decreasing or static. The intrinsic rate of increase is related to the basic reproductive rate R_0, such that:

$$r = \frac{\log(\text{average number of offspring per individual})}{\text{generation time}} = \frac{\ln R_0}{T}$$

The parameter r is usually applied to a closed population, where no immigration or emigration occurs, and represents the difference between instantaneous birth and death rates per individual.

Density-independent population growth

Most populations reproduce over a protracted period and have overlapping generations, that is individuals of different ages are present in the population at any one time. This situation is best described by a continuous population model and expressed in terms of the rate of change in population numbers at time t:

$$\text{Rate of change of population size at time } t = \text{Intrinsic rate of increase} \times \text{Population size}$$

$$\frac{dN}{dt} = rN$$

At any point on the curve (*Fig. 3a*) the population increase is equivalent to the gradient of the tangent to the curve. The equation above gives the gradient of the tangent at a value of t.

Density-dependent growth – the logistic equation

The growth of a simple population in a confined space, where resources are not unlimited is simply described by a graph that always looks sigmoid (Greek for 'S'-shaped) (*Fig. 3b*). In the early stages resources are abundant, the death rate is minimal and reproduction can take place as fast as possible allowing the individuals to attain their intrinsic rate of increase.

The population increases geometrically until an upper limit is approached. This upper limit, or saturation value, is a constant for a particular set of conditions in a particular habitat and is called the **carrying capacity (K)**. The

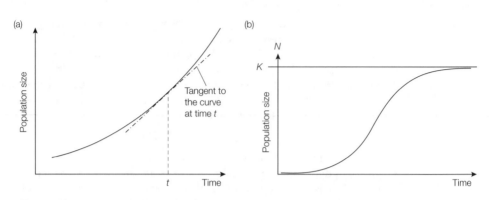

Fig. 3.   (a) A geometrically increasing population; (b) a sigmoid population growth curve, where the population growth rate slows as the population approaches the environmental carrying capacity, K.

population growth rate declines to zero as the population becomes more crowded and the population size stabilizes at the maximum that the environment can support, reaching an equilibrium population density. Competition among individuals of the same species has a dampening effect on population growth (see Topic I1).

Density dependence is occurring (see Topic H3), resulting in a decline in r with increasing density. This is in contrast with the density-independent case (above) where r remains constant. The sigmoid curve can be explained and described by multiplying the equation for density-independent growth by a density-dependent factor, to give the **logistic equation**:

$$\begin{array}{l}\text{Rate of change of}\\\text{population size at time } t\end{array} = \begin{array}{l}\text{Intrinsic rate}\\\text{of increase}\end{array} \times \begin{array}{l}\text{Population}\\\text{size}\end{array} \times \begin{array}{l}\text{Density dependent}\\\text{factor}\end{array}$$

$$\frac{dN}{dt} = rN\left(1 - \frac{N}{K}\right)$$

When the population size N is small, as in the early stages of population growth, the quantity N/K is also small and the dampening effect is negligible because $(1 - (N/K))$ is close to 1. The population increase is essentially given by rN and is geometric. However, as N grows large, the dampening effect increases, until when $N = K$, the quantity $(1 - (N/K))$ becomes $(1 - (K/K))$, which is zero.

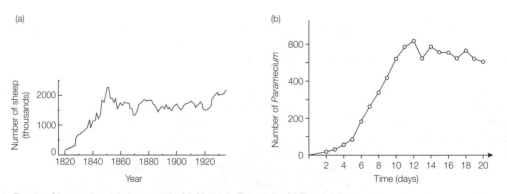

Fig. 4.   Observed population growth. (a) Sheep in Tasmania; (b) Paramecium.

At this point the population growth rate is zero and the population reaches equilibrium at a stable and constant size.

Real examples of population growth (*Fig. 4*) show some fluctuation in numbers as a result of fluctuating environmental conditions. Note also that both populations slightly overshoot the equilibrium population density, this probably occurs because of a delay in the effect of density on r, which is not incorporated into the simple logistic model above.

H3 DENSITY AND DENSITY DEPENDENCE

Key Notes

Density

Density is defined as the number of individuals per unit area, per volume or per unit of habitat. Total counts may be obtained for some plants and large or conspicuous animals. For many animals only a proportion of the individuals present in a given area will be detected and density must be estimated.

Estimating density – mark release recapture

Estimates of absolute density can be made with this method which involves trapping, marking trapped individuals, releasing them and trapping a second time. The proportion of marked to unmarked individuals in the traps will be the same as the proportion of marked to unmarked in the whole population, assuming a random proportion of the population is trapped. Since the total number of marked animals is known an estimate of the total population can be obtained from the following equation:

$$\frac{\text{Number trapped}}{\text{Number recaptured}} \times \text{Number originally marked} = \text{Total number in population}$$

Density dependence

Population parameters such as birth and death rates which vary with density are said to be density dependent. Birth and death rates that do not change with density are density independent. In the absence of immigration, a population will continue to increase in number unless either the *per capita* birth rate or death rate is density dependent.

Equilibrium population density

The equilibrium population density occurs when the *per capita* death rate exactly balances the *per capita* birth rate such that the density is neither increasing nor decreasing. The equilibrium population density is equivalent to the carrying capacity K.

Compensation and density dependence

There are three types of density dependence: overcompensating, undercompensating and exactly compensating. If a decline in numbers due to density dependence does not outweigh or balance the initial increase in numbers, density dependence is undercompensating. Overcompensating density dependence occurs where the effect of increased density more than outweighs the initial augmentation. If the decline in numbers exactly balances the initial increase in density, density dependence is said to be exactly compensating.

Related topics

Populations and population structure (H1)

Natality, mortality and population growth (H2)

Population dynamics – fluctuations, cycles and chaos (H4)

The nature of competition (I1)

The nature of predation (J1)

Density

Density is defined as the number of individuals per unit area, per volume or per unit of habitat. Densities will vary enormously, for example from hundreds of thousands m^{-2} for soil arthropods to a few km^{-2} for large mammals such as deer. Techniques for estimating density depend on the ease with which the species can be counted in its natural habitat. Total counts may be possible for some plants and large or conspicuous animals such as birds or butterflies. An estimate of population size can be made based on the density in a sample. For example, the number of beetles in a $0.1\ m^2$ sample can be extrapolated to give the absolute population size for the whole area. For many elusive and less conspicuous animals only a proportion of the individuals present will be sighted or trapped making total counts impossible, so density must be estimated.

For some species the best sampling technique does not estimate numbers per area but per sampling effort – giving the **relative density**, e.g. moths caught per light trap per hour or fish caught per rod per day.

Estimating density – mark release recapture

Very numerous, small organisms such as plankton or invertebrates may be sifted out of their habitat medium and counted. Less numerous, more motile organisms must be counted, but how is this done if they are rarely seen? Such organisms must be trapped, marked (e.g. with dye or by fur clipping) and then released and an estimate of the number present is made from the success of the trapping. After releasing the marked animals the ecologists knows that there is a definite number of marked individuals mixed in with the wild population. If traps are set again some of the animals caught will have been trapped before and will bear a mark. The proportion of marked to unmarked individuals in the traps will be the same as the proportion of marked to unmarked in the whole population, assuming the traps take a random proportion of the population. Since the total number of marked animals is known an estimate of the total population can be obtained from the following equation:

$$\frac{\text{Number trapped}}{\text{Number recaptured}} \times \text{Number originally marked} = \text{Total number in population}$$

Unfortunately, the assumption of randomness of the trapping often does not hold true; animals can become 'trap happy' or conversely avoid traps having been trapped once. Careless trapping techniques may elevate mortality among the marked individuals. Ingenious trapping programmes and statistical methods have been devised to help overcome these problems, allowing reasonable estimates of minimum population size to be obtained.

Density dependence

Many population parameters will vary with density. As habitats become more crowded the death rate is likely to increase and the birth rate to decrease because of competition and scarcity of resources. Birth rates may increase with density and death rates may, in some circumstances, decrease. These rates are said to be **density dependent**. Birth and death rates that do not change with density are **density independent**. In the absence of immigration, a population will continue to increase in numbers unless either the *per capita* birth rate or death rate is density dependent.

For example, increasing the egg density of experimental populations of the fruit fly, *Drosophila melanogaster*, reduces both survival to adulthood and fecundity (*Fig. 1a*). Crowding in the great tit, *Parus major*, has a similar effect on clutch size and the number of pairs producing a second brood (*Fig. 1b*).

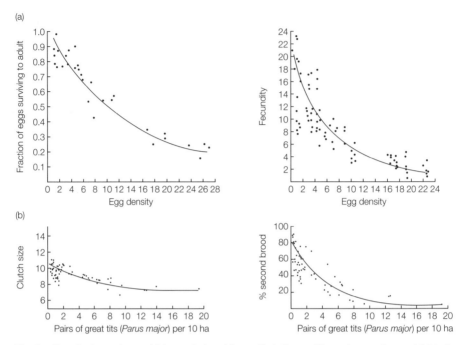

Fig. 1. Density dependence; (a) in survival and fecundity in Drosophila melanogaster, *and (b) in fecundity in the great tit,* Parus major.

Whilst resource limitation will cause negative density-dependent responses at high densities, population growth is not necessarily maximal at very low densities. Any process which gives advantage to individuals in a group (see Topic N1), such as antipredator behavior, increased foraging efficiency, and increased success of mate location or gamete transfer will lead to positive density dependence (until resource limitation counteracts these benefits) and hence maximal population growth at intermediate density – the **Allee effect**.

There is increasing awareness that the Allee effect is an important component of many populations. The vulnerability of small populations to extinction is exacerbated by this effect (e.g. the passenger pigeon, see Topic A2), and hence understanding the nature and importance of positive density-dependent processes is critical in conservation and harvesting.

Equilibrium population density

The equilibrium population density occurs when the *per capita* death rate exactly balances the *per capita* birth rate such that the density is neither increasing nor decreasing. Population density can only reach this equilibrium if birth and/or death rates are density dependent (*Fig. 2*). If both are density independent, population density will fluctuate and not reach equilibrium. In reality there is not one birth or death rate for a given density, but a range of values dependent on environmental conditions. *Figure 2d* shows how mortality and birth rates can balance over a range of densities to give a carrying capacity range rather than a single value.

The equilibrium population density is equivalent to the carrying capacity *K*. Density dependence explains the S-shaped, or sigmoidal pattern of population growth seen in Topic H2. As density increases, birth and/or death rates adjust, reducing the rate of increase and slowing population growth until the equilibrium density/carrying capacity is reached.

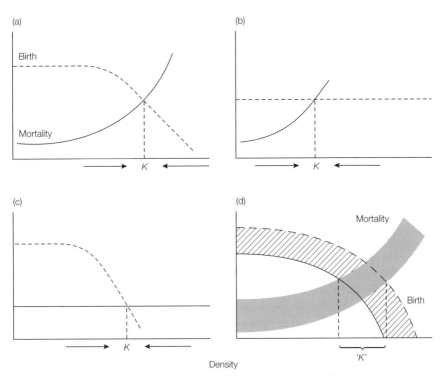

Fig. 2. *Density-dependent birth and mortality balance to give the equilibrium density, or carrying capacity K. (a) Birth and mortality both density dependent; (b) density-independent birth; (c) density-independent death; (d) as for (a) but incorporating environmental variation in birth and death rates. Reprinted from* Ecology, *3rd edn, Begon et al., 1996, with permission from Blackwell Science.*

Compensation and density dependence

It is possible to classify three types of density dependence, depending on the effect of an increasing starting density of individuals competing for a fixed resource: **exactly compensating**, **undercompensating** and **overcompensating** (*Fig. 3*).

With exact compensation, the number of survivors will be constant, regardless of the starting density – in other words, density-dependent processes exactly compensate for the variation in starting number. If overcompensating density dependence is operating, the number of survivors falls as the starting density rises. If density dependence is undercompensating, the number of survivors rises as the starting density rises (the density-dependent processes are failing to match the population to the available resources). These patterns may themselves vary with density, so at low densities density dependence may

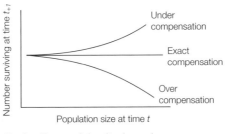

Fig. 3. *Forms of density dependence.*

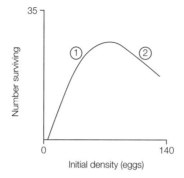

Fig. 4.   Density dependence in the flour beetle Tribolium confusum. In region 1 there is
undercompensating density-dependent mortality. In region 2 density dependence is
overcompensating.

exert little effect, and as density rises density dependence becomes first undercompensating then overcompensating. This pattern is observed in data from a
series of experiments in flour beetles (Fig. 4).

Exactly compensating density dependence would occur if the increase in
mortality due to density exactly balanced the excess in initial density over the
carrying capacity. This pattern is found in the field grasshopper, Chorthippus
brunneus (Fig. 5). The adult numbers are very consistent from year to year
despite the large variation in egg and nymphal numbers over the years.

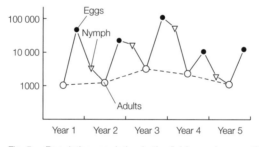

Fig. 5.   Population regulation in the field grasshopper, Chorthippus brunneus.

H4 POPULATION DYNAMICS – FLUCTUATIONS, CYCLES AND CHAOS

Key Notes

Expanding and contracting populations	Most real populations are not at their constant equilibrium density for very long, but are dynamic and changing. Populations may be expanding or contracting because of changes in environmental conditions or because of changes to their biotic environment.
Population fluctuations	Populations may fluctuate for a number of reasons: (i) a time lag between a change in density and its effect on the population size, or delayed density dependence. The population can overshoot the carrying capacity and then show gradually diminishing, dampened oscillations before eventually stabilizing at equilibrium. This delayed density dependence may also produce cycles in predator and prey abundance; (ii) overcompensating density dependence. This can lead to dampened oscillations, stable limit cycles (regular cycles that do not damp down) or chaotic fluctuations that appear random; (iii) environmental stochasticity. This is a nondeterministic, unpredictable variation in the environmental conditions, resulting in a changing equilibrium density.
Chaos	The mathematical definition of chaos is quite distinct from the colloquial use of this term. A chaotic system is driven by a deterministic process, it is not random. The outcome of chaos depends on the precise value of the starting conditions. As perfect accuracy is impossible to attain, chaotic systems are effectively unpredictable. It is currently unclear to what extent observed fluctuations in population sizes are influenced by chaotic processes. Some analyses suggest that measles outbreaks and some insect populations may display chaotic dynamics.
Related topics	Natality, mortality and population growth (H2) Predator behavior and prey response (J2) Density and density dependence (H3)

Expanding and contracting populations

Most real populations are not at a fixed equilibrium density for very long, if at all, but are dynamic and changing. Populations may be expanding or contracting because of changes to environmental conditions or because of predation or interspecific competition. The decline in whale populations due to extreme (human) predation is shown in Topic T2, *Fig. 1*.

Population fluctuations

Populations may fluctuate around their carrying capacity for a number of reasons:

(i) a time lag between a change in density and its effect on the population size, or **delayed** density dependence;

(ii) **overcompensating** density dependence (see Topic H3 for an explanation of density dependence);

(iii) environmental stochasticity.

Delayed density dependence

Fluctuations can be easily induced in theoretical populations by introducing a delay between a change in density and density having an impact on the birth or death rate. The population can overshoot the carrying capacity and then show gradually diminishing, damped oscillations before eventually stabilizing at equilibrium.

In some cases the cycles can persist. The larch bud moth occurs in forests in Switzerland. The larvae emerge in the spring simultaneously with the flushing of the larch. Feeding has an effect on the physiology of the larch, reducing the needle size and their food quality *in the following year* (*Fig. 1*). High larval density results in poor host quality for consecutive years, causing the larch bud moth population to crash. The low larval numbers allow the larch to recover which in turn leads to an increase in the number of larvae in response to improving food quality.

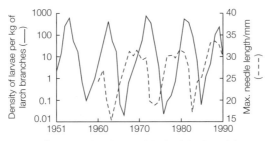

Fig. 1.   Cycles of abundance of the larch bud moth in response to larch quality (needle length).

This delayed density dependence also produces cycles in predator and prey abundance (*Fig. 2*). A classic example of a predator–prey cycle is illustrated by the Canadian lynx (*Lynx canadensis*) and the snowshoe hare (*Lepus americanus*) (*Fig. 3*). In this case, high lynx numbers depress the snowshoe hare population. This in turn causes a reduction in the number of lynx in subsequent years, allowing the hare population to rise again, resulting in a roughly 10-year cycle. However, as with the larch bud moth, the plants the hares eat also influence this cycle. As the hare numbers increase, the food quality of the plant leaf tissue decreases, which decreases the hare's reproductive potential. Thus, snowshoe hare–lynx population cycling is best thought of as the result of three interacting components: plants, hares and lynx.

Overcompensating density dependence

Density dependence is only stabilizing under certain conditions. Where there is no overcompensating density dependence the population will approach the carrying capacity with no oscillations. As density dependence becomes more overcompensating, damped oscillations and then population cycles will occur (*Figs. 4a* and *b*). These **stable limit cycles** have a fixed interval between each cycle and do not diminish in amplitude over time. Extreme overcompensation

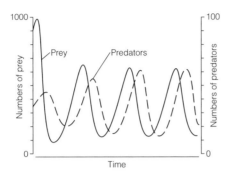

Fig. 2.   Theoretical predator–prey oscillations.

in combination with high reproductive rates can lead to chaotic fluctuations with no fixed interval or amplitude (*Fig. 4c*). Chaotic dynamics appears to be random but is actually deterministic in that its occurrence can be predicted (see below).

Environmental stochasticity
The carrying capacity of the environment can fluctuate in response to environmental conditions such as the weather. As a result, many real populations experience unpredictable fluctuations in numbers in response to good years and bad years. These populations are said to experience environmental stochasticity, are regulated by environmental conditions and are not predictable, deterministic density-dependent processes.

Small, short-lived organisms are more likely to show dramatic fluctuations in numbers than large, long-lived organisms which are more tolerant of environmental variation. Algae are small, short-lived and able to reproduce rapidly making them sensitive to environmental changes. The algal population fluctuations in *Fig. 5* are largely driven by temperature variation and consequent changes in nutrient availablity.

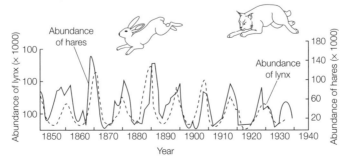

Fig. 3.   Cycles in the numbers of a predator (Canadian lynx) and its prey (snowshoe hare) over 90 years. From Ecology of a Changing Planet, *M.B. Bush, 1997. Reprinted with permission from Simon and Schuster, Upper Saddle River, NJ.*

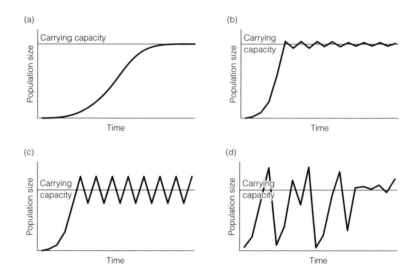

Fig. 4.   Alternative population dynamics: (a) monotonic increase; (b) damped oscillations; (c) stable limit cycles; (d) chaos.

Chaos

One might expect that in a constant environment, without any biotic or abiotic disturbance, population dynamics would be simple and predictable. However, even in an environment that is wholly predictable and constant, population parameters can interact in such a way so as to result in chaotic dynamics. A chaotic system may look random, but actually has no noise and is driven entirely by a deterministic process. The outcome of chaos depends on the precise value of the starting conditions. As perfect accuracy is impossible to attain, chaotic systems are effectively unpredictable. Populations may follow very different trajectories because of very minor differences in their starting conditions, as illustrated in *Fig. 6.*

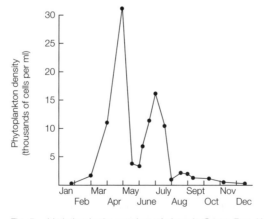

Fig. 5.   Variation in the number of algae in Green Bay, Wisconsin in response to changes in the environment.

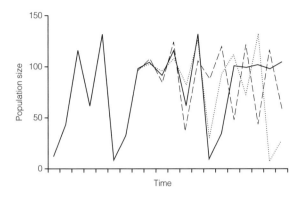

Fig. 6.   An illustration of the dependence of chaotic systems upon initial conditions. The three lines here are all derived from the logistic equation (see Topic H2) with identical values for the rate of increase r and the carrying capacity K, but vary minutely in the starting population size (starting sizes vary by 0.1%).

The discovery of chaos in theoretical models has led ecologists to speculate that the kind of erratic population trajectories commonly observed in nature, which are commonly assumed to be determined by a complex interaction of abiotic and biotic factors, might in fact be driven by chaotic processes. However, the mathematical techniques to discern chaos require longer time series than most ecology studies are able to provide, and questionable assumptions have to be made to apply them. It has also been pointed out that populations possessing attributes likely to give rise to chaotic dynamics are likely to become extinct (as chaos will occasionally push the population to very low densities), and thus will be uncommon. Evidence of evolution of systems away from violent fluctuation towards stability is found in many parasite–host systems (e.g. the myxoma virus in European rabbits, see Topic K2).

Nevertheless, analysis of large datasets of the incidence of measles in New York, USA and Liverpool, England suggests that some measles outbreaks may be chaotic. The abundance of voles, lemmings and forest insects fluctuate dramatically and demonstrate some characteristics of chaotic dynamics. The prevalence of chaos in natural populations is still unclear, but it is likely that chaos, if it does occur in natural populations, is not a major force in ecology.

I1 THE NATURE OF COMPETITION

Key Notes

Classifying interactions	Interactions between individuals and species can be classified on the basis of the effects and the mechanism of the interaction. The key interspecific interactions are competition, predation, parasitism and mutualism, whilst the main intraspecific interactions are competition, cannibalism and altruism.
Competition	Competition is an interaction among individuals utilizing a limited resource, resulting in reduced fitness in the competing individuals. Competition occurs both between species utilizing a shared resource (interspecific competition) and among members of a species (intraspecific competition). The niche of an individual or species (the conditions under which it is found, the resources it utilizes and the time it occurs there) is critical in determining the degree of competition with other species or individuals. Large niche overlap generally results in intense competition.
Intraspecific competition	As individuals are quite similar in their resource requirements, such competition may be particularly intense. Intraspecific competition is a major force in ecology and is responsible for phenomena such as dispersal and territoriality, as well as being the primary cause of population regulation via density-dependent processes.
Interspecific competition	Interspecific competition occurs between two species using the same limited resource. Very few species can escape from the effects of other species competing for the same resource.
Exploitation competition	There are two ways in which competition can operate. In exploitation competition, individuals only interact indirectly, by depleting the resource in short supply. Reduced fitness occurs due to a shortfall in resource availability.
Interference competition	In interference competition, individuals interact directly, most obviously, in the case of some animal species, by fighting, but also by producing toxins (e.g. plant allelopathy). Fitness reduction in the 'loser' in such interactions may be due to the interference (e.g. injuries or death) as well as the lack of resource access.
Asymmetry in the effects of competition	Competition often unevenly affects competitors, such that the cost for one individual is far greater than for another. It is common for competition to kill the losers, either via exploitation or interference.
The Lotka–Volterra interspecific competition model	The Lotka–Volterra model of interspecific competition is a simple mathematical model representing the interaction between two competing species. It makes three main simplifying assumptions: (i) only two species are involved in the interaction; (ii) each species has a population growth pattern described by the logistic equation; (iii) an individual of one species is perceived as having a defined competitive equivalence to a given number of members of the other species. This simple model makes valuable predictions as to when coexistence or competitive exclusion occur.
Related topics	The niche (B3) Resource partitioning (I3) Intraspecific competition (I2)

Classifying interactions

Interactions between individuals and species can be classified on the basis of the effects and the mechanism of the interaction. The key interspecific interactions are **competition, predation, parasitism** and **mutualism**, whilst the main intraspecific interactions are **competition, cannibalism** and **altruism** (*Table 1*). **Parasitoidism** is a type of parasitism, also known as hyperparasitism, which occurs in some insect species (mainly wasps and flies), in which the parasitoid lays eggs in or on the body of the host, and usually results in the death of the host.

Table 1. A classification of interactions between individuals and species. Note that by these definitions, herbivores may be either predators (e.g. wildebeest) or parasites (e.g. aphids)

	Interaction between two species (interspecific)	Interaction between members of a species (intraspecific)
Use of the same limiting resource, with resulting fitness loss	Competition	Competition
Consumption of all or part of another individual	Predation	Cannibalism
Individuals live in close association with mutual benefit	Mutualism	Altruism or mutualism
Individuals live in close association, to cost of host	Parasitism	Parasitism[a]

[a] Intraspecific parasitism is relatively rare, and may be difficult to distinguish from mutualism, especially if individuals are related.

Occasionally, interactions between species occur where one species exhibits no effect, whilst the other either benefits (**commensalism**) or is adversely affected (**amensualism**). Hermit crabs often carry hydroids on the mollusc shell they occupy, which benefits the hydroid by transporting it to feeding sites, but bears no obvious cost to the crab – an apparent example of commensalism. Examples of amensualism may be provided by species that produce toxins (e.g. soil-living fungi producing antibiotics) regardless of whether the species which suffers a deleterious effect is present or not. It may be convenient to classify interactions on the basis of whether effects are positive (+), negative (–) or neutral (0) (*Table 2*).

Table 2. A classification of interspecific interactions by effect

Type of interaction	Response of species A	Response of species B
Competition	–	–
Predation	+	–
Parasitism	+	–
Neutral	0	0
Amensualism	0	–
Commensalism	0	+

Competition

Competition is an interaction among individuals utilizing a limited **resource** (e.g. food, space etc. – see Topic B1) resulting in reduced fitness in the competing individuals. Competition occurs both between species utilizing a shared resource, and among members of a species.

The **niche** of an individual or species (the conditions under which it is found, the resources it utilizes and the time it occurs there) is critical in determining the degree of competition with other species or individuals (see Topic B3). The greater the degree of niche overlap, generally the more intense the degree of competition (see 'Limiting similarity' in Topic I3).

Intraspecific competition

Competition occurring between members of the same species is defined as **intraspecific competition**. It is likely to be intense because individuals will tend to share requirements for the same resources. Although there may be age differences in resource requirements (e.g. young bream, *Abramia brama*, a fresh-water fish, feed on small pelagic zooplankton, whilst adults feed on large benthic invertebrates) or sex differences (e.g. as the male of most birds of prey is much smaller than the female, the preferred prey size differs between the sexes), the general degree of overlap in resource use means that intraspecific competition is a major force in ecology. By depressing the fitness of individuals in crowded populations, it both influences fundamental processes like fecundity and mortality, and hence regulates population size, and leads to behavioral adaptations to overcome or cope with competition, such as **dispersal** and **territoriality**.

Interspecific competition

Where two species utilize the same limited resource, interspecific competition will occur. Examples of interspecific competition are given below.

(i) Protozoans *Paramecium aurelia* and *P. caudatum*. When cultured separately in yeast medium, *P. aurelia* was found to have a faster rate of increase than *P. caudatum*. When both species were added to the same culture vessel, *P. aurelia* dominated the mixture and eventually *P. caudatum* died out (*Fig. 1*).

(ii) Diatoms *Asterionella formosa* and *Synedra ulna*. Diatoms are unicellular algae with a characteristic, often complex, silicate capsule. All diatoms therefore have a requirement for silicate. When both species were grown separately in a culture to which silicate was constantly added, they both thrived. However, *Synedra* reduced the silicate concentration to a lower level than *Asterionella*. Therefore, when the two species were cultured together, *Asterionella* was rapidly excluded, as the silicate was reduced to a level which it could not access (*Fig. 2*).

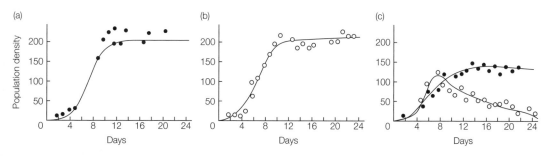

Fig. 1. Competition between two species of Paramecium. *(a)* P. aurelia *alone; (b)* P. candatum *alone; (c) both species together.*

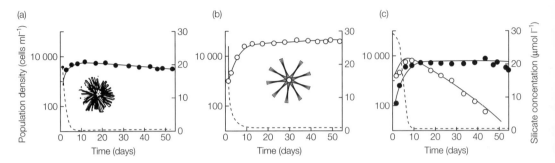

Fig. 2. Competition between diatom species. (a) Synedra *alone; (b)* Asterionella *alone; (c) both species together. Dotted line, silicate concentration. Redrawn from Tilman* et al., Limnol. Oceanog. **26**, *1020–1033, 1981.*

(iii) Darwin's finches, *Geospiza fortis* and *Geospiza scandens*. On the small island of Isla Daphne in the Galapagos, there was a drought in the late 1970s which reduced the production of seed (the food of both *G. fortis* and *G. scandens*) dramatically. Both species survived the drought, but they changed their diets, with *G. fortis* concentrating on small cactus seeds and *G. scandens* selecting larger seeds. This is an example of competition resulting in coexistence via a **niche shift**.

(iv) Barnacles *Balanus balanoides* and *Chthamalus stellatus*. In northwest Europe, these two species commonly co-occur on the same rocky shore, but with most adult *Chthamalus* found higher up the shore, and adult *Balanus* further down the shore. Young *Chthamalus* usually settle on the lower shore but apparently do not survive. In an experiment where young *Chthamalus* on the lower shore were protected from being smothered by *Balanus* individuals, they survived and grew well. However, in the upper zone, *Chthamalus* do not have to compete as *Balanus* cannot survive the desiccation. Thus, the observed distribution pattern is due to a mixture of competition and environmental tolerance.

(v) Invasive weeds. Exotic plant species occasionally become troublesome weeds because of their vigorous competitive ability, allowing them to exclude a broad spectrum of native plant species in a particular habitat. Examples are (i) the purple loosestrife (*Lythrum salicaria*), a European wetland species which has invaded wetlands throughout temperate North America; (ii) the Japanese knotweed, (*Fallopia japonica*) which is a vigorous riverside weed in Europe; (iii) water hyacinth, (*Veronica beccabunga*), a native of Southeast Asia, which has become a problem blocking drainage channels and waterways in Europe and the Americas.

Exploitation competition

Competition can operate in two ways, either solely via the depletion of the limiting resource – (**exploitation competition**) whereby individuals only interact indirectly, or by direct interaction between competing individuals (interference competition). In exploitation competition reduced fitness occurs due to a shortfall in resource availability, as occurs in the tadpoles of *Rana tigrina* (*Fig. 3*).

Interference competition

Competing individuals may interact directly. The most obvious instances of such **interference competition** are animals which fight over territories or food items. Lions may arrive at the kill of a smaller carnivore and displace it. Physical

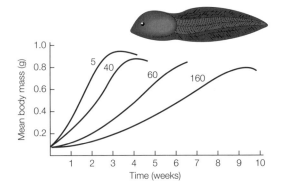

Fig. 3.   The tadpole Rana tigrina *grows more slowly as density increases from five individuals per tank to 160 due to the reduced food availability, an example of exploitation competition. Redrawn from* Ecology and Field Biology, *5th edn, R.L. Smith, 1996, Addison Wesley Longman.*

conflict occurs at smaller scales too – many ichneumonid wasp species possess large mandibles as larvae, which they use to fight to the death with any other larvae they find inside their caterpillar hosts.

Interference may occur via the use of toxins by competitors. A braconid wasp parasitic on aphids produces a toxin as it hatches from the egg, which kills all other parasitoid eggs. In plant species, the process of competing via toxin production is termed **allelopathy**. There are many examples, including

(i)   the black walnut (*Juglans nigra*) of North America, which suppresses plant growth up to 25 m from the trunk, killing many plants outright. A root extract contains the chemical quinone which kills alfalfa and tomato plants;

(ii) the Californian chapparal shrub, *Salvia leucophylla* produces volatile turpenes which appear to suppress competitors in the field. Cucumber seedlings grown in an experimental chamber alongside 2 g of *Salvia* leaves produced only 8% of the stem extension of control plants grown without *Salvia* leaves;

(iii) In the reedmace, *Typha latifolia*, intraspecific allelopathy occurs, such that the centre of stand dies back.

Some competing species may reciprocally predate upon each other, as happens in the grain beetles *Tribolium confusum* and *T. castaneum*.

Competing predators may interfere with each other by disturbing the prey, making them harder to catch. This occurs in wading birds in estuary mudflats searching for invertebrates such as worms, isopods and shrimps. This may force some individuals away from the richest sites, and onto more marginal environments.

Fitness reduction in the 'loser' in interference interactions may be due to injuries, possibly resulting in death, as well as the lack of resource access.

Asymmetry in the effects of competition

Both intraspecific and interspecific competition often affect the competitors very unevenly, such that the cost for one individual is far greater than for another. It is common for competition to kill the losers, either via exploitation (by depriving them of resources) or interference (by directly injuring or poisoning them). Examples of **competitive asymmetry** appear to greatly outnumber examples of symmetrical outcomes.

Resource distribution **within a species** may be fairly **even**, so that most individuals get a similar amount of food. Under high densities, no individual gets enough to survive, and the population crashes. This scenario may be classified as **scramble competition**, and has been observed in the sheep blowfly, *Lucilia cuprina*. Alternatively, in some species high densities result in a very **uneven** distribution of resources, classified as **contest competition**, so that some individuals grow and reproduce whilst others die. It may be noticed that there is a clear link between contest competition and interference competition, and between scramble and exploitation competition, but these terms are not fully interchangeable; contest/scramble refers to the **evenness** of resource distribution whilst exploitative/interference refers to the mechanism.

The Lotka–Volterra interspecific competition model

The Lotka–Volterra model of interspecific competition is a development of the logistic equation of population growth (see Topic H2):

$$\frac{dN}{dt} = rN \left(1 - \frac{N}{K}\right)$$

where N is the population density, K the carrying capacity and r the intrinsic rate of increase of a species. As there are two species, we have two sets of these variables: $N_1, N_2, K_1, K_2, r_1, r_2$. To apply this model to two competing species, it is necessary to be able to translate the number of individuals of one species that are competitively equivalent to the one member of the other species. This is the **competitive coefficient,** and is denoted α. There are two competitive coefficients: (i) the perceived competitive equivalence of species 2 on a member of species 1 (α_{12}), and (ii) the perceived competitive equivalence of species 1 on a member of species 2 (α_{21}).

If each member of species 2 exerts five times the competitive effect of a member of species 1 on species 1 (which may occur if say, species 2 are bigger and hence consume more resources), then α_{12} will be 5.0. Thus, we can translate the number of individuals of species 2 into species-1 equivalents by multiplying N_2 by α_{12}. The maximum number of individuals of species 1 the environment can support is of course the carrying capacity K_1. But as more and

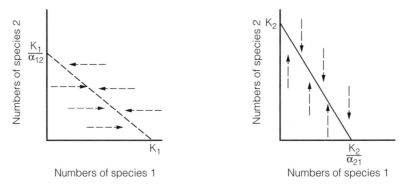

Fig. 4.   *The zero growth isoclines from the Lotka–Volterra competition equations. (a) The species 1 zero growth isocline. Species 1 increases to the left of the line (the environment is not saturated) and decreases to the right of the line (the environment is over-saturated). (b) The species 2 zero growth isocline. Species 2 increases below the line (the environment is not saturated) and decreases above it (the environment is over-saturated). The arrows indicate the change in population size for a range of different values of N_1 and N_2.*

more members of species 2 are present, this maximum will fall, until when there are K_1/α_{12} members of species 2 present, the environment will be saturated. Thus, it is possible to represent the numbers of species 1 the environment can support for all values of the numbers of both species – this is the **zero net growth isocline** for species 1, shown graphically in *Fig. 4a*. A reciprocal effect of the number of individuals of species 1 on species 2 occurs, as shown in *Fig. 4b*.

By adding together both of the zero net growth isoclines defined in *Fig. 4*, we can find the different outcomes of the model (*Fig. 5*). Four different outcomes

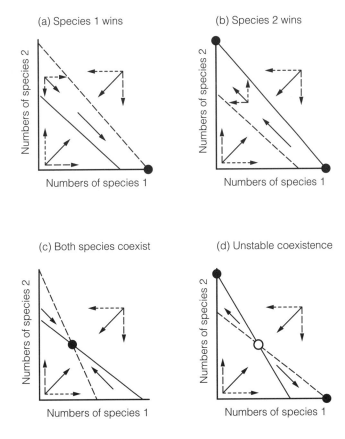

Summary

	Outcome
One species is not limited by its competitor and exerts strong **interspecific** competition	(a), (b) One species wins
Both species have intense **intraspecific** competition (possibly due to niche differentiation)	(c) Coexistence
Both species have intense **interspecific** competition (possibly due to co-operation within a species)	(d) One species wins, but the outcome depends on the starting densities (unstable coexistence may temporarily occur)

Fig. 5. *The four alternate outcomes of the Lotka–Volterra model. The dotted arrowed lines represent the trajectories of species 1, the dashed arrowed lines represent the trajectories of species 2 and the solid arrowed lines the joint trajectory.*

arise from this simple model, based on the relative values of the species' carrying capacities K_1 and K_2 and their competitive equivalents K_1/α_{12} and K_2/α_{21}. In the simplest cases, as shown in *Fig. 5a* and *b*, one species will win if it is a superior competitor at all values of N_1 and N_2. In these cases, the weaker competitor is eliminated. This occurs when intraspecific competition within the dominant species exceeds the interspecific effects of the competitor *and* the interspecific competitive impact on the inferior species exceeds the intraspecific effects of that species on itself. If both species exert more intense intraspecific than interspecific competition, then stable coexistence occurs, as shown in *Fig. 5c*. This may happen if competitors have only partial niche overlap. In the final case, where both species have more intense interspecific than intraspecific competition (which may occur in cooperative animals which may act aggressively towards competitive species), the outcome is more complex. There are two alternative stable states, in which the species numerically dominant will exclude the other (*Fig. 5d*). If equivalent densities balance, an unstable coexistence may temporarily occur, but in real populations this will not persist and one or other species will be excluded.

12   INTRASPECIFIC COMPETITION

Key Notes

Density dependence	Density dependence describes the relationship between fitness and population size. A key population regulatory factor is negative density dependence, where declining fitness occurs as population density increases within a species due to intraspecific competition.
Dispersal	Organisms can respond to high levels of intraspecific competition by dispersing away from the area of high population density. Even in species that are sessile for most of the lifecycle, there is a mobile dispersal stage. Dispersal is often undertaken by the younger members of a population, whilst in many mammal species males disperse more than females do.
Territoriality	In many animal species (including insects, birds and mammals) individuals or groups compete for areas of space. There is active interference between individuals to maintain the territory boundaries. Territoriality gives a benefit to the territory-holder, such that the costs of defending the territory against intruders are outweighed by advantages such as increased food supplies, increased mating success and reduced predation risk.
Self-thinning	Sessile organisms, including plants, cannot escape competition by movement, and therefore the losers in the competitive battle die. In a group of plants of the same age, this results in fewer individuals of larger size surviving. This process is described as 'self-thinning'. Self-thinning results in a relationship between density and individual plant mass, which typically has a slope of $-\frac{3}{2}$ on a log–log plot. This relationship is known as Yoda' s $-\frac{3}{2}$ law.
Related topics	The niche (B3) The nature of competition (I1) Density and density dependence (H3) Resource partitioning (I3)

Density dependence

Density dependence describes the relationship between fitness and population size. A key population regulatory factor is **negative density dependence**, where **declining fitness** occurs as **population density increases** within a species due to intraspecific competition. As the number of individuals rises, mortality rates rise, birthrates and growth rates fall. All of these effects are due to intraspecific competition, and are discussed in detail in Topic H3.

Dispersal

If competition is intense, individuals may **disperse** away from the area of high population density. Even in species that are sessile for most of the lifecycle, there is usually a mobile dispersal stage. Thus, most plant species produce seeds which may be carried by the wind (e.g. the 'helicopter' seeds of maples and sycamores), on the fur of animals (e.g. cleavers, *Galium aparine*), in the gut of fruit-feeding birds and mammals (e.g. blackberries, *Rubus fructicosus*) or on water currents (e.g. the coconut palm). Similarly, most sessile marine organisms have a mobile larval stage. As well as permitting possible escape, dispersal provides a mechanism by which individuals avoid the costs of **inbreeding**. Dispersal is often a risky activity as it involves the location of a suitable habitat in which population densities are low. Therefore, it might be expected that species found in 'secure' stable environments might be less willing to disperse than those in 'dangerous' ephemeral habitats. This pattern is found in British water beetles, in which only 25% of species from permanent habitats are able to fly, compared to 64% of those from temporary habitats. The tendency to disperse may also depend on the species' competitive ability. 'Supertramp' bird species are those like the pigeon *Macropygia mackinlayi*, which is quick to colonize Pacific islands, but is easily ousted by more competitive bird species. This **trade-off** between competitive ability and dispersal ability is widespread, and is found intraspecifically too. Winged aphids allocate resources to wing muscle and fat, and have lower fecundity compared to wingless individuals. Many plant species, including *Galinsoga parviflora* produce seeds of two different sizes: the larger having higher competitive ability, the smaller having greater dispersal potential.

Because dispersal is risky, life history theory dictates that it should be undertaken by individuals with a relatively low **reproductive value**, in other words by the younger members of a population. This pattern is commonly found in plants, sessile marine animals, birds and small mammals. Some species produce special dispersal morphs. In aphids, many species can produce winged forms when the population density is high and then revert to wingless morphs, which have higher fecundity rates, when a new plant is colonized. In some mammal species, such as lions and bushbabies, there is a sex bias in dispersal, where males leave the family group but their sisters remain with their mother.

Territoriality

Individuals or groups in many animal species (including insects, birds and mammals) compete for areas of space, known as territories. Active interference between individuals maintains the territory boundaries. Commonly, an individual or mating pair defend a territory, but social animals (e.g. gray wolves) may defend a common territory.

There are different types of territories. Some animals (e.g. many songbirds) establish general purpose territories, in which almost all activities occur, including feeding, mating and rearing of the young. In contrast, some territories are more specialized. The most extreme form of specialization are **leks**, closely aggregated male territories occurring for the purposes of breeding, which are often short-lived. Lekking occurs in bird (e.g. black grouse), amphibian (e.g. natterjack toad) and mammal species (e.g. fallow deer).

Territoriality gives a benefit to the territory-holder, such that the costs of defending the territory against intruders are outweighed by the advantages. In great tits (*Parus major*) territory-holders have improved winter food availability, are more attractive to mates, provide more food to their young and have reduced predation risk. Female red-winged blackbirds (*Agelaius phoeniceus*) choose males on the basis of territory quality. In ground squirrels, pairs with large territories are less likely to cannibalize their young. In dragonflies, males with better quality territories get more copulations. In many species, territory sites are in short supply, and a proportion ('floaters' or 'satellites') of the population cannot secure these benefits. For example, some male black grouse fail to establish a lek territory. These satellite individuals are rarely able to successfully mate females.

Territory size is determined by the balance of costs and benefits. Both will increase as territory size increases, but at some critical size the costs of maintaining the territory will outweigh any benefits. If resources are abundant, the optimal territory size may be smaller than if resources are scarce. Such a pattern is observed in the golden-winged sunbird (*Nectarina reichenowi*), a nectar-feeding hummingbird, in which territories vary considerably in size and flower density, but each territory provides an equivalent nectar supply.

Self-thinning

In an even-aged group of sessile organisms, competing individuals cannot escape, and typically competition results in the survival of fewer individuals of larger and larger size. This process is described as **self-thinning**. Self-thinning results in a relationship between density and individual mass, which typically has a slope of $-\frac{3}{2}$ on a log–log plot (*Fig. 1*). This relationship is known as **Yoda's $-\frac{3}{2}$ law** and has been found to be approximated by numerous plant species as well as sessile animals like barnacles and mussels. This pattern indicates that in a growing, self-thinning population, weight increases more rapidly than density decreases. Whilst there is variation among species as to the precise value of this slope, the general pattern is due to the simple proportionality between volume (which is directly proportional to weight) and area, and does not invoke special biological processes.

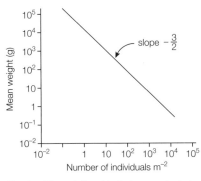

Fig. 1.   The relationship between plant density and plant size, demonstrating Yoda's $-\frac{3}{2}$ law.

I3   RESOURCE PARTITIONING

Key Notes

Competitive exclusion	If two species compete in a stable environment, there are two possible outcomes: (i) one species is excluded, or (ii) both species coexist. The competitive exclusion principle states that coexistence can only occur in a stable environment if the species niches are differentiated.
Limiting similarity	How much niche differentiation is needed for species to coexist? This critical threshold of differentiation in resource utilization is termed the limiting similarity.
Competitive release	In the absence of a competitor, a species may expand its niche. Examples of such competitive release include ground doves in New Guinea and gerbils in Israel.
Character displacement	When realized niches contract under the influence of competition, morphological changes may follow as adaptations to the new resource spectrum. Such character displacement is found in the ant *Veromessor pergandei* and in Darwin's finches, *Geospiza fortis* and *G. fuliginosa*.
Spatial and temporal heterogeneity	Environments are patchworks of habitats of varying qualities and differing resource levels. Some patches are temporary and/or unpredictable in occurrence. In time and space, the environment is heterogeneous. This means that competitive 'battles' may not be completed before the environment changes and the balance of play alters. Therefore, species that coexist in the real world may do so because of environmental heterogeneity rather than niche differentiation.
Apparent competition	If a predator attacks two prey species, then each prey species may adversely affect the other, by increasing the local predator population. Therefore, the interaction between the two prey species is exactly as if they were competing, yet they may utilize entirely different resources. This phenomena is known as apparent competition.
Enemy-free space	If apparent competition occurs between two species, they may try to escape the impact of the predator by escaping from the vicinity of the other prey species into so-called 'enemy-free space'. Therefore, niche differentiation may occur, but mediated by a predator rather than a resource shortage.
Related topics	The niche (B3) Intraspecific competition (I2) The nature of competition (I1)

Competitive exclusion

Where two species compete in a stable, homogeneous environment, either one species will win and the other be excluded, or both species will manage to coexist. The **competitive exclusion** principle states that coexistence can only occur in a stable, homogeneous environment if the species niches are differentiated, because if two species had identical requirements one would dominate and outcompete the other.

Limiting similarity

How much niche differentiation is needed for species to coexist? This critical threshold of differentiation in resource utilization is termed the '**limiting similarity**'. The limit in the similarity of competing species is caused by a balance between (i) the intensity of intraspecific competition (which a narrow niche width intensifies), and (ii) the intensity of interspecific competition (which a broad niche width intensifies) (*Fig. 1*). However, theory is currently unable to predict where this balance might lie in real world examples.

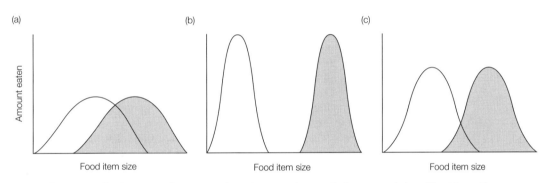

(a) (b) (c)

Amount eaten

Food item size Food item size Food item size

Fig. 1.   Resource utilization curves for two species, where species vary in the range of size of food items they can utilize. In (a) there is strong interspecific competition and weak intraspecific competition; in (b) there is strong intraspecific competition and no interspecific competition; and in (c) the curves overlap only slightly, so interspecific competition is relatively weak, yet the range of resource utilized by each species is quite wide, so intraspecific competition is also relatively weak.

Evolution may act to reduce the degree of competition between species – thus current patterns of resource utilization are a result of competition over time, even though little or no competitive interactions are currently observed. This phenomenon is known as '**the ghost of competition past**'.

Competitive release

In the absence of a competitor, a species may expand its **realized niche** (see Topic B3). Examples of such **competitive release** can be regarded as evidence of competition acting in the field.

In northern Israel two gerbil species, *Gerbillus allenbyi* and *Meriones tristrami*, overlap in range. Where they overlap, *G. allenbyi* only is found on non-sandy soils, whereas where *G. allenbyi* occurs alone it occupies sandy and non-sandy soils alike. It would appear that in the absence of *M. tristrami*, *G. allenbyi* is able to expand its realized niche.

Competitive release also appears to occur in ground doves in New Guinea. On the main island of New Guinea, three species of ground doves are found, each in a different habitat (coastal shrub, second-growth forest and rainforest). On small islands (which tend to be species-poor, see Topic Q2) possessing only one ground dove species, that species is able to exploit all three habitat types. It can be concluded that competitive exclusion results in niche differentiation in New Guinea, but competitive release occurs on the small islands.

Character displacement

Occasionally, niche contraction due to competition results in **morphological character change**, known as **character displacement**. (It should be noted that behavioral and physiological changes also occur as a result of niche shifts.)

In the harvester ant *Veromessor pergandei*, the variability in mandible size is inversely correlated with the number of competing seed-eating ants. This suggests that as the competition from other ant species increases, *V. pergandei* becomes more specialized and concentrates on a smaller range of seed sizes.

On the Galapagos, when the two Darwin's finches *Geospiza fortis* and *G. fuliginosa* occur on islands alone, they have similar beak sizes, whereas when they occur together, *G. fuliginosa* invariably has a markedly narrower beak than *G. fortis*.

Spatial and temporal heterogeneity

Natural environments are **not constant or homogeneous** in structure, but are patchworks of habitats which vary extensively in quality and resource levels, in both time and space. Some patches are temporary and/or unpredictable in occurrence. In **time and space**, the environment is **highly heterogeneous**. Therefore, competitive 'battles' between individuals or species may not be completed before the environment changes and the balance of play alters in favor of the previously weaker competitor. Heterogeneity thus plays a very important role in fostering ecological diversity.

The constantly changing balance of play between competitors has been suggested as the explanation of **'the paradox of the plankton'** – the persistence of large numbers of plankton species in the structurally simple habitat of the upper layers of the sea. The constantly changing environment, diurnally and seasonally, with changing levels of temperature, light, oxygen and nutrients, may exclude any species equilibrium being reached.

Gaps may occur unpredictably in many habitats, caused by extreme weather, or the death, by predation or some other cause, of a dominant competitor. The individuals which first fill these gaps are often species which are weak competitors but good colonizers. In due course, they will be outcompeted by stronger competitors. A classic example of this is provided by the sea palm (*Postelsia palmaeformis*) and the mussel *Mytilis californicus* on the coast of Washington, USA. The sea palm is an annual species, which must therefore recolonize bare rock every year. In heavily storm-damaged sites, with frequent gap formation, both species coexist, whilst relatively undisturbed sites are dominated by mussels.

In a similar fashion, ephemeral patches of habitat, such as carrion, dung, desert ponds and fungi are too unpredictable for any individual or species to be able to locate many patches. As a result, weaker competitors are able to persist. In a Finnish study on carrion flies, 50 patches of carrion produced a total of nine species, yet each patch on average produced only 2.7 species. Thus, many pairwise interactions between species occurred only rarely. The situation whereby competitive interactions are curtailed by the limited co-occurrence of competitors in a patchy environment has been termed the **probability refuge**.

The best competitor in a confrontation may depend on which individual was at the site first – the **priority** effect. This applies to both animals and plants. In many territorial animals, there is an owner-advantage, where an owner will tend to fight harder to retain a territory than a competitor will be willing to fight to gain it. In plants, the first species to develop may out-compete a later arrival at a site, even if the former is a weaker competitor in a head-to-head confrontation. Thus, repeated colonization in an unpredictable environment favors coexistence.

Apparent competition

If a predator attacks two prey species, then each prey species may adversely affect the other, by increasing the local predator population. Therefore, the interaction between the two prey species is exactly as if they were competing, yet they may utilize entirely different resources. This phenomenon is known as apparent competition.

Enemy-free space

If apparent competition caused by a common predator occurs between two species, they may escape the impact of the predator by escaping from the vicinity of the other prey species – into so-called 'enemy-free space'. Such differentiation in habitat use between species occurs in Californian rocky reefs, where gastropods (*Tegula* spp.) and a clam (*Chama arcana*) are predated by both lobsters and octopuses. In habitat dominated by large boulders (with many crevices) there are many clams, whilst in sites with little crevice space gastropods dominate. Transfer of clams to the gastropod-dominated sites resulted in increased predation and a reduction in gastropod densities. This pattern thus appears to be maintained by predator activity rather than resource utilization. Therefore, niche differentiation may be mediated by a predator rather than a resource shortage.

J1 THE NATURE OF PREDATION

Key Notes

Defining predation
Predation can be defined as the consumption of all or part of another individual (the prey). This wide definition thus encompasses: (i) 'true predators', which kill their prey soon after attacking them; (ii) grazers, which consume only part of a prey individual; (iii) parasites, which live in very close association with a single prey individual (the host), often inside the host's tissues.

Carnivores and herbivores
Predators can be categorized as (i) herbivores which consume plant tissue, (ii) carnivores which feed on animal tissue and (iii) omnivores which feed on both. The difference between animals and plants as prey types requires different physiological and behavioral adaptations, and has lead to repeated evolutionary divergence between carnivorous and herbivorous lineages.

Generalists and specialists
Predators vary in the number of species of prey they will feed on, with some species being specialists, whilst others are more generalist. Generally, parasites tend to be more specialist than true predators and herbivores tend to be more specialist than carnivores.

The impact of predators on prey population size
Do predators and parasites regulate the population size of their prey? This is not as simple a question as it may appear. There are two main issues: (i) the effect of any one predator may only be a small component of the total mortality causes affecting a prey species, so removal of the predator will have only a minor effect; (ii) predation may kill animals which were going to die anyway, so there will be no impact on the final prey population size. However, in a number of cases there is clear evidence that predators have a considerable impact on prey numbers.

Lotka–Volterra predator–prey model
The Lotka–Volterra predator–prey model is a simple mathematical model representing the interaction between predators and their prey. It makes three simplifying assumptions: (i) there is only one predator and one prey species involved in the interaction; (ii) prey numbers increase if the number of predators falls below a threshold and decrease if there are more predators; and (iii) predator numbers increase if the number of prey rises above a threshold and decrease if there are fewer prey. This simple model makes an interesting prediction: predator and prey populations will tend to cycle, as is observed in natural predator–prey dynamics.

Related topics

Natality, mortality and population growth (H2)

Population dynamics – fluctuations, cycles and chaos (H4)

Predator behavior and prey response (J2)

The nature of parasitism (K1)

The dynamics of parasitism (K2)

Defining predation

As mentioned in Topic I1, **predation** can be defined as the consumption of all or part of another individual (the prey). This wide definition thus encompasses (i) '**true predators**', which kill their prey soon after attacking them; (ii) **grazers**, which kill prey gradually (or not at all), and consume only part of the individual; (iii) **parasites**, which live in very close association with a single prey individual (the host), often inside the host's tissues. Parasites pose a special set of ecological and evolutionary problems, and are treated separately in Section K.

Carnivores and herbivores

Predators may also be divided into plant tissue-eating **herbivores** and animal tissue-eating **carnivores**, or **omnivores** which consume both. Whilst both prey types defend themselves by **structural** devices (for example the thick shell of a coconut or a tortoise), plants primarily use **chemical defences** whilst animals are able to adopt a variety of **behavioral strategies** (e.g. fight, run, hide). Vertebrates generally lack protective toxins, although some fish and amphibians are exceptions (e.g. puffer fish and the cane toad *Bufo marinus*). The different strategies needed to deal with, on the one hand, immobile but chemically defended prey compared to mobile and behaviorally complex but palatable prey have lead to an evolutionary divergence between carnivores and herbivores, which is particularly marked in eutherian mammals, where one order (Carnivora) of the 12 major orders contains all the carnivores (cats, dogs, bears, hyenas etc.) which feed on other mammals. Among the other orders, three specialize in insect predation (bats, shrews and anteaters) and six are wholly or almost wholly herbivorous (elephants, manatees, horses, ruminants, rabbits, rodents).

Generalists and specialists

Predators vary in their diet breadth, with some species being extremely selective **specialists**, concentrating exclusively on one prey type, whilst others are more **generalist**, being able to feed on a number of prey species. Herbivores tend to be more specialized than carnivores (either **monophagous** (feeding on one prey type) or **oligophagous** (few prey types)), concentrating on a narrow range of related plant species with similar defence chemicals, although generalist (or **polyphagous**) species feed on a range of plant species but avoid the more toxic parts and individuals. **Parasites** of both plants and animals tend to be specialists; thus, for example, most species of aphids (plant parasites) are highly restricted in diet – around 80% of the approximately 550 British aphid species are restricted to one host genus. Similarly, each of the 13 primate hosts (including *Homo sapiens*) of the nematode pinworm (*Enterobius* sp.) parasite is infested by a specific specialized pinworm species. In contrast, larger carnivores and herbivores generally have a wider diet width, thus most herbivorous mammals are relatively polyphagous. There are exceptions to these patterns – both monophagous mammals (e.g. the giant panda, a bamboo specialist, and the koala, a eucalypt specialist) and polyphagous parasites (e.g. the peach-potato aphid, which can attack over 500 plant species).

The impact of predators on prey population size

Do predators and parasites regulate the population size of their prey? This is not a trivial question. There are two main issues:

(i) the effect of any one predator may only be a small component of the total mortality causes affecting a prey species, so removal of the predator will have only a minor effect. Thus, in an experiment where ants were excluded from bracken fronds, the majority of bracken herbivores showed no significant change in numbers, indicating ant predation was swamped by other mortality factors;

(ii) predation removes an excess of animals above the number that the environment can sustainably support, so there will be no impact on the final prey population size. This is illustrated by the increase in sparrowhawk (*Accipiter nisus*) predation on great tits (*Parus major*) in Wytham Wood, England, following the banning of toxic pesticides. Mortality due to hawk predation increased from less than 1% to over 30%, yet there was no net reduction in great tit numbers, possibly because a scarcity of nest-holes is the key factor limiting population size in this species. An artificial example of the same process is demonstrated by the shooting of woodpigeons (*Columba palumbus*), which reduced overwintering mortality (presumably by reducing competition at a time when food was scarce), with no net impact on pigeon numbers.

However, there are numerous demonstrations of a marked impact of predators on prey numbers. Perhaps the most marked are the multiple extinctions caused by predators introduced onto tropical islands. For example, on the Pacific island of Guam, the introduced brown tree snake caused extinction or considerable reduction in the numbers of 10 native bird species. In such instances the prey are greatly disadvantaged because they have no evolutionary history of predation, and have evolved no appropriate responses. However, impacts can also be strong when prey have had long-term exposure to predators (*Table 2*).

Table 2. Some examples of the impact of predation upon prey numbers

Experiment	Result
Aphid predators (ladybirds, hoverflies and lacewings) removed from goldenrod plants	Aphid numbers increased by 30%
Duck predators (foxes, racoons, badgers and striped skunks) removed from nesting areas	Duck nesting densities increase by 300%, nest success increased by 50%
Foxes removed from some areas, small carnivores removed from others	Jackrabbit numbers increased by 300% in absence of foxes, but unchanged by removal of small carnivores
Dingo control and exclusion in Australia	Increase in numbers of more than 10 medium-sized mammals, and outbreaks of previously feral pigs.

When predator populations are not limited primarily by prey numbers, but by other factors, such as the availability of nesting sites or territories, then these predators are less likely to regulate their prey numbers. Thus, the effect of lion predation on migratory herbivores (buffalo, wildebeest, zebra) in Africa is constrained by the sedentary territoriality of the lions.

Lotka–Volterra predator–prey model

A variety of models have been developed to explore predator–prey dynamics. The **Lotka–Volterra predator–prey model** is a simple yet valuable example. It makes three simplifying assumptions: (i) there is only one predator and one prey species involved in the interaction; (ii) prey numbers increase if the number of predators falls below a threshold and decrease if there are more predators; and (iii) predator numbers increase if the number of prey rises above a threshold and decrease if there are fewer prey. As shown in *Fig. 1*, these two assumptions give rise to a very interesting prediction: the **predator and prey populations cycle in abundance**. *Figure 1a* shows the prey zero growth isocline, the critical predator density above which the prey population falls as death from predation exceeds the birthrate. In *Fig. 1b*, the predator zero growth isocline is shown. This is the critical prey density below which the predator population falls due to starvation. Combining these two isoclines combines the change in prey numbers and predator numbers to give the joint instantaneous change in predators and prey (*Fig. 1c*). A cyclical pattern will be followed, almost regardless of the starting numbers of predators and prey (providing both are above zero), as rising numbers of prey are followed by a rise in predator numbers, which in turn reduces the number of prey leading to a fall in predator numbers. Prey numbers are then able to rise and the cycle begins again. (The cyclical pattern illustrated in *Fig. 1* represents a single possible outcome – different starting numbers would result in cycles of different magnitudes.) Despite the simplicity of this model, the pattern shows clear similarities to the cycling observed in the snowshoe hare/lynx example shown in Topic H4.

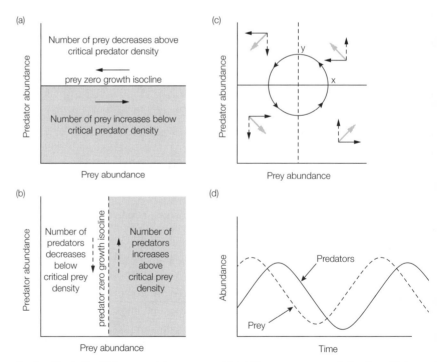

Fig. 1. The Lotka–Volterra predator–prey model. (a) The prey zero growth isocline; (b) the predator zero growth isocline; (c) both isoclines are combined to give the joint instantaneous change in predators and prey (solid arrows, change in prey numbers; dashed arrows, change in predator numbers). Note that the peak prey abundance occurs at x whilst the peak predator abundance occurs at y; (d) the coupled cycles of the abundances of predator and prey in (c) plotted against time.

J2 PREDATOR BEHAVIOR AND PREY RESPONSE

Key Notes

Profitability of prey	Given a choice between two potential prey types, a predator which is optimizing its effort should choose the most profitable prey. Evidence from common shore crabs and pied wagtails demonstrates that prey of a size which return the greatest energy reward per unit time are preferred over smaller and larger individuals.
Switching between prey types	Predators may alter or 'switch' their preference for a particular prey species depending on the abundance of that species. When this occurs, common prey are consumed superproportionately whilst less common prey are largely ignored.
The effect of prey density – functional responses	It is generally expected that at high densities of prey, a predator's consumption rate will increase and then flatten out as prey saturation occurs. This relationship is termed the functional response and may adopt different patterns, which can be stereotyped into three classes: functional responses I, II and III.
Searching and handling	To obtain food, a predator must first search for its prey and then 'handle' (catch, process and eat) it. Diet width can be regarded as being determined by a balance between a generalist strategy of searching for a wide variey of prey (relatively easy) and a specialist strategy of searching for one type of prey and handling that very efficiently. Optimal foraging theory assumes that evolution will have optimized predator behavior to maximize the rate of energy gain and makes predictions about how we should expect predators to balance searching and handling.
Heterogeneity and prey refuges	Predator–prey experiments in the laboratory indicate that in a simple environment, either (i) predators are able to consume all prey individuals, or (ii) the predator population becomes extinct and the prey survives. If, however, the habitat is more complex some prey individuals may be able to escape from predation in prey refuges and coexistence between predators and prey may occur. In corollary with the role of habitat patchiness in maintaining coexistence between competing species (see Topic I1), environmental heterogeneity is likely to be of critical importance in allowing predators and prey to coexist.
The ideal free distribution	Predators do not solely respond to the distribution and density of prey – they may also respond to the distribution of competing predators. Predators will tend to aggregate in the most profitable patches, but predator crowding will reduce the patch profitability until it is better to move to another less crowded patch. The ideal free distribution theory suggests that predators should move among sites until profitability is equal.

Plant defense	Plants defend themselves from predation in two main ways: (i) toxicity and unpalatablity, and (ii) defensive structures. There is a vast variety of chemical ammunition found in the plant kingdom used to defend plants against attacks from predators and parasites. These secondary compounds may either be directly toxic or they may reduce the food value of the plant, for example, by reducing the availablity of the leaf tissue protein to the animal gut. Defensive structures exist on a variety of scales, from small hairs on the leaf surface which may trap insects and other invertebrates, to large spines which deter mammalian herbivores. Both the levels of secondary compounds and the size of defensive structures may be elevated or 'induced' in plants that have suffered defoliation.
Related topics	The nature of competition (I1) The nature of predation (J1)

Profitability of prey

If a predator has a choice of a number of prey types, then if it is optimizing its effort it should choose the most **profitable prey**. The most profitable prey is not necessarily the larger individuals but rather those individuals which offer the **highest energy reward per unit time**. Commonly, it is observed that large prey require extensive handling times and hence predators prefer some intermediate between these and small prey, which offer low energy rewards. Shore crabs (*Carcinus maenas*) prefer intermediate-sized mussels (*Mytilus edulis*) which give the greatest energy gain per unit time. Similarly, pied wagtails (*Motacilla alba*) tend to select those scatophagid flies which give the greatest profitability (*Fig. 1*).

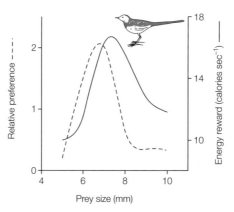

Fig. 1.   The relative preference of pied wagtails to different sizes of fly prey compared to the energy reward per unit time. Wagtails prefer the most profitable prey.

Switching between prey types

Predators which may consume a number of prey species may partially specialize on the commonest prey type available at any one time and concentrate their attacks on this more common prey. If the relative abundance of prey species then change such that another species becomes the commonest, this species will now be the focus of more attacks. This process is known as **predator switching**. When guppies were offered varying mixtures of water fleas and tubificid worms, they displayed a clear switching response, superproportionately consuming the more abundant prey (*Fig.* 2). Interestingly, when the diet comprised 50% tubificids and 50% waterfleas, the guppy population showed no bias in diet preference but individual guppies specialized on one of these two prey types. This tendency to specialize suggests that there are benefits to the predator – possibly searching efficiency is increased if the predator has a single 'search image'.

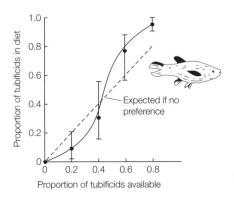

Fig. 2.   *Switching in a predator. Guppies superproportionately feed on the commoner prey type when offered water fleas and tubificid worms in varying ratios. The straight line indicates the expected proportion eaten if prey were selected randomly. Reproduced from Murdoch, W.W. et al., Ecology* **56**, *1094–1105, 1975, with permission from the Ecological Society of America.*

The effect of prey density – functional responses

As prey density increases, a predator's consumption rate will increase until it is consuming at its maximum possible rate. This relationship is the **functional response**. A functional response may be a direct linear relationship with predator consumption increasing at a constant rate as prey density increases, until the predator's processing ability is saturated and the consumption rate plateaus. This is a **Type I functional response** (*Fig.* 3). Alternatively, the predator's processing ability gradually becomes saturated as it consumes prey, and as a result the rate of increase of the consumption rate gradually reduces with prey density – a **Type II functional response**. If predator response is low at low prey densities due to low hunting efficiency or lack of a 'search image', and thereafter is curtailed by processing saturation, a **Type III functional response** is observed.

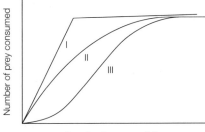

Fig. 3.   *The function responses of predators to increasing prey density. Functional responses of Types I, II and III are shown.*

Searching and handling

If evolution has optimized predator behavior to maximize the rate of energy gain, it is possible to predict how predators should behave. This is the basis of **optimal foraging theory**. Three separate components have to be taken into account to calculate the rate of energy gain: (i) the energy content of the prey item; (ii) the **search time** – the time to locate the prey; (iii) the **handling time** – the time between locating the prey and ingesting it, which includes such activities as processing (such as gnawing a nut or unrolling a hedgehog) and chasing and capture of elusive mobile prey.

The optimal foraging model makes **predictions** (i) that no predator can afford to consume **unprofitable** prey, regardless of how common they are; (ii) **search-intensive** predators should be **generalists** (otherwise they will spend an unprofitable length of time searching); (iii) **handling–intensive** predators should be **specialists** (to reduce the handling time to a profitable length); (iv) **productive** environments favor **specialism** (because search is easy), whilst **unproductive** environments (with high search costs) favor **generalists**.

The following are examples of fulfillment of these predictions.

- When great tits were offered small and large pieces of mealworm prey at high densities they ignored unprofitable small prey and specialized on large prey at high densities, but at low prey densities they generalized and accepted both types.
- Observation of search-intensive predators such as insectivorous birds which 'glean' leaves indicates that they are unspecialized, and thus minimize handling times.
- Handling-intensive predators are found to be specialists, such as the dormouse – a hazelnut specialist – or lions, which specialize on wildebeest and kudu.
- Blue herons in the productive waters of Florida are specialists, whilst those in the unproductive Adirondacks are more search-intensive and are generalists.

Heterogeneity and prey refuges

In a **simple environment**, a predator species may either (i) consume all prey and then become extinct, or (ii) fail to locate enough prey to avoid starvation, resulting in extinction of the predator but persistence of the prey. Both outcomes occurred in experiments between two protozoans, the predatory *Didinium* and the prey *Paramecium*, in glass culture vessels.

- *Didinium* was able to wipe out the population of *Paramecium* within three days when the culture vessel had no sediment on the bottom. The *Didinium* population then became extinct.
- When sediment covered the bottom of the vessel, *Paramecium* individuals buried in the sediment were able to escape *Didinium* predation. *Didinium* subsequently became extinct through starvation and the *Paramecium* population expanded.

Work on two mite species, the herbivore *Eotetranychus* (which feeds on the outer skin of oranges) and the predatory *Typhlodromus* demonstrates how **heterogeneity can allow coexistence** between predator and prey.

- In the first experiment, oranges were placed on a tray, mixed with rubber balls. A *Eotetranychus* population was established and then some predators added. The predators rapidly consumed all the prey and then became extinct themselves.

● In the next experiment, the 'microcosm' of the orange tray was made more heterogeneous. Vaseline barriers were added, which the mites could not cross, and 'launching posts' were added – small upright sticks which *Eotetranychus* could use to jump off and disperse, using silken threads to catch the air currents to drift to a new site. This complexity created a mosaic of patches. Within each occupied patch at any time, either the predators were 'winning' (by consuming all prey individuals) or the prey were winning (the population expanding in the absence of predators), or predators were 'losing' if they were in a patch devoid of prey. At the level of the whole microcosm, this **complexity resulted in long-term coexistence** between *Eotetranychus* and *Typhlodromus*.

A **prey refuge** exists when predators are unable to drive prey to extinction. In the case of both *Paramecium* and *Eotetranychus* above, a **physical hiding place** exists (the medium sediment and the oranges behind Vaseline barriers, respectively) which predators could not access. Prey refuges may take other forms, including (i) **differential habitat occupation** by predators and prey, (ii) **predator behavior** of avoiding low prey densities (either migrating or switching to other prey), (iii) **prey polymorphism** – visible variation (e.g. snail banding patterns) in the prey may allow the rarer forms to escape predation, and (iv) **superpredation**, where the predators themselves may be subject to intense predation in particular locations or at particular times.

The ideal free distribution

Predators optimizing energy intake will tend to aggregate in the most profitable patches, where prey density is highest, but predator crowding will reduce the patch profitability due to **interference** until it is better to move to another lower prey density, but less crowded, patch. The **ideal free distribution** theory suggests that predators should move among sites until profitability is equal at all sites. If profitability was higher at one site, the theory predicts that predators would redistribute themselves. This assumes that predators are 'ideal' in their assessment of the quality of patches and 'free' to move from patch to patch. An experimental demonstration of this process was performed in a laboratory **stickleback** population. Six fish were placed in a tank and food added at either end, with one end offering **high food density** and the other **low food density**. The distribution of the fish accurately reflected the variation in the food density.

Plant defense

There are two main ways that plants defend themselves from predation: (i) **toxicity and unpalatablity**, and (ii) **defensive structures**. Within the plant kingdom, many thousands of toxic **secondary compounds** have been recorded, for example cardiac glycosides in milkweeds, cyanide in white clover, nicotine in tobacco, mustard oils in cabbages. Some secondary compounds are not toxic but reduce the food value of the plant. Thus, tannins in the mature leaves of many woody plants bind to proteins, making them inaccessible to the predator's gut. Similarly, tomato plants produce protease inhibitors, which inhibit the protease enzymes in the herbivore's gut. The levels of secondary compounds may be elevated in plants which have been defoliated by herbivores. This **induction** of defense suggests an optimization of resource allocation – resources are only spent on defense when the costs are exceeded by the benefits. Thus, defoliation of 25% of the canopy of oak (*Quercus robur*) trees resulted in greatly increased mortality in leaf-mining moth larvae in the remaining leaves.

Defensive structures exist on a variety of scales, from small hairs (often hooked or bearing sticky exudates) on the leaf surface, which may trap insects and other invertebrates, to large hooks, barbs and spines (e.g. nettle, (*Urtica dioca*), dog rose (*Rosa canina*), holly, *Ilex aquifolium* and *Acacia* species) which primarily deter mammalian herbivores. The size and prevalence of such defensive structures may also be induced in defoliated plants.

K1 THE NATURE OF PARASITISM

Key Notes

The diversity of parasites

Parasites are a subgroup of predators (which consume the tissue of another living organism) that live in close association with their host. Parasites can be classified into two broad groups: (i) microparasites, which multiply within, or on the surface of, the host, and (ii) macroparasites, which grow in or on the host, but do not multiply. The main microparasites are viruses, bacteria, fungi and protozoans. Helminth worms and insects are important macroparasites. A large group of insect macroparasites are parasitoids, which lay eggs in or on the body of their insect host, and usually cause the death of the host.

Modes of transmission

Parasite transmission can be either horizontal (among members of a population) or, less commonly, vertical (passed from mother to offspring). Horizontal transmission may either be direct or indirect, mediated by a vector (e.g. a mosquito) or an alternate host.

Host response to disease

In vertebrates, infection by microparasites results in a strong immunological response. There are two components to this response: (i) the cellular immune response, where specialized cells directly attack pathogen cells, and (ii) the B-cell immune response, which gives rise to antibodies. After the first infection by a pathogen, immunological memory creates a rapid response to future attacks by the same organism, resulting in immunity. Invertebrates and plants also may suppress infections, but by less sophisticated, and usually less specific mechanisms. The loss of potential hosts which immunity engenders accentuates the boom-and-bust strategy of microparasites (see Topic K2).

Complex life cycles

Many parasites obligately switch between two or three host species in the course of their life cycle. Three different explanations have been advanced to explain such complex life cycles: (i) alternate hosts are vectors which have been attacked by the parasite (this cannot apply to immobile hosts which cannot be vectors); (ii) optimal habitat use is occurring, where different species are the optimal resource in different seasons, or for different life cycle stages; (iii) the pattern is due to evolutionary constraint, as parasites become highly adapted to one host for part of their life cycle, they are unable to leave it even though the rewards are higher elsewhere.

Social parasites

A completely different form of relationship is found between 'social parasites' and their hosts. Social parasites gain benefit from their animal hosts not by feeding on their tissues but by coercing them to provide food or other benefits. Such relationships are found in cuckoos which lay their eggs in the nests of other bird species, which then undertake the rearing of the young, and in some ant species which coerce the workers of another species to provision their brood.

Related topics

Natality, mortality and population growth (H2)

The dynamics of parasitism (K2)
The nature of predation (J1)

The diversity of parasites

Parasites can be classified into two broad groups: (i) **microparasites**, which multiply within, or on the surface of, the host, and (ii) **macroparasites**, which grow in or on the host, but do not multiply. The main microparasites are **viruses, bacteria, fungi** and **protozoans**. A selection of human and plant examples is given in *Table 1*. Macroparasites of plants and animals are dominated by invertebrates. In animals, **helminth worms** are particularly important (*Table 2*), whilst **insects** are the main macroparasites of plants (particularly butterfly and moth larvae and beetles), though other plants (e.g. mistletoes) may be important. Note that the size of the parasite is not always the determinant of whether it behaves as a microparasite or a macroparasite. Thus, aphids are microparasites of plants (reproducing on the surface of the plant) whilst fungi may be macroparasites of insects and plants, not reproducing until the host is killed.

Parasitoids (also known as hyperparasites) comprise a large group of insect macroparasites (mainly wasps and flies) which lay eggs in or on the body of their insect host and usually cause the death of the host.

Most parasites are **biotrophs**, only surviving on living tissue, but some (such as the sheep blowfly, *Lucilia cuprina*, and the plant fungal disease *Pythium*, which causes 'damping-off') continue to live on the host after causing its death. These are **necrotrophs**.

Table 1. A selection of microparasitic diseases

Microparasite	Human host	Plant hosts
Viruses	Measles HIV (the cause of AIDS) Influenza	Barley dwarf yellows virus Cauliflower mosaic virus
Bacteria	Thyphoid Scarlet fever Streptococcus	Fireblight Willow watermark disease
Fungi	Ringworm	Cabbage clubroot Rusts Potato blight
Protozoan	Trypanosomes (sleeping sickness) Plasmodium (malaria)	

Table 2. A selection of animal macroparasites

Class	Order	
Helminth worms	Tapeworms Schistosomes Flukes Monogeneans Nematodes	e.g. ox tapeworm in humans e.g. bilharzia in humans e.g. liver fluke in cattle fish gill parasites e.g. elephantiasis in humans
Insects	Fleas Lice Fly and wasp parasitiods	e.g. cat flea e.g. swift louse macroparasites on insects
Arachnids	Ticks	

Modes of transmission

Parasite transmission can be either **horizontal** (among members of a population) or, less commonly, **vertical** (passed from mother to offspring). Horizontal transmission may either be **direct** or **indirect**, mediated by a vector (e.g. a mosquito) or an alternate host. Occasionally the major route of transmission is 'accidental' acquisition via another species. Some examples of human parasites using alternative modes of transmission are given in *Table 3*.

Table 3. *Modes of transmission in selected human parasites*

Mode of transmission	Examples in humans
Vertical	HIV, rubella
Horizontal Direct	
close contact	Measles, common cold
sexual	HIV, syphilis
water contamination	Poliovirus, cholera
Indirect	Malaria (mosquito), sleeping sickness (tsetse fly)
Other species	Rabies, plague

Host response to disease

In vertebrates, infection by microparasites results in a strong immunological response, which has two distinct components: (i) **the cellular immune response**, where phagocytic cells (e.g. white blood cells – T lymphocytes) attack and engulf pathogen cells, and (ii) **the B-cell immune response**, which is based on the production of specific proteins (or **antibodies**) which bind to the surface of the pathogen by B lymphocytes. If the same pathogen (or **antigen**) is encountered subsequently, immunological memory results in rapid production of the specific antibody, giving rise to immunity.

Invertebrates and plants are also able to suppress infections, but by less sophisticated, and usually less specific mechanisms. Immunity results in a loss of potential hosts for microparasites, which accentuates their boom-and-bust strategy (see Topic K2 and Topic M1, *r*- and *K*-selection).

Behavioral strategies may also be important in reducing parasite levels. Many vertebrates exhibit **preening** behavior, which effectively removes ectoparasites. Denying chickens the opportunity to preen by debeaking changed the louse load from less than 50 per individual to 1600 per individual. Caribou avoid intense mosquito attacks by **migration** to higher altitudes in the summer months.

Plants and lower animals may also exhibit elevated immunity after parasite attack, but without the sophisticated specificity of vertebrates. For instance, in tobacco plants the infection of one leaf with tobacco mosaic virus will result in an increase in the levels of defensive chemicals throughout the plant, which will increase resistance to a wide range of pathogens. Herbivory may elicit similar responses.

Plants also have a further response to pathogens – **localized cell death**. When tobacco leaves are infected with tobacco mosaic virus, the plant responds by killing the cells in the locality, thus depriving the parasite of its food source. Similarly, leaves of the black poplar, *Populus nigra*, parasitized with aphid galls are shed long before unparasitized leaves.

Complex life cycles

Many parasite species **obligately switch** between two or even three host species in the course of their life cycle (*Fig. 1*, *Table 4*). Often, there are a suite of different

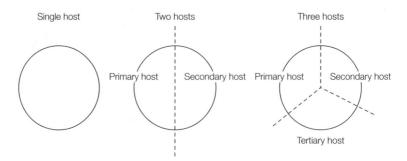

Fig. 1.   Complex life cycles in parasites.

morphs associated with the different hosts. In most cases, sex only occurs on the primary host, and if reproduction occurs on the other host(s), it is asexual.

Why do parasites commonly have such complex life cycles? Three different explanations have been advanced.

(i)   Some alternate hosts were initially used simply as **vectors** which have subsequently been attacked by the parasite to increase its reproductive output. However, this cannot apply to cases where the alternate hosts are sessile.

(ii)   Complex life cycles are a result of **optimal habitat use**. Thus, it has been suggested that host-alternating aphids use the hosts on which phloem sap quality (on which aphids feed) is highest when plants are growing most rapidly, which is in trees in the spring and herbaceous plants in the summer.

(iii)   Alternatively, complex life cycles may be due to an **evolutionary constraint**, as parasites become highly adapted to one host for the sexual part of their life cycle and are unable to leave the primary host even though the rewards are higher elsewhere. This explanation may account for a few cases, but it seems unlikely that the general widespread pattern is explained by such maladaptation.

Table 4.   Examples of complex life cycles in parasites. Two examples of single host use are provided for comparison

		Host		
Parasite	One host	Primary	Secondary	Tertiary
Measles	Human			
Apple sawfly	Apple			
Bilharzia		Human	Freshwater snail	
Ox tapeworm		Human	Cattle	
Black stem rust		Barberry	Wheat	
Trypanosoma brucei (sleeping sickness)		Man	Tsetse fly	
Peach-potato aphid		Peach tree	Potato	
Potato leaf roll virus		Potato	Peach-potato aphid	
Tapeworm (Diphyllobothrium latum)		Human	Copepod	Fish

Social parasites   'Social parasites' do not feed on the tissues of their hosts like true parasites, but gain benefit from their animal hosts by coercing them to provide food or other benefits. One such example is **brood parasitism** in birds. **Intraspecific brood parasitism** is found in a number of species, and is particularly common in ducks, including the goldeneye (*Bucephala clangula*). After the parasitic female has laid some eggs in a foreign nest, the host female typically responds by reducing the number of eggs she subsequently lays. **Interspecific brood parasites** include the cuckoo (*Cucuclus canorus*) in Europe and the brown cowbird (*Molothrus ater*) in North America, which lay eggs in the nests of other bird species. The cuckoo lays an egg in the nest of its host and removes one of the original clutch, thus conserving the clutch size. (For further discussion of the cuckoo, see Host-parasite coevolution in Topic K2.) Social parasitism is also prevalent among **ant and wasp** species. Some species, such as the ant *Lasius regina*, have a worker caste and are able to rear their own broods but may coerce other species to undertake this task, whilst other, obligate parasites, are workerless and depend entirely on other species to rear their young. In either case, colony takeover usually occurs by the parasite queen invading the nest and killing or dominating the resident queen. The resident workers continue to supply food and services to the brood, to which the parasite queen adds her own eggs.

K2 THE DYNAMICS OF PARASITISM

Key Notes

Parasite–host dynamics	Vertebrate hosts commonly acquire immunity to parasites. These immune hosts are no longer available to the parasites, and the size of the susceptible population is reduced. This results in a fall in the incidence of disease. However, as new susceptible hosts enter the population (for example, by birth), the disease will increase in incidence again. Thus, there is a tendency of diseases to cycle, rising as the number of new susceptibles increases and falling as the level of immunity rises. As the reproductive rate of a disease falls, the response of disease incidence to the influx of susceptibles is slower and the dynamics change from clear, synchronous cycles with a short period between peaks to reduced synchrony and longer periodicity until finally no cycles are observable.
Host–parasite evolution	The close association between parasites and their hosts often results in evolutionary interactions, or coevolution. Coevolution may give rise to defence mechanisms in the host and routes to overcome these defences in the parasite – so called 'arms races'. Not all coevolution between host and parasite is escalatory, however. For example, coevolution may lead to the reduction in the virulence of a parasite.
A model of microparasite disease	The reproductive rate of a microparasite (R_p) can be expressed as the number of new cases that will arise from an infected host: $R_p = \beta Sd$. If R_p is less than one, then the disease incidence is falling in the host population, whilst if R_p is greater than one, the level of disease is increasing. This model can predict (i) the evolution of reduced virulence when host deaths occur rapidly, (ii) altered host behavior to maximize parasite fitness, (iii) that there is a threshold density of hosts for a given parasite, below which the parasite will not survive, and (iv) that diseases with short periods of infectivity should not persist in small populations.
Heterogeneity in parasite populations	Whilst some diseases are stable, others have evolved 'escapes' from host defences by variation. The 'childhood' diseases of humans – measles, mumps, chickenpox etc. are largely restricted to children because they are antigenically stable. In contrast, parasites such as influenza, rhinoviruses (the cause of the common cold) and *Salmonella* have multitudes of strains which are constantly being added to. Hosts therefore have a constantly varying army of attacking parasites to contend with.
Heterogeneity in host populations	Individuals within a host population are very rarely equally at risk of being successfully attacked by a given parasite. The age, behavior, state of health, proximity of the infected individuals, and, of particular importance, the genetic predisposition of an individual will all influence the outcome.

Parasites as a reason for sex – the 'Red Queen'	The costs of sexual reproduction are high compared to asexual reproduction, as males produce no offspring, so population growth is slow. Therefore, there must be some balancing benefit. It has been suggested that the main benefit of sex is to produce genetic variation to overcome the ubiquitous and dynamic attack of parasites.
Related topics	Natality, mortality and population growth (H2) The nature of parasitism (K1) The nature of predation (J1) Sex in ecology (N2)

Parasite-host dynamics

Many microparasite diseases in vertebrate hosts give rise to **immunity**. This reduces the size of the susceptible population, resulting in a fall in disease incidence. However, as new susceptibles enter the population (for example, by births), the disease will increase in incidence again. Thus, there is a tendency of such diseases to **cycle**, rising as the number of new susceptibles increases and falling as the level of immunity rises (*Fig. 1*).

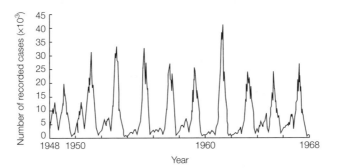

Fig. 1.   Cycling of the incidence of measles in England and Wales over a 20-year period.

As the **reproductive rate** of a disease falls, the response of disease incidence to the influx of susceptibles is slower, and the dynamics change from **clear, synchronous cycles** with a short period between peaks to **reduced synchrony** and longer periodicity until, in the case of diseases with low rates of increase and limited immunity, **no cycles** are observable (*Table 1*).

As with true predators (see Topic H4, *Fig. 3* and Topic J1, *Fig. 1d*), parasites may cause the populations of their hosts to cycle. A laboratory population of the azuki bean weevil, *Callosobruchus chinensis*, cycles when infested with the parasitoid wasp, *Heterospilus prosopidus*. Cycling is observed in Scottish red grouse (*Lagopus lagopus*) populations parasitized by a nematode, but not in unparasitized populations (*Fig. 2*).

Table 1.   Human disease dynamics vary with the reproductive rate of the parasite

Reproductive rate of parasite	Parasite	Cycle period
High	Measles (virus)	2 years
	Mumps (virus)	3 years
	Rubella (virus)	4–5 years
Low	Sleeping sickness (protozoan)	No detectable cycles

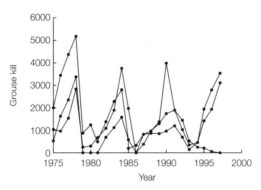

Fig. 2. Cycling in three red grouse (Lagopus lagopus) *populations parasitized by the nematode* Trichostrongylus tenuis *in southwest Scotland. Unparasitized populations show no cycling. Reprinted with permission from* Nature **390***, p. 547, 1997, Macmillan Magazines Limited.*

Host–parasite evolution

The close association between parasites and their hosts commonly gives rise to reciprocal evolutionary selection pressures, in which evolutionary changes in the host in response to the parasite give rise to evolutionary changes in the parasite. This is a type of **coevolution** (which may occur between any interacting members of a community). Coevolution may cause an **'arms race'**, which escalates as the increase in defense mechanisms employed by the host is countered by the development of the means to overcome these defenses in the parasite.

A simple example of host–parasite coevolution is found in the plant *Glycine clandestina* and its fungal parasite, the rust, *Phakospora pachyrhizi*, in which individual genes for virulence in the parasite have a complementary gene in the host conferring resistance. This is known as **gene-for-gene** coevolution and appears to be largely restricted to plants and their fungal pathogens. In most coevolutionary interactions, many genes are involved.

However, not all coevolution between host and parasite is an escalatory arms race. As explained below, coevolution may lead to the reduction in the virulence of a parasite.

A rather different example of coevolution comes from the **social parasite** (see Topic K1), the cuckoo, *Cuculus canorus*. The cuckoo has several common host species, in particular the reed warbler, meadow pipit, pied wagtail and dunnock. Each of these hosts is parasitized by a different race of cuckoo, the female of which lays an egg which **mimics** that of its host in color and pattern (with the marked exception of the dunnock egg) (*Fig. 3*). Hosts are generally likely to evict eggs which are unlike their own, which provides a clear benefit to the evolution of a mimic egg. Some apparently potential host species are particularly acute in their discriminatory powers and reject 100% of strange eggs. It has been suggested that these are species which have 'won' the evolutionary arms race against the cuckoo. The dunnock, on the other hand, appears to have rather low powers of discrimination, possibly because the association with the cuckoo has only occurred relatively recently, and the proportion of the dunnock population affected by cuckoo parasitism is relatively small, reducing the evolutionary impact of parasitism.

A model of microparasite disease

The **basic reproductive rate of a microparasite** is most easily expressed as the number of new cases that will arise from one infected host. The three components are (i) the **transmission rate** (β), (ii) the number of **susceptible hosts** (S),

Cuckoo egg

Host egg

Reed warbler   Meadow pipit   Pied wagtail   Dunnock

Fig. 3.   The eggs of cuckoos mimic those of their hosts, except the dunnock.

and (iii) the **duration** of the infectious period (d). The microparasite's basic reproductive rate (R_p) is defined as:

$$R_p = \beta S d$$

If R_p is less than one, then the disease incidence is falling in the host population, whilst if R_p is greater than one, the disease incidence is rising.

This model produces a number of simple predictions.

- For the disease to **persist**, neither β , S nor d should be small; in other words, transmission should be effective, the susceptible population must not be too small and the infectious period must not be too short.
- If hosts are rapidly killed by their parasites, this will reduce the infectious period d, (unless the parasite is a necrotroph) and will reduce the reproductive rate of the parasite. For this reason, new diseases may **evolve reduced virulence** as they maximize their R_p (see the myxomatosis example below). It is important to note that this can happen as a consequence of the parasite optimizing its strategy. It does not follow that evolution inevitably leads to a benign relationship between parasite and host.
- If R_p is low, the parasite will be selected to increase its virulence. This may occur by **manipulation of the host's behavior**. For example, (i) sneezing is a response which very effectively disperses common cold viruses, (ii) parasitized aphids and bumble bees will migrate to sites where the parasitoid wasps can safely develop and emerge.
- The critical size below which the number of susceptible hosts must not fall (the **threshold density**) can be expressed in terms of the equation above. As the disease will not persist if R_p is less than one, the equation can be stated in terms of the threshold density of susceptible hosts (S_T):

$$S_T = \frac{1}{\beta d}$$

This predicts that for a parasite to persist in small host populations, it should (i) be highly infectious, and (ii) give rise to long periods of infectiousness (and be unlikely to kill their hosts).

These predictions are confirmed in observations of natural and human populations:

(i)   highly infectious diseases (such as those sexually transmitted) and diseases with long periods of infectivity (such as protozoan diseases in mammals) are found in small populations, and

(ii) diseases which have short periods of infectivity are not found in small populations. For example, measles in humans requires a population of around 0.5 million to persist.

Vaccination affects R_p by reducing the number of susceptible individuals, S. Thus, as the rate of disease spread is reduced, it is not just the vaccinated individuals which benefit. This effect is known as the **herd immunity.**

Heterogeneity in parasite populations

Whilst some diseases are stable, others have evolved 'escapes' from host defences by variation. The childhood diseases of humans – measles, mumps, chickenpox etc. – are largely restricted to children because they are **antigenically stable**. Once an individual has been infected by such a parasite, the resulting antibody defence will ensure no further disease occurs throughout the individual's life. In contrast, parasites such as influenza, rhinoviruses (the cause of the common cold) and *Salmonella* have multitudes of **strains** which are constantly being added to. Hosts therefore have a constantly varying army of attacking parasites to contend with. Sometimes, a novel strain will overwhelm a population, as occurred in the influenza epidemic of 1919 which killed over 20 million people.

Variation in parasites does not always favor the most virulent, as the example of the **myxoma virus** demonstrates. This virus, originally only found in the South American jungle rabbit, and which causes **myxomatosis in the European rabbit**, was introduced to both Australia and the UK in the early 1950s. A variety of strains exist, which have been classified into five grades (I–V) on the basis of their virulence, with I being the most virulent and V the least. In the UK in 1953, 95% of the parasite population comprised high-virulence grade I type. By 1962, less than 5% of the myxoma population was grade I, and the dominant type was grade III. A very similar pattern was recorded in Australia (*Fig. 4*).

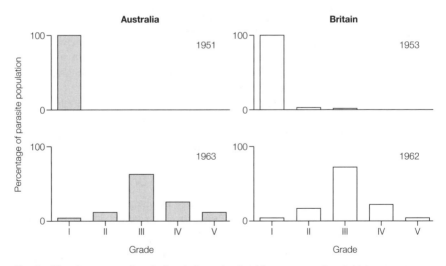

Fig. 4. The change over time in the virulence levels of the myxoma virus (which causes myxomatosis in rabbits) in Britain and Australia. Virulence is highest in the Grade I strains and lowest in the Grade V strains.

Heterogeneity in host populations   Individuals within a host population are very rarely equally at risk of being successfully attacked by a given parasite. The **age, behavior, state of health and genetic predisposition** of an individual will all influence the outcome, as will the proximity of the infected individuals. As the disabling effects of microparasitism often depend on the disruption of processes at the cellular level, small genetical changes to cellular proteins may alter the ability of the parasite to disable its host. Genetic variation within a host population thus provides a route by which some individuals can avoid disease. An example of such variation is found between mouse strains (where, within each strain there is little genetic variation) in resistance to nematode (*Trichuris muris*) infection. After infection with 400 nematode eggs, the level of infection generally reduced over time as a result of immunological defences, but there were large differences between mouse strains in this response, demonstrating **genetic variation in immunity** (*Fig. 5*). Selection experiments on mice to increase resistance to this nematode resulted in the mean percentage resistance rising from 72% to 100%.

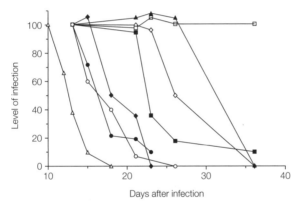

Fig. 5. The response of eight strains of mice to inoculation with eggs of the nematode Trichuris muris.

Another commonly cited example, from humans, is the maintenance of the gene for **sickle cell hemoglobin**. The gene alters hemoglobin so that it polymerizes at low oxygen concentrations, distorting ('sickling') the red blood cell and killing it. Individuals with two copies of this gene (homozygotes) usually die at an early age due to the debilitating 'sickle cell disease', which results in feebleness and heart, kidney, spleen and brain damage. Heterozygotes (possessing one copy of the gene) suffer from 'sickle cell trait' which results in milder symptoms of anaemia. Despite these costs, the gene is maintained in human populations exposed to the **malarial parasite *Plasmodium falciparum*,** as it confers resistance to malaria. In individuals with sickle cell trait, infected red blood cells bind to capillary walls in tissues where oxygen concentrations are low, which results in sickling and death of the red blood cells and their parasite load. In West Africa, where *P. falciparum* is abundant, the sickle cell gene is common, possessed by around 24% of the population. It is also found (in lower levels, as the advantage of malarial resistance is irrelevant) in communities in Europe and North America of West African origin.

Parasites as a reason for sex – the 'Red Queen'

Genetic variation is prevalent in all large populations that sexually reproduce (see Topic O1). The **costs of sexual reproduction** are high compared to asexual reproduction, as males produce no offspring, so population growth is slow. Therefore, there must be some balancing benefit. It has been suggested that the **main benefit of sex is to overcome parasite attack** by the production of genetic variation. If a species is constantly being attacked by new parasites, then no matter how well adapted a species is at any one time, the quality of the environment will be constantly declining as the parasite burden grows. Sex, due to the effects of recombination which creates new combinations of genes, may help to overcome this problem by creating new phenotypes which pose new problems for parasites. This idea of the need for continuous evolutionary change has been termed the '**Red Queen**' effect, from the character in Lewis Carroll's *Through the Looking Glass*. The Red Queen says to Alice 'it takes all the running you can do, to keep in the same place'.

L1 MUTUALISM

Key Notes

Mutualism and symbiosis	Mutualism is a positive reciprocal relationship between two individuals of different species which results in increased fitness for both parties. Mutualism may be symbiotic, in which the organisms live together in close physical association.
Obligate mutualism	Some mutualisms, such as lichens, are permanent pairings in which one or both partners cannot lead an independent life. Most symbioses are obligate, as are some nonsymbiotic mutualisms, such as those formed by fungus-farming ants.
Facultative mutualism	The majority of mutualisms are nonobligatory and opportunistic. They may be diffuse, involving a varying mixture of species, as occurs between many pollinators and their plants.
Pollination	An outcrossing plant needs to transfer its pollen to the stigma of a conspecific plant, and receive pollen from a conspecific. Some plant species rely on the wind to achieve this, which can work acceptably well if plants grow in large homogeneous stands of few species, as occurs in grasslands and pine forests. However, in most species of flowering dicotyledonous plants, insects, birds bats or small mammals are employed to transfer pollen from plant to plant, usually in exchange for either nectar or pollen itself as a foodsource
Seed dispersal	Large seeds cannot be effectively wind-dispersed, and unless dispersal by water (as occurs in the coconut palm) occurs, such plants are dependent on animals for dispersal. Rodents, bats, birds and ants are all important seed dispersers.
Symbiotic mutualists within animal tissues and cells	A number of animal species rely to some extent on mutualisms with mutualists which reside within their bodies. Ruminants (deer and cattle) possess a multi-chambered stomach in which bacterial and protozoan fermentation take place. In some termites, which feed on wood, the necessary breakdown enzymes are provided by bacterial mutualists. Intracellular bacterial symbionts which transform amino acids occur in a number of insect groups, including aphids and cockroaches.
Defensive mutualisms	Some mutualisms provide one partner with a defense against predators or competitors. Examples of such defensive mutualisms are found between some grasses and alkaloid-producing fungi, and between many plant species and ants.
Mutualism and evolution	Mutualism may have arisen from parasite–host and predator–prey relationships, or between closely coexisting species with no cooperation or mutual benefit. Evolutionary changes in both partners (coevolution) have then resulted in both partners benefiting from the relationship, although it is

possible for mutualisms to 'deteriorate' into unbalanced exploitation of one partner to the benefit of the other – parasitism. Mutualistic interactions have been central to a number of important steps in the evolution of multicellular organisms. Many of the cell organelles of higher organisms, including mitochondria and chloroplasts are believed to be derived from symbiotic bacteria.

Related topics   The nature of competition (I1)   The nature of parasitism (K1)

Mutualism and symbiosis

Mutualism is a reciprocal relationship between two individuals of different species which results in **increased fitness** for both parties. **Symbiotic** mutualism occurs when the organisms live together in close physical association, whereas **nonsymbiotic** mutualisms involve species which do not live together.

Obligate mutualism

Obligate mutualisms involve **permanent pairings** in which one or both partners cannot lead an independent life. This is the case with the fungus–algae symbiosis called **lichens**. A lichen consists of a mat of fungal hyphae within which a thin layer of cells of photosynthetic algae or cyanobacteria is embedded. The algae gain the protection from drought and solar radiation from the fungus whilst the fungus receives the products of photosynthesis from the algae. Thus lichens can thrive in exposed and extreme environments in which neither fungus nor algae could. Another example of a symbiotic mutualism comes from **mycorrhizae**, the intimate association of fungal hyphae with the roots of many species of higher plants. The fungi aid plants in the uptake of nutrients (particularly phosphate) whilst receiving nutrients from the plant. Mycorrhizal associations are particularly important in poor soil types, and it is now common practice to inoculate shrubs and trees with mycorrhizae to aid establishment. **Ectomycorrhizae**, which occur on trees, form a mantle of fungus around the tips of rootlets, whilst **endomycorrhizae**, which are associated with a wide range of wild and crop plants, invade the root cells. **Corals** consist of a symbiotic mutualism of corraline **anthozoans** with photosynthetic **dinoflagellate algae** which provide the anthozoans with sugars *(Fig. 1)*. The anthozoans are carnivorous filter-feeders, but zooplankton provide only about 10% of their daily energy requirements, the majority being provided by their algal partners.

Most symbioses are obligate, as are some nonsymbiotic mutualisms such as those formed by **fungus–farming ants**, in which neither ant nor fungus can survive without the other. Another close mutualism occurs between **cleaner fish** and cleaner shrimps and the **'customer' fish,** from which they remove parasites and dead skin. Removal of cleaner fish from a reef site in the Bahamas caused a rapid onset of skin disease, and increased mortalities.

Facultative mutualism

The majority of mutualisms are **facultative**, where the partners may coexist without a reliance on each other, and are only mutualists **opportunistically**. Often, such relationships do not involve tight **pairwise** relationships between two species, but are **diffuse**, involving a varying mixture of species. Thus, honey bees visit many different species of flowering plants in the course of the season, and many of these plants will be visited by a number of insect pollinators. The relationship between plants and **nitrogen fixing bacteria,** such as occurs between legumes (peas) and *Rhizobium*, is facultative. In nitrogen-poor soils,

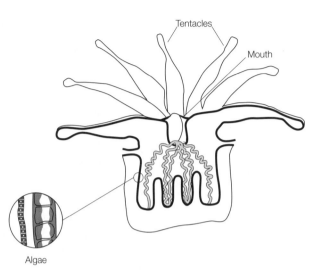

Fig. 1.   The anatomy of a coral polyp showing the location of the symbiotic photosynthetic algae.

legumes benefit greatly from the nitrogen-fixing activities of the bacteria which form nodules in the plant roots. However, when the soil nitrogen levels are higher, the plants can thrive in the absence of the bacteria.

Pollination   To exchange genes with the rest of the population (see Topic N2 for the advantages of sex), an **outcrossing** plant needs to transfer its **pollen** to the **stigma** of a conspecific plant, and receive pollen from a conspecific. **Wind-pollination** can achieve this in species that grow in large **homogeneous** stands of few species, as occurs in **grasslands** and **pine forests**. However, in most species of flowering dicots, **pollinators**, which may be **insects, birds, bats or small mammals** are employed to transfer pollen from plant to plant. Normally, the pollinators benefit by receipt of either **nectar** (a sugary syrup usually enriched with amino acids), oil or pollen itself as a food source. Some plant–pollinator relationships involve a tight **pairwise** interaction in which both species depend on the other, as occurs in *Yucca* cacti and **yucca moths,** and **fig** trees and **fig wasps**. Male **Euglossine bees** are **orchid pollinators** which receive no food reward, but complex chemicals which the males transform into sex pheromones. However, most plant–pollinator relationships are more **diffuse** than this, with each pollinator using a range of plants from which nectar or pollen is harvested, which changes through

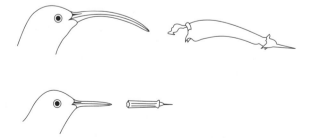

Fig. 2.   Two hummingbird morphologies and the flower shapes they specialize on. Redrawn from Coevolution, *D.J. Futuyma and Slatkin (eds), 1983, Sinauer Associates Inc.*

the season as the suite of available flowers changes. Often there is a degree of specialization, as is apparent in the two **hummingbird** types shown in *Fig. 2*, which respectively specialize on flowers with long and short corollas.

Seed dispersal

Air currents can effectively disperse very small seeds, but the only options for large seeds are either dispersal by **water currents** (as occurs in the coconut palm) or by animals. **Rodents, bats, birds** and **ants** are all important seed dispersers. Some specialist seed dispersers are **seed predators**, which consume seeds but aid dispersal by dropping or storing and losing seeds. Although this seed loss may be accidental, the relationship is still mutually beneficial to both parties. Other dispersers include **frugivores** which consume **fleshy fruit** but excrete or dispose of the seed. In tropical forests up to 75% of the tree species produce fleshy fruits whose seeds are animal-dispersed. Plants have evolved these energy-rich fruits as 'pay' to encourage the attention of frugivores. Frugivore–plant relationships are usually **diffuse**, a number of different animal species consuming the fruits of a given plant. It may be noted that some animal-dispersed seeds (such as burrs which are trapped in the fur of mammals) confer no benefit to the animal and are not mutualistic.

Defensive mutualisms

Some mutualisms provide one partner with a defense against predators or competitors. A number of **grass species**, including the common perennial ryegrass, *Lolium perenne*, have a mutualistic relationship with **Claviciptacae fungi**, in which the fungi grow either within the plant tissue or on the leaf surface and produce **alkaloids** which are powerful toxins, conferring protection to the grass from both grazers and seed predators. **Ant–plant** mutualisms are very widespread. Many plants provide the ants with a food source from specialized glands called **extrafloral nectaries** on the leaves or stems which secrete protein- and sugar-rich fluid. In many **acacia** tree species, the ants also gain physical protection, living in the hollow thorns of the trees. The ants provide their hosts with a strong defense against herbivores, vigorously attacking any intruder, and in some cases also limit competition for the plant by removing the surrounding vegetation. Demonstration of the protective effect of ants is shown by experimental removal of ants from acacia and *Tachygali* trees, which subsequently suffered greatly elevated levels of herbivory. In acacia species which are facultative mutualists with ants, individual plants which have no ants have high levels of protective **secondary chemicals** (see Topic J2), whilst those which harbor ants have very low levels of investment in these compounds. This further indicates the value of ants as herbivore deterrents.

Symbiotic mutualists within animal tissues and cells

Mutualists which reside inside the **guts** or **cells** of their animal partners are common. **Ruminants** (deer and cattle) possess a **multi-chambered stomach** in which bacterial and protozoan fermentation take place. In some **termites**, which feed on wood, the necessary breakdown enzymes, **cellulases**, are provided by bacterial mutualists which live in specialized gut structures. Termites maximize the efficiency of energy extraction from their food by eating their feces – **refecation**. Some termites also possess bacteria capable of **fixing atmospheric nitrogen** (as occurs in legumes and other plants), a valuable benefit as wood is very nitrogen-poor. Intracellular **bacterial symbionts** which aid nitrogen metabolism by synthesizing essential amino acids occur in a number of insect groups, including aphids and cockroaches. These bacterial symbionts have closely **coevolved** with their hosts.

Mutualism and evolution

The evolution of mutualism may have occurred in different situations from either **parasite–host** or **predator–prey** relationships, or between **closely co-existing species** with no cooperation or mutual benefit. Insect pollination, for example, is likely to have started by the theft of pollen by insects from wind-pollinated flowers. Evolutionary changes in both partners (**coevolution**) have then resulted in both partners benefiting from the relationship. Thus, in the plant–pollinator relationship, the advantages of enhanced pollination success produced flowers that attracted insects (bright colours, scents, nectar). Nevertheless, it is possible for mutualisms to 'deteriorate' into unbalanced exploitation of one partner to the benefit of the other – **parasitism**. For example, many **orchids** offer **no reward** at all for their pollinator, but instead induce them to alight on the flower by using scent, shape and color patterning to mimic female insects (especially bees and wasps). This is an example of a mutualistic relationship which has changed evolutionarily to a parasitic one.

It is now widely accepted that the **origins of the eucaryotic cell** have involved the acquisition of procaryotic symbionts. Both **mitochondria** (the cell 'power-house' in all higher organisms where oxidative respiration occurs, resulting in the production of **ATP**) and **chloroplasts** (the light-capturing and harvesting units found in all plants) have their origins as free-living procaryotes, and contain circular DNA genomes and other characteristics of bacteria.

M1 LIFE HISTORY

Key Notes

Life history variation	An organism's life history is its lifetime pattern of growth and reproduction. The observed variation in the life history patterns of different species is vast. Some species live for hundreds or even thousands of years, some grow to vast sizes whilst other are minute. Some produce many, tiny offspring whilst others have few, relatively large offspring. How these differences evolve is a key question in ecology.
Energy allocation and trade-offs	It is not possible for an organism to reach a large adult size shortly after birth, produce many large offspring and live to a great age. Energy allocated to one aspect of life history cannot be spent on another, so 'trade-offs' between different traits are inevitable.
r- and K-selection	Two distinct life history strategies are described by r- and K-selection theory, which states that r-selected species are adapted to maximize the rate of increase of the population size, whilst K-selected species are adapted to be competitive. Thus, r-selected species have rapid development, small adults, many and small offspring and a short generation time. In contrast, K-selected species have slow development, large adult size, few, large offspring and long generation times. Although some species fit this theoretical dichotomy, many do not, and this theory is now out-dated as better models have wider predictive powers.
Reproductive values	The reproductive value (RV_x) of an individual of age x is the number of offspring that individual is expected to produce in the immediate future plus those expected over its remaining lifespan. The reproductive value of an individual inevitably rises after birth and falls towards old age. The difference in reproductive values between individuals provides a powerful predictor of life history strategy.
Habitat classification	In addition to the concepts of r- and K-selection, there a variety of schemes which classify habitats in an attempt to discern a pattern linking habitats and life histories. Habitats may be classified on the basis of the balance of benefits between growth and reproduction, into 'high-cost-of-reproduction' and 'low-cost-of-reproduction'. An alternative approach, 'bet-hedging', considers the impact of the habitat on the relative variability of mortality or fecundity for different growth stages and uses this to predict optimal life history strategies. Grime's 'CSR triangle' classifies habitats from a perspective of plant life history, using the degree of habitat disturbance (or stability) and its severity to plants.
Diapause and dormancy	Many organisms undergo a delay in development at some stage in their life cycle, such as seed diapause or implantation delay in a red deer embryo. This strategy is an adaptive response to avoid unfavorable conditions (such as giving birth to young deer or germinating in the midwinter). Individuals may

also enter periods of reduced metabolic activity such as long-term hibernation or short-term torpor when the benefits of foraging or other activities are limited.

Migration

Organisms may also avoid local harsh conditions by moving to another location. Migration is directional movement, such as the autumn flight of swallows from Europe to Africa. Dispersal, in contrast, is a nondirectional movement away from the birth or breeding site. There are three categories of migration, depending on whether an individual makes either (i) repeated return trips, (ii) a single return trip, or (iii) a one-way trip.

Complex life cycles

Many plants, fungi and animals have complex life cycles, in which either individuals adopt radically different morphological forms (e.g. caterpillar/butterfly, tadpole/frog) or generations differ radically from one another in a predictable fashion (e.g. haploid/diploid alteration of generation in plants, sexual/asexual generations in rusts and aphids). Why are these complex strategies adopted? In some species, different stages are mainly devoted to either growth or dispersal, whilst in others different habitats are optimally utilized in different stages.

Senescence

Why do individuals deteriorate as they grow old, with resultant reduced fecundity and vigor? There are two levels of answer to this question. At the mechanistic level, senescence is caused by the breakdown in cellular machinery to the effects of toxins and natural radiation. However, this cannot be the complete story – the onset of senescence varies hugely between species. This suggests evolution affects or determines senescence. Evolutionary models of senescence suggest that either (i) a mutation deleteriously affecting older individuals will be selected against more weakly than one affecting younger individuals, or (ii) there are genes which benefit early reproduction which have deleterious effects later in the lifespan.

Related topics	Populations and population structure (H1)	The nature of competition (I1) The nature of parasitism (K1)

Life history variation

The **life history** of an organism is its lifetime pattern of **growth** and **reproduction**. Key components of life histories are **size, growth rates, reproduction** and **longevity**. There is **enormous variation** in the life history patterns between species. Some species live for hundreds or even thousands of years (such as the yew, *Taxus baccata*); some, like blue whales or Californian redwoods (*Sequoia sempervirens*), grow to vast **sizes** whilst others are minute. Some, such as fungi or pelagic fish produce **many, tiny offspring**, whilst others (for example the pipistrelle bat in which the two offspring comprise 50% of the mother's post-birth weight) have **few, relatively large** offspring. A key question in ecology is asking how these differences evolve.

Energy allocation and trade-offs

The perfect **hypothetical organism** will have every trait required to maximize reproductive output – it will reach a large adult size shortly after birth, produce many large offspring and live to a great age. However, such '**Darwinian demons**' do not exist, because it is not possible to maximize every life history trait in this way. Energy allocated to one aspect of life history cannot be spent on another, so 'trade-offs' between different traits are inevitable. A trade-off between growth and reproductive output is found in many temperate tree species, including Douglas fir (*Fig. 1*). This **cost of reproduction** is apparent from a wide variety of organisms.

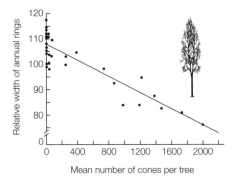

Fig. 1. A trade-off between growth and reproductive output in Douglas fir Pseudotsuga menziesii.

In red deer, the mortality of milk hinds (those with calves) is significantly higher than that of yeld hinds (which have no calves), demonstrating that survival trade-offs with fecundity. Nonreproductive females of the woodlouse (or pill-bug) *Armadillium vulgare* allocate three times as much resources to growth than reproductive females. Many perennial garden plants will exhibit enhanced survival and future flower production if they are stopped from allocating resources to reproduction by removal of seed-heads. Within reproduction, there are choices on **energy allocation** available to the organism. Resources may be allocated to one 'big bang' bout of reproduction – **semelparity** (as occurs in Atlantic salmon, and the bamboo plant), or spread over time more evenly – **iteroparity** (found in most large mammals and in trees). The same energy allocation can produce either **many small offspring** or **fewer, larger offspring**.

r- and K-selection The concept of *r*- and *K*-selection theory, suggests that **r-selected species** have evolved in **unstable environments** and thus maximize **population growth** (the *r* of the logistic equation – see Topic H1), or evolved in **stable** habitats near the **carrying capacity** (*K* of the logistic equation) and are thus adapted to be **competitive**. Thus, *r*-selected species have all the characters which maximize population growth: **rapid development, small adults, many** and **small offspring, a high energy allocation to reproduction** and a **short generation time**. In contrast, *K*-selected species have features which maximize competitive ability: **slow development, large adult size, few, large offspring, a low energy allocation to reproduction** and **long generation times**. Making broad comparisons across taxa, the general pattern supports these kinds of life history differences – for example, woodland trees and large mammals have many *K*-selected characters whilst annual plants and insects have some *r*-selected characters, although close scrutiny suggests the fit is less than perfect. More robust evidence supporting the *r/K* dichotomy is provided by two species of **reed mace** (*Typha* sp.) from Texas and North Dakota respectively. The North Dakota species, *T. angustifolia* experiences high winter mortality and low competition compared to the

Texan *T. domingensis*. As predicted by the *r*/*K* scheme, *T. angustifolia* matures earlier (44 days compared with 70 days), reaches a lower height (162 cm compared with 186 cm) and produces more fruits (41 per plant compared with eight per plant) than *T. domingensis*. However, there are many examples of cases which do not fit the *r*/*K* dichotomy – one study found only 50% of examples agreed with the predictions. For example, aphids have among the highest population growth rates of all animals of comparable size (suggesting they are *r*-selected) and yet give birth to relatively large offspring (a *K*-selected trait). The *r*/*K* theory is now not generally regarded as wrong, but rather of being a **special case**, encompassed by better models with wider predictive powers.

Reproductive values

All organisms have to compromise between the energy allocated to **current reproduction** and that which is allocated to survival and thus **future reproduction**. The **reproductive value (RV_x)** of an individual of **age *x*** is the number of offspring that individual is expected to produce in the immediate future (**the current reproductive output**) plus those expected over its remaining lifespan (**future reproductive output**). Evolution is expected to maximize the total number of offspring an individual passes to the next generation in other words, the **reproductive value at birth**. Thus, reproductive values provide an evolutionarily relevant way of comparing different life histories. If future life expectancy is low, allocation to current reproduction should be high, whilst if the remaining expected lifespan is long, allocation to current reproduction should be lower. The reproductive value of an individual inevitably **rises** after birth and **falls** towards old age (*Fig. 2*).

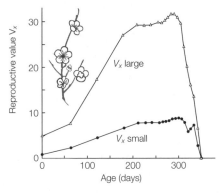

Fig. 2. Changing reproductive values with age in large and small Phlox drummondi.

The difference in reproductive values between individuals provides a powerful predictor of life history strategy. An example of this is provided by the winkles *Littorina saxatilis* (shore snails) which adopt different life histories in two different habitats, (i) narrow **crevice habitats** between immobile rock-faces and (ii) on the surface of mobile **boulders**. The crevice population lives in a very different environment to the boulder population – crevices are highly sheltered environments which protect snails from waves and predators, but are limited in space and give rise to intense competition, whilst moving boulders can crush small snails, making this a risky habitat. These differences result in life history divergence – the crevice population have thin shells, are smaller, mature at a small size, have high allocation to reproduction and produce **few**,

large offspring. In contrast, the boulder population have thick shells, are large, mature at a large size, have low allocation to reproduction and produce **many, small offspring**. Note that both morphs fit the r/K dichotomy (see above), with the crevice form showing r-selected characters and the boulder form showing K-selected characters with the exception of the number and size of offspring, **which are reversed**. These patterns can be explained by comparing the reproductive values of different sized snails in the two habitats (*Fig. 3*). The boulders favor large adults with strong shells (requiring more allocation to growth and less to reproduction, in keeping with the K-selected strategy), but small individuals are randomly killed by moving boulders, so the optimal solution is to produce many offspring, which are then necessarily small, in keeping with the r-selected strategy. The crevices, on the other hand, favor fast growth to a small adult size (crevices are too small to accommodate large adults) and more allocation to reproduction, in line with r-selected strategy, but the intense competition for space favors large young, a K-selected trait.

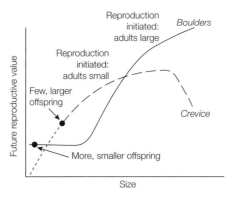

Fig. 3. The relationship between future reproductive value and snail size in boulder and crevice habitats. This pattern predicts the observed difference in life history traits between the two habitats.

Habitat classification

There are a variety of schemes which classify habitats in an attempt to discern a pattern linking habitats and life histories, in addition to the concepts of r- and K-selection described above. Such classification schemes must be able to classify all habitats, and do so from the point of view of the organism in question. Whether a habitat is homogeneous or heterogeneous, or benign or severe, will differ according the perspective of the organism concerned.

Habitats can be divided into those which result in a **high cost of reproduction** (high-CR) and those which have a **low cost of reproduction** (low-CR). In high-CR habitats (where there is intense competition, or intense predation on small adults) any reduced growth that results from reproduction will have a high cost on future reproduction. Species in such habitats would thus be expected to start reproduction relatively late in life, after reaching a moderate size. In contrast, in low-CR habitats (where competition is weak, large individuals are subject to intense predation or mortality rates are very high and randomly targeted), there will be few advantages to putting off reproduction.

'Bet-hedging' theory contrasts the variability of habitats in terms of the impacts upon different parts of the life history (birthrates, juvenile mortality rates, adult mortality rates, etc.). If adult mortality rates are relatively constant compared to juvenile mortality rates, adults may be expected to 'hedge their bets' and spread the births of their offspring out over a long period (i.e. be **iteroparous**), whereas if juvenile mortality rates are lower than adult mortality rates, the allocation to reproduction should be high, and the offspring produced all at once (**semelparity**).

Grime's CSR triangle is a three-way classification of **plant life histories** which has somewhat broader applicability than the r/K dichotomy. This classification has two axes, one describing the **habitat disturbance (or stability),** the other its **severity** to an average plant. Three types of environment are potential habitats for plants: (i) **low severity, low disturbance** (ii) **low severity, high disturbance;** and (iii) **high severity, low disturbance.** (High severity, high disturbance habitats (such as active volcanoes and highly mobile sand dunes) are uninhabitable.) Each of these three habitats favors particular life history strategies (*Fig. 4*). **Low severity, low disturbance** habitats will favor life histories which maximize **competitive** (C) ability among adults. **Low severity, high disturbance** habitats will favor a high reproductive rate, which is characteristic of weedy species – the **ruderal** (R) strategy. **High severity, low disturbance** habitats, such as deserts, favor a **stress-tolerant** (S) strategy which allocates, resources to storage at the expense of competitive or reproductive ability.

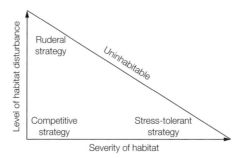

Fig. 4. Grimes' CSR classification of habitats and plant life histories.

Diapause and dormancy

If current conditions are harsh, and the future is expected to be more agreeable, it may be adaptive for an organism to enter into **dormancy**, where development is **temporarily suspended**. Dormancy may occur just once, as is commonly observed with **plant seeds**, or may be repeated, as occurs with many temperate and arctic mammals during the winter. Dormancy in insects, termed **diapause,** is common. In the common field grasshopper, *Chorthippus parallelus*, the egg stage is resistant to sub-zero conditions which would kill other stages. The egg determines that winter is over by monitoring the temperature – only after 90 days below 4°C does development continue. Thus diapause allows the grasshoppers to 'migrate' in time from autumn to spring. In many temperate mammals, such as red deer, young are born at the optimum time for survival and food availability by a **delay in implantation** of the **embryo**. Breaking seed dormancy in plants often requires a combination of conditions (temperature, water, oxygen). If these do not occur, the seed may remain in the soil for some time, as part of a **seed bank**. Some seeds, if kept in the correct conditions, may survive for hundreds or even thousands of years, as in the case of wheat *'Triticum'* and the weed, *Chenopodium*, taken from archaeological excavations.

In the minute, moss-dwelling animals, the tardigrades, a special form of dormancy called **cryptobiosis** may occur at any point in development, and animals may remain in this state for many years.

Some birds and mammals save energy by allowing their body temperatures to temporarily fall to near ambient, during periods when they are inactive. Such **torpor** may occur as part of a daily cycle, as occurs in species of humming-birds, bats and mice, or may last for longer periods. Deep torpor in response to cold conditions is called **hibernation,** and it is usually characterized by a reduction in heart rate, and total metabolism, with a core body temperature of below 10°C. Hibernating mammals, such as hedgehogs (*Erinaceus europeaus*) and woodchucks (*Marmota monax*) generally feed heavily in late summer to build up a fat reserve which will provide energy during winter months. Some species which spend less time in torpor, such as the grey squirrel (*Sciurus carolinensis*) and the chipmonk (*Tamias striatus*) lay up stores of food. In bears, such as the grizzly *Ursus arctos*, hibernation involves a cessation of activity and a reduction in metabolic demands by about 50%, but the body temperature is only dropped slightly, from 37°C to 35°C. Bears may remain in hibernation for up to six months. This remarkable feat is not simply achieved by huge fat reserves, but also by the ability to convert toxic urea, which is normally excreted in the urine, into amino acids.

Some species of birds and mammals may respond to prolonged periods of very high temperatures in deserts and similar habitats by entering a summer dormancy similar to hibernation, termed aestivation (or **estivation**).

Migration

Organisms may avoid local unfavorable conditions by moving to another location by migrating. There is a clear similarity between dormancy, which allows the organism to 'move' in time through an unfavorable period, and migration, which is movement in space to a more favorable location. **Migration** is **directional movement,** such as the autumn flight of swallows (*Hirundo rustica*) from Europe to Africa. In contrast, **dispersal** is a **nondirectional movement** away from the birth or breeding site. Dispersal may be viewed as a process evolved to avoid intraspecific and intersibling competition, and to avoid inbreeding, which is discussed further in Topic I2.

Migration may occur on a variety of timescales, from return trips which are repeated on diurnal or tidal cycles to annual or less frequent cycles. Migrations may involve remarkable levels of investment. For example, in the Arctic tern (*Sterna paradisaea*) the annual round trip, from the breeding grounds in the Arctic to the summering location of the Antarctic shelf, totals about 20 000 miles (32 000 km). It is possible to classify migration in relation to the frequency of migration in an individual lifecycle. There are three categories of migration, depending on whether an individual makes (i) **repeated return trips**, (ii) **a single return trip**, or (iii) a **one-way trip** (*Table 1*). Repeated return trips are probably the commonest form of migration, occurring as a part of the daily or seasonal behavior patterns of many species. Many migrant species make only one complete return trip in their lifetime, however. In such species, growth occurs away from the habitat in which they are born, and which they return to, to breed once before dying. Eels and migratory salmon are both examples of this. A less common type of migration is the one-way trip, in which one generation moves in one direction, and a subsequent generation makes the return trip. The most remarkable instance of this is found in the monarch butterfly, *Danaus plexippus*, which migrates south from Canada and the northern USA to a few

Table 1. Patterns of migration

Migrant species	Habitat 1	Habitat 2
Repeated return trip		
Marine zooplankton	Sea surface (night)	Deeper water (day)
Bats	Roosting sites (day)	Feeding areas (night)
Many birds	Feeding areas (day)	Roosting sites (night)
Frogs, toads	Water (breeding period)	Land
Caribou (reindeer)	Tundra (summer)	Forest (winter)
One return trip		
Eels (*Anguilla anguilla* in Europe, *A. rostrata* in N. America)	Ponds and rivers in Europe and North America (growth)	Sargasso Sea (breed)
Pacific salmon	American rivers (breed)	Pacific Ocean (growth)
Butterflies, dragonflies, etc.	Larval habitat	Adult habitat
One-way trip		
Monarch butterfly	Mexico	Northern USA and Canada
Red admiral butterfly, painted lady butterfly	Southern Europe	Britain

restricted sites in Mexico in the autumn. These individuals initiate the next generation in the far south of the USA. Succeeding generations move gradually northwards until the late summer comes and the southern migration occurs.

Complex life cycles

In many species, complex life cycles occur, in which either **individuals** adopt radically different **morphological forms** or **generations** differ radically from one another in a predictable fashion. Morphological transformation in individual life histories is known as **metamorphosis,** as occurs in the holometabolous **insects** (e.g. beetles, butterflies and moths, flies) in which the larval form is completely different from the adult (e.g. caterpillar/butterfly) and in **amphibians** (e.g. tadpole/frog). **Intergenerational** change may also involve morphological transformations, as occurs in many 'host-alternating' **aphids** and fungal **rusts** (which alternate a sexually reproducing generation on one plant host with a number of asexual generations on another host) and in many animal parasites which move between hosts (see Topic K1). In **plants** (most obviously in ferns and mosses) this intergenerational change involves the chromosomal complement change from haploid to diploid – 'the alteration of **generations'.** Why have these complex strategies evolved? It has been suggested that complex life cycles are unstable and maladaptive as organisms will have to make evolutionary compromises in adapting to different environments. Several hypotheses have been put forward to account for adaptive advantages to life cycle complexity. A trade-off between **dispersal and growth** has been proposed to account for the life cycles of many marine invertebrates such as barnacles, which are adapted for dispersal as larvae but do not grow significantly until they metamorphose into a sessile form. The same argument may be applied, with the roles of adult and larva reversed, to butterflies and caterpillars. A more generally applicable suggestion is that complex life cycles represent an **optimization in habitat utilization.** As habitats change seasonally, or the requirements of an individual change with growth, the optimal strategy may change. Thus, frogs switch from herbivory as larvae to carnivory as adults and host-alternating aphids feed on fast-growing woody plants in the spring and then switch to herbaceous plants when the woody plants cease growth in the early summer.

Senescence

The **deterioration** of individuals as they grow old, with resultant **reduced fecundity**, **vigor** and **survival** is inevitable, although deterioration occurs in a matter of days in some species and only after hundreds of years in others. Why does this occur? There are two levels of answer to this question. At the **mechanistic level**, senescence is caused by the breakdown in cellular machinery due to the effects of chemical **toxins** such as highly reactive free radicals and natural radiation. However, this cannot be the complete story as the onset of senescence varies hugely between species, suggesting **evolution** affects or determines senescence. There are two competing evolutionary models of senescence: (i) **mutation-accumulation** and (ii) **antagonistic pleiotropy**. The **mutation-accumulation** model states that a mutation deleteriously affecting older individuals will be selected against more weakly than one affecting younger individuals, as a gene which acts early will reduce the reproductive output of an individual much more markedly. Thus, the population will have early-acting 'bad genes' removed efficiently by selection, but late-acting ones will be more persistent. In contrast, **antagonistic pleiotropy** occurs where genes which benefit early reproduction have deleterious effects later in the lifespan. Some evidence for antagonistic pleiotropy comes from red deer (*Cervus elaphus*) in which a gene associated with early breeding and high fecundity is also associated with lower survival. In *Drosophila melanogaster*, selection for enhanced survival results in the depression of early fecundity (supporting antagonistic pleiotropy); but reversed selection of these lines for increased early fecundity does not reverse an increased resistance to stress, suggesting that deleterious genes which made individuals susceptible to stress were purged by the first bout of artificial selection, supporting the mutation-accumulation model. It thus seems likely that both processes occur. Natural selection for increased longevity in a species may thus have effects on a range of life history parameters.

N1 SOCIAL GROUPS, COOPERATION AND ALTRUISM

Key Notes

Cooperation	In many animal species, individuals cooperate for their mutual benefit. This cooperation can be temporary e.g. (for hunting or offspring care), or permanent, lasting the lifespan of the individual (e.g. an ant colony, or a pair of mute swans).
Grouping – benefits	Many species of animals form groups, such as flocks of starlings, shoals of herring or prides of lions. Being in a group gives an animal some advantages: avoidance of predation, location of food and catching elusive or large prey.
Grouping – costs	There are also costs of belonging to a group: increased competition for food, increased conspicuousness to predators, increased risk of disease.
Altruism	Altruism occurs when an individual causes an increase in the fitness of another individual of the same species at a cost to its own survival or offspring production. An example is the sterile worker in a eusocial ant society, which sacrifices its opportunity of reproducing and instead supports its mother's reproduction. This appears paradoxical, as it implies that the altruist is actively reducing its fitness. However, if the altruist is related to the beneficiary, this strategy may result in more of the altruist's genes passing to the next generation.
Eusocial society	A select group of animal species display an extreme form of sociality, in which there is a reproductive division of labor: some individuals (usually called 'workers') forego reproduction, whilst others are fertile ('sexuals' such as queens). This type of society is only common among a group of insects in the order Hymenoptera equipped with a sting – ants, bees and wasps.
Related topics	Intraspecific competition (I2) Sex in ecology (N2)

Cooperation Cooperation is a commonly observed phenomenon among animal species, whereby individuals act for the mutual benefit of each other. Cooperation is often temporary or transitory, as in the example of offspring care, but may be permanent, lasting the lifespan of the individual (e.g. an ant colony, or a pair of mute swans) the commonest general kind of cooperation involves the provision of nutrition and defense of the young. This may take the form of one cooperative donor (usually the mother), two donors (where both parents are involved) or many donors (in eusocial species such as ants and cooperatively breeding birds such as the pied kingfisher).

**Grouping –
benefits**

The fact that so many species of animals live in social groups, such as flocks of geese or herds of wildebeest, suggests that there are some fundamental advantages to this strategy. There are two fundamental types of advantage: avoiding predators and getting food. Individuals in a group can avoid predation through three routes: (i) **increased total vigilance**; (ii) **the dilution effect or 'selfish herd'**; (iii) **group defense.** As a group becomes larger, each individual may spend less time being vigilant (and is able to spend time on, say, foraging) and yet the total vigilance of the group can continue to rise. Thus, goshawks are highly successful at catching single wood pigeons, with an 80% attack success rate, but this success rate drops as the flock size increases, falling to less than 10% for wood pigeons in flocks of 50 or more birds (*Fig. 1*). This change is apparently due to the increased distance from the hawk at which larger flocks take flight.

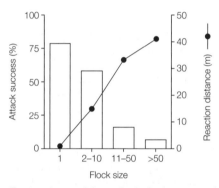

Fig. 1.   As woodpigeon flock size increases, the reaction distance to a goshawk increases and goshawk attack success declines.

   The dilution effect of a group is simple to comprehend – if a lion is going to attack a group of gazelles, the probability of any individual being killed, given that the lion will only be able to kill one gazelle per attack, is inversely proportional to the number of individuals in the herd (i.e. 1/10 if there are ten gazelles, 1/100 if there are 100). However, this effect may be offset by the redistribution of predators to match the clustering of prey (see Topic J2, **ideal free distribution**). It nevertheless appears to be important in some situations, especially where the opportunities for escape or defense are limited, as occurs in water skaters when predated from below by a small marine fish, *Sardinops sagax*. The fish attack groups of water skaters independently of the group size, so the mortality rate is higher for individuals in smaller groups. Further secondary support for the selfish herd comes from the fact that in many species, the animals on the edge of the group, which are those most exposed to predation, are individuals of low social rank, such as newcomers or immature individuals. **Group defense** occurs in many aggregations of prey species. It is common for small birds to mob birds of prey or corvids (crows) which threaten to predate them or their eggs, whilst musk ox adults form an outward-facing circle around the juvenile members of the group when threatened by a pack of wolves. The advantages of belonging to a group in accruing food are two-fold: **locating food-rich sites** and, for predators, **catching elusive prey**.

Grouping – costs   Being in a group has its disadvantages, particularly **increased competition for food, increased conspicuousness to predators** and **increased risk of disease.**

Cliff swallows (*Hirundo pyrrhonota*) in Nebraska form colonies of up to 3000 pairs and the level of **parasitism** with the swallow bug increases as colony size increases (*Fig. 2*). This parasitism reduces the growth rate of the swallow nestlings. If bugs are eliminated by fumigation, the nestling growth rate is higher in the largest colonies (indicating enhanced foraging efficiency in large groups), whereas in unfumigated nests there is no clear relationship between colony size and nestling growth rate.

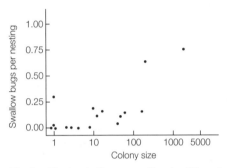

Fig. 2.   *The level of bug parasitism in cliff swallow colonies increases as colony size increases. Reproduced from Brown and Brown,* Ecology **67***, 1206–1218, 1986, with permission from the Ecological Society of America.*

Grey wolves in North America form packs of up to 12 animals, yet a recent study shows that the amount of food caught per individual *declines* linearly as the pack size increases, even if the prey are large (e.g. moose). In other words, single wolves or pairs would be better off. Why then do wolves aggregate? As a pack generally comprises a mating pair and a group of younger related animals (offspring and siblings), it is likely that the dominant individuals gain benefit by increasing their **inclusive fitness** (see Altruism below) by elevating the fitness of their relatives (who share some of their genes).

Altruism   Altruistic behavior, whereby an individual appears to sacrifice fitness to benefit another, seems paradoxical on first impression. Why should worker caste ants (see below) sacrifice the ability to reproduce? Why should aphids and wasps produce 'soldier' larvae which defend the colony but do not develop into adults? The solution lies in the idea of **inclusive fitness.** An individual's fitness is high if it passes many copies of its genes on to future generations. One way of doing this is to leave many offspring, but gene copies are shared amongst an individual's relatives. Supporting the reproductive output of a relative may be an effective way for an individual to increase its fitness. As close relatives share more genes than distant ones, we would expect a balance between the cost to the donor and the relatedness of the recipient. This idea is captured by **Hamilton's Rule** which says that the fitness benefit to the donor must be greater than the fitness cost to the donor (C). The benefit to the donor is given by the total fitness benefit (B) multiplied by the **relatedness** (r) of the recipient to the donor. Thus, for altruism to occur, $rB > C$. Thus altruism is expected to be uncommon except among highly related individuals. Clear examples of sacrifice of fitness, such as the sterile worker castes in termites, do involve highly related individuals. Where there is **apparent altruistic** behavior among unrelated individuals, as occurs in the pied kingfisher, closer observation may reveal a different picture. Unrelated males will help a breeding pair to provide their young with food, for which they gain no direct fitness benefit. However, when the breeding male dies, the female often pairs with the unrelated helper. There is, therefore, a clear payback for the helping behavior.

Eusocial society   A few animal species display an extreme form of sociality, in which altruistic behavior is apparent by a division of labor: some individuals (usually called 'workers') forego reproduction, whilst others are fertile ('sexuals' such as **queens**). **Eusociality** is found in ants, bees, wasps (all in the order **Hymenoptera**), termites, aphids and one mammalian species – the naked mole rat.

Ants, bees and wasps, like all Hymenopteran insects, have an unusual chromosomal make-up, know as **haplodiploid**. The females are diploid, that is they have a paternal and a maternal set of chromosomes. However, males are haploid – they are the product of unfertilized eggs, and hence have no paternal chromosome set. This unusual situation means that full sisters (sharing both parents) are more closely related to one another than to their mother, as they have all of their father's genes and half of their mother's, and hence have a relatedness of 0.75. It has been suggested that this high relatedness amongst sisters might partially account for the frequency of altruistic behavior and eusociality found in Hymenoptera, although other factors must be important too. In particular (i) in the modified ovipositor which female (but not male) ants, bees and wasps possess – **the venomous sting** which can defend the group, and (ii) the tendency to build **nests**. Naked mole rats, termites and aphids are all **diploid**, and eusociality in these species appears to be primarily determined by a **high relatedness** amongst individuals.

N2 SEX IN ECOLOGY

Key Notes

Sex	Sex is a widespread phenomenon, found in plants and animals, which has a number of ecological consequences. Sex is evolutionarily important as it generates genetic variation, and ecologically important because the different sexes behave differently. The difference between sexes arises from the asymmetry in the energy invested in offspring development.
The costs of inbreeding	When closely related individuals interbreed, 'inbreeding' occurs, which gives rise to (i) inbreeding depression (which causes a reduction in offspring fitness) and (ii) increased genetic homozygosity (which reduces the possibility of evolutionary adaptation to a changing environment, and increases the possibility of extinction). Many species have mechanisms that prevent or curtail inbreeding.
Self-fertilization	Self-fertilization involves male and female gametes produced by the same, hermaphrodite individual. Self-compatibility is a means by which sessile and less mobile organisms' such as plants, that are unable to actively seek mates, insure against failure to cross-fertilize. A single species may switch from self-fertilization to cross-fertilization depending on the environment and level of isolation.
Sexual versus asexual reproduction	Sexual reproduction recombines the genomes of the parents producing genetically variable gametes and offspring. This maintenance of high levels of genetic variation enhances disease resistance and allows adaptation to unpredictable environmental conditions. In contrast, asexual reproduction does not generate genetic variation but can be seen as an adaptation to predictable environmental conditions.
Mating systems	The social structure of matings in animals, and particularly in vertebrates, is an important ecological parameter. Monogamy is where one male and one female form a pair bond, either for a breeding season, or until one dies. Where one male mates with a number of females, polygyny occurs, whilst the opposite situation (which is rarer) where one female has a mating group of a number of males is known as polyandry.
Sex ratio	The sex ratio is the ratio of males to females in a population, and usually is close to 1 male:1 female. However, unequal costs, crowding and mate competition can bias the sex ratio.
Sexual selection	Sexual selection causes such features as the peacock's tail, the elaborate song of the male reed warbler and the antlers of the male stag. Sexual selection is caused by differences in reproductive success due to competition for mates.
Related topics	Populations and population structure (H1) Genetic variation (O1)

Sex   Sexual reproduction is characterized by the **meiotic** subdivision of a **diploid** nucleus which rearranges the distribution of genetic material in the chromosomes, and the production of **haploid** gametes. In species in which males and females have differentiated, the males produce small gametes (**sperm** and pollen) whilst the females produces large gametes (**ova**). In many species the female makes a markedly higher energy investment in offspring than the male does. In some cases (e.g. fallow deer) males provide only sperm, whilst the mother gives extensive energy provision, both in the uterus and after birth, by lactation. This difference in energy allocation gives rise to differences between the sexes (see Sexual selection p. 161).

The costs of   **Inbreeding** occurs when closely related individuals, sharing many of the same
inbreeding   genes, interbreed. It results in two deleterious effects: a loss of genetic variation within individuals (increased **homozygosity**) and a loss of fitness in the offspring, a phenomenon called **inbreeding depression**.
                          Genetic **homozygosity** is the proportion of individuals in a population that are homozygous at a particular genetic locus. Homozygosity represents a loss of potential for adaptation to changing environments, an increased susceptibility to disease and a greater probability of extinction (see Topic V1). Homozygous populations are more susceptible to decimation through disease as there will be little variation for disease resistance. The classic example is the Irish potato famine in the 1840s caused by the devastation of the staple crop through *Phytophora infestans* infection. Once established the potato blight was able to spread through the entire crop (of Western Scotland as well as Ireland) because a single variety of potato was cultivated.
                          Inbreeding tends to result in a reduction in offspring fitness. The effects of this **inbreeding depression** are attributed to the increased frequency of offspring homozygous for **deleterious recessive alleles.** These are lethal or disadvantageous alleles (see Topic O1) that are only expressed when they are present in the homozygous form. Homozygotes will result from mating between individuals each carrying a single copy of the deleterious allele, which is more likely to occur where there is breeding among related individuals.

                          Captive animals
                          Inbreeding depression is apparent in **artificially maintained populations** such as laboratory rats and pedigree dogs. It is also a serious problem for captive breeding attempts in zoos who often have only a small population of related animals from which to breed. A study of the offspring of zoo animals, including the Indian elephant, zebra, giraffe, reindeer and Pere David's deer, showed those born to a female mated to a related male had a lower chance of survival than those born to females mated to unrelated males.

                          Wild animals
                          Inbreeding in wild animals is rare because they have many **behavioral mechanisms** which prevent it. These include kin recognition and male biased dispersal, both of which prevent mating between close relatives. Whether inbreeding depression has a significant effect in natural populations is a contentious issue as it has proved very difficult to detect. It has been observed in small, isolated populations of the adder, *Viperus berus*, in southern Sweden. Litter sizes were smaller and the proportion of stillborn and deformed offspring was higher than in nonisolated populations. Introducing males from other

areas into the isolated population sharply reduced the incidence of inviable offspring – good evidence that the low fitness was not caused by some environmental factor.

The primary reason why it can be difficult to observe the effects of inbreeding depression in the field is that the underlying deleterious alleles will tend to be eliminated from the population by natural selection. As the frequency of these alleles decreases the chance of homozygotes being produced on mating also decreases. Therefore, it may be that inbreeding depression can only be detected immediately following population decline or isolation. This is probably the case for the northern elephant seal (see Topic O1), which has very low levels of genetic variation but with no obvious ill affects.

Self-fertilization

Self-fertilization involves male and female gametes produced by the same individual. Plants and animals with the ability to produce both male and female gametes are **hermaphrodite**. Not all hermaphrodites are self-fertile. Some plants are self-compatible but are predominantly cross-fertilized, for example *Primula vera*. Self-compatibility can be viewed as an insurance measure against failure to cross-fertilize. At the other extreme, some plants have flowers which do not open at all (are cleistogamous) and reproduce solely through self-fertilization. Self-fertilization and hermaphroditism can be advantageous in marginal habitats where population densities are very low and mates rarely encountered. Clearly, sessile organisms such as plants, that are unable to actively seek mates, benefit from being able to produce both male and female gametes and from the potential to self-fertilize.

A single species may adopt one or more fertilization strategies. *Viola* produces insect pollinated flowers in the spring and closed, cleistogamous flowers in summer in response to changing day length. This may be an adaptation to decreasing visibility to pollinators or decreasing pollination success as the season progresses. The snail, *Rumina*, typically fertilizes itself in nature. In contrast, the white-lipped land snail, *Triodopsis albolabris*, only self-fertilizes if it is kept in isolation for several months and then produces offspring that are less fit than those produced through cross-fertilization.

Sexual versus asexual reproduction

Sexual reproduction shuffles, or **recombines**, the genome of the parents resulting in genetically variable gametes which in turn result in genetically variable offspring. This generation of genetic novelty maintains high levels of genetic variation in the population on which natural selection can act, ensuring disease resistance and adaptation to unpredictable environmental conditions. In contrast, asexual reproduction generates little or no genetic novelty (depending on the mechanism employed). The asexual seeds produced by the blackberry, *Rubus fructicosus*, and the dandelion, *Taraxacum officinale*, differ genetically from their parent to some small degree whilst aphid daughters are genetically identical to their mother. Asexual reproduction allows the rapid colonization of a habitat by a single well-adapted genotype and can be seen as an adaptation to predictable, less risky conditions.

The advantages of sexual reproduction
Recombination occurs during meiosis when the genotypes of each parent is rearranged such that each gamete, while containing half of each parental genotype, carries a unique combination of parental genes. Further novelty is introduced when these genetically varied gametes come together to yield genetically

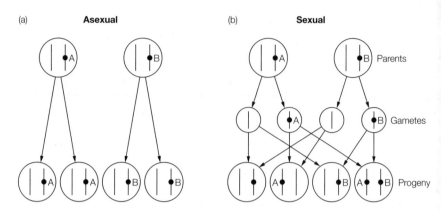

Fig. 1. *The genetic consequences of (a) asexual, and (b) sexual reproduction. Asexual reproduction produces progeny that are exact replicas of the parents, carrying the same genes (A, B). Sexual reproduction produces genetically variable progeny that carry genes from the two parents in different combinations.*

variable progeny carrying genes from the two parents in different combinations (*Fig. 1*).

If we suppose that two genes A, and B, are deleterious we can see that, unlike asexual reproduction, sexual reproduction has the capacity to eliminate them from the progeny. If A and B were advantageous, sexual reproduction can give rise to a new combination containing both genes.

Sexual reproduction may represent a means by which an individual 'keeps its options open' and ensures the survival of some offspring. The genetic shuffling inherent in sexual reproduction may have another primary function, namely to ensure the individual stays one step ahead of continuously adapting disease organisms, thus preventing loss of all offspring to a specialized pathogen or parasite.

Is asexual reproduction advantageous?

In addition to bringing together favorable gene combinations, sexual reproduction will tend to break up existing advantageous combinations. This is referred to as the cost of recombination. Asexual reproduction produces progeny that are wholly exact replicas of the parent which may be particularly appropriate in predictable, favorable environments.

Aphids have complex life cycles involving both sexual and asexual phases. The sycamore aphid, *Drepanosiphon platinoides,* has a female-only parthenogenetic stage to its life cycle in spring and summer (*Fig. 2*) where founder females give birth to live female offspring asexually. This mode corresponds to the period when food supply in the form of sycamore sap is abundant and increasing. In contrast, in autumn, when food supply is declining and weather conditions deteriorate, males are produced and the aphid enters a sexual stage, shuffling the genome to produce eggs that, in theory, maximize the chance that a founder female's offspring will survive the winter and contribute to the next generation. Aquatic blue-green algae enter into a sexual phase when nutrient levels are low, producing cysts which lie dormant at the bottom of the pond or lake until conditions trigger growth. In this case sex, appears to be partly a mechanism for avoiding unfavorable conditions.

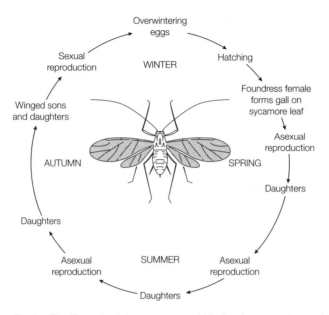

*Fig. 2. The life cycle of the sycamore aphid, showing asexual reproduction in the spring and summer and sexual reproduction in the autumn. Adapted from Johnson, Nature **332**, 726–728, 1988, with permission from Macmillan Magazines Limited.*

Mating systems

The social structure of the mating group determines the **mating** system in operation. **Monogamy** occurs where one male and one female form a pair bond, either for a breeding season, or until one dies. Monogamy is common among birds, where 90% of species are monogamous, but less so amongst mammals where **polygyny** is much more common, where one male mates with several or many females. Relatively rare in any group is **polyandry**, where one female is the centre of a mating group with several males. This spectrum of mating patterns can be viewed partly as a reflection of the environmental productivity. Where a female can successfully raise offspring on her own, a male can mate with other females to improve his mating success. However, if the environment is less productive, it may require two parents to raise the young, in which case a male should stay with his partner. In very harsh environments, it may be necessary to have more than two parents, and in this case polyandry may be the most successful strategy. This pattern is supported by observations – polyandry is found in birds of the sub-arctic tundra such as dotterel, and in human societies in the harshest areas of the Himalayas, whilst in very productive reedbeds, reed warbler populations are polygynous.

Sex ratio

The sex ratio is usually expressed as the **relative number of males to females in a population**. Thus, an equal number of males and females may be expressed as a **ratio of 1:1**. (Alternatively, it may be expressed as the proportion of the total population comprised of males. An equal number of males and females gives a proportion of 0.5.) Populations of most organisms tend towards a 1:1 sex ratio. The evolutionary reason for this is known as **Fisher's sex ratio theory**, which can be illustrated as follows. If males were rare relative to females, each male would mate with many females and produce many offspring. The male fitness would thus be higher than the female fitness. If the opposite situation occurred, and females were rare, then female fitness would exceed male fitness. Thus we would expect any sex ratio bias to be corrected by evolution as a mother's fitness would be higher if she biased production of offspring to the rarer sex. This is an example of **rare type advantage**. Fisher's sex ratio theory suggests that there should be **equal investment** in either sex as a result of rare type advantage.

However, there are a variety of **exceptions** to equal sex ratios.

Unequal costs

If an individual of one sex is **more costly** to the mother than an individual of the other sex, equal investment in either sex will result in a greater number of the cheaper sex. An example of this is found in the solitary bee, *Anthophora abrupta*, in which females are 58% heavier than males. If mothers invested equally in males and females, we would expect there to be 58% more males hatching from eggs, in other words a sex ratio of 1:1.58. The observed sex ratio is 1:1.63, very close to that predicted. Another example is given by the general bias in mammals in the **sex ratio at birth** towards males, coupled with higher male juvenile mortality than female juvenile mortality. For example, the sex ratio of fetuses of the moose (*Alces alces*) in Canada is slightly biased towards males, a sex ratio of 1.13:1. However, the adult population is strongly female biased. The bias in the allocation of resources to males at birth does not mean that males are necessarily gaining more resources. The elevated mortality levels of males reduces the chance of males gaining postnatal maternal investment such as lactation and protection, and hence reduces the mean level of maternal investment in sons.

Local resource competition

In some species the sex ratio is biased at birth because females can maximize their reproductive success by producing unequal numbers of male and female offspring. In many primates, sons disperse from their mothers' home range while daughters stay put. In crowded conditions **local resource competition** between females is intense and it is better to produce sons who will move away from the home range and may stand a higher chance of breeding successfully (*Fig. 3*).

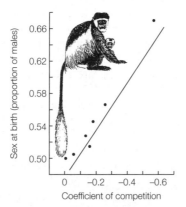

Fig. 3. The relationship between sex ratio at birth in primate species and competition between females. Each point on the graph is the average for a primate genus.

Local mate competition
Where mating commonly occurs among siblings, it is wasteful for a mother to produce as many sons as she does daughters, and the sex ratio is thus female-biased. This is known as **local mate competition**, and is found in a number of invertebrate species. In *Adactylidium* mites the sex ratio is 1 male to 6–9 females. Bizarrely, the male mates with his sisters in the womb and then dies before he is born.

Sexual selection

Males and females often differ by much more than simply the structure of their reproductive organs, but also in behavior, size and many morphological characters. These **secondary sexual traits**, such as the peacock's tail, the elaborate song of the male reed warbler and the antlers of the male stag are a product of **sexual selection** (*Fig. 4*). Sexual selection is caused by differences in reproductive success due to competition for mates. The greater the differential in investment in offspring between the sexes, the more intense competition will be amongst the low-investment sex (usually the males) for access to the high-investment sex (usually the females), and the more choosy the high-investment sex should be to get the best of what is on offer from the low-investment sex. Put slightly over-simply, males should thus be **aggressive** and females **choosy**. Sexual selection 'therefore' can arise in two ways – through competition between members of one sex for mates (**intrasexual selection**) or via preference for particular traits in the other sex (**intersexual selection**) (or via both routes). Intrasexual selection can account for the development of fighting 'equipment' such as antlers, horns, tusks and enlarged canines in male mammals. Intersexual selection, on the other hand, provides an explanation for apparently useless articles such as the **extravagant** tails or head plumes of male birds of paradise or peacocks. How do preferences for such traits arise in females? The **handicap** theory suggests that possession of a good quality large

Fig. 4.   Sexual dimorphism in a hummingbird Spathura underwoodi, *in which the brightly colored and long-tailed male contrasts markedly with the dull female (from Darwin, 1871).*

tail (or other extravagant character) indicates that the owner must have **good genes** as a weak individual could not bear the energy cost and enhanced predation susceptibility of an extravagant trait. Alternatively, **Fisher's runaway** model suggests that such male **epigamic** traits can be initiated by arbitrary female choice and will continue to evolve if the genes coding for choice (e.g. for large tails) in females also code for the character (e.g. tail size) in males.

01 GENETIC VARIATION

Key Notes

Genes and alleles	Genes are pieces of DNA which contain the genetic code necessary to produce a specific protein. A gene can have many alleles and individuals can have a homozygous or heterozygous genotype. The phenotype expressed depends on whether alleles are dominant, recessive or codominant. The total set of genes and alleles present in a population is known as the gene pool.
Polymorphism	The term polymorphism refers to the presence of alleles in a population. A population or species may be polymorphic for color, as in snails, or for some biochemical function, as in plant toxicity. Some polymorphisms are maintained by natural selection, but others appear to result from the effect of many genes controlling the polymorphic character.
Measuring genetic variation	Genetic variation within populations and species can be estimated directly from the DNA or from protein (allozyme) variation. Proteins or DNA fragments are separated by gel electrophoresis to give bands that can be scored to determine the genotype of individuals. Detailed investigation of the DNA of individuals yields a unique genetic fingerprint, useful for establishing paternity.
Genetic drift	Genetic drift is a random change in gene frequency arising through chance alone. It tends to be more apparent in small populations. Allele frequencies 'drift' away from their starting values, increasing, decreasing or fluctuating up and down. It occurs because there is an element of chance in which individuals and which gametes will produce offspring and which individuals will die before reproducing. Random changes in allele frequency can lead to fixation and the progressive loss of genetic variation from the population.
Genetic bottleneck	When a population undergoes an abrupt contraction in numbers, this is accompanied by a change in gene frequencies and a decline in the total genetic variation. Genetic drift in the small population during the bottleneck results in the loss of genetic variation. Although population numbers may recover, genetic variation will remain low for many generations.
Founder effect	This term describes the establishment of a new population by one or a few individual(s), resulting in low levels of genetic variation and often a disproportionate number of alleles that are rare in the parent population. In human populations the high incidence of certain genetic diseases can be traced to founder effects.
Related topics	Adaptation (B1) Rare species, habitat loss and extinction (V1)

Genes and alleles   Genetic information is carried on **chromosomes** which consist of DNA. In **diploid** organisms, chromosomes come in matched pairs. One of these **homologous** chromosomes is inherited from the mother and one from the father. Each chromosome carries the units of heredity called **genes**, pieces of DNA that code for a particular protein. The gene is a paired structure, consisting of two **alleles**, one from each homologous chromosome. These alleles can be identical or they can be different alternatives. The site alleles occupy on a chromosome is known as a **locus** (loci is the plural). A diploid individual has two alleles (identical or different) at each locus. An individual with two identical alleles at a locus is said to be **homozygous** at that locus, if the alleles are different the individual is **heterozygous**. The genetic make-up of an individual is known as its **genotype**.

When an individual is heterozygous, the resulting **phenotype** or expression may be some intermediate state between the two homozygotes. In this case both alleles are expressed and they are said to be **codominant**. Frequently only one is expressed in the phenotype, in this case the expressed allele is **dominant** over the other allele. The nondominant, or **recessive**, allele carried by the individual does not influence the phenotype. It can, of course, be passed on to the individual's offspring and may then be expressed provided it is not inherited with another dominant allele. Many characters do not have this simple genetic control but are influenced by many genes at different loci and even on different chromosomes. These characters are said to be **polygenic**. Height in humans is a polygenic character.

The total set of genes and alleles present in a population is known as the **gene pool**. Populations that are separated from one another and do not interbreed will have separate gene pools. Depending on their environmental conditions and their history the gene pools of populations even within a species can be quite different.

Polymorphism   The presence of a number of different alleles resulting in more than one phenotype (e.g. flower color) in a population is termed a **polymorphism**. Seashore molluscs such as winkles, dog whelks and clams coexist in different shell color forms. Some polymorphisms appear to be maintained by natural selection; in the land snail, *Cepaea nemoralis,* shell color and banding is at least partly related to camouflage and the temperature regime of the local habitat. There is also evidence that predators of *C. nemoralis* preferentially choose the more common shell types, giving a **rare type advantage** (also known as frequency-dependent selection) which will tend to favor maintenance of a polymorphism. Polymorphisms in plant toxicity are widespread. Clover and vetches vary in their ability to produce cyanide when their leaves are broken. Individuals of the ponderosa pine, *Pinus ponderosa,* vary vastly in the number of the black pineleaf scale insects they carry, probably because of differences in the type and concentration of plant monoterpenes.

However, the reason behind many polymorphisms is unknown. Often the genetic basis has not been investigated, but where it has been studied the coexistence of distinct phenotypes is not attributable to natural selection but to the effect of several different genes or groups of genes acting in combination on the polymorphic character.

Measuring   The estimation of genetic variation within populations and species usually
genetic variation involves extracting either the proteins, known as **allozymes**, associated with particular alleles or the DNA containing the alleles themselves. Allozymes or

fragments of DNA are separated according to size and electrical charge by **gel electrophoresis** (*Fig. 1a*). Samples of DNA or protein extract are applied to a gel medium to which an electric current is applied. The charge is carried through the gel by an aqueous buffer which also provides the optimum pH for separation.

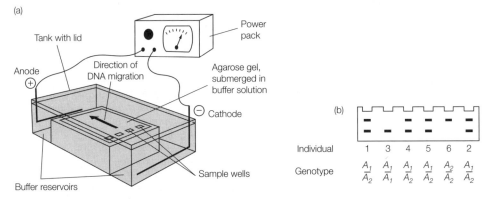

Fig. 1. (a) Agarose gel electrophoresis of DNA,(b) an example of part of the resulting electrophoresis gel (enlarged) showing two restriction fragments. From Practical Skills in Biology, Jones *et al., 1994. Reprinted by permission of Addison Wesley Longman Ltd.*

Before DNA fragments can be separated on a gel, the DNA is exposed to an enzyme which latches on to specific places, or **restriction sites,** and cuts the DNA at that point. If an individual does not have the specific restriction site at which the enzyme binds the DNA will not be cut, hence the size and number of fragments provides information about the individual's DNA. Scoring this variation among individuals on the gel provides an estimate of the overall level of genetic variation present in the species or population. *Figure 1b* shows a number of individuals carrying different combinations of two fragments A_1 and A_2.

Detailed investigation of a piece of DNA from a single individual can be undertaken to reveal a 'genetic fingerprint'. The level of genetic variation is such that each individual will have a unique fingerprint allowing it to be identified. The separation of small fragments of DNA produces the characteristic 'bar-code' pattern. This technique is useful in forensic studies for matching body samples to individuals and for establishing paternity in humans and other animals. For example, genetic fingerprinting has revealed that dunnock females, who appear monogamous from field observations, bear a significant number of offspring that are not fathered by their attendant male. It seems that females indulge in swift and opportunistic copulations with other males in order to increase the genetic variation in their offspring (see Topic N2).

Genetic drift

Unlike adaptation (see Topic B1) genetic drift is a change in allele frequency in a population that cannot be ascribed to natural selection. Evolution is a change in allele/gene frequencies over time that occurs as less adapted alleles are replaced with better adapted ones.

Genetic drift is a random change in gene frequency arising through chance, and tends to be more apparent in small populations. The allele frequencies in a population experiencing significant genetic drift can be observed to 'drift' away from their starting values over time. Because the changes are at random and not under the influence of natural selection, frequencies will alter in a nondirectional way, increasing, decreasing or fluctuating up and down.

Drift occurs because of the influence of chance on which genes are transmitted from one generation to the next. Not all individuals mate and not all of the gametes produced by an individual contribute to reproduction. Those gametes that do result in offspring may not be representative, in terms of the frequencies of the alleles they carry, of the parents. In a large population these so-called sampling effects will tend to cancel each other out; if the offspring of one pair of parents under-represents an allele it is likely that others will over-represent it. This is less likely to be the case in a small population, and the result will be a different allele frequency in the offspring population than in the parents. These random changes can cause alleles to become lost from the population completely, such that only one allele remains at that locus. Successive **fixation** events such as these will lead to the progressive loss of genetic variation from the population (*Fig. 2*).

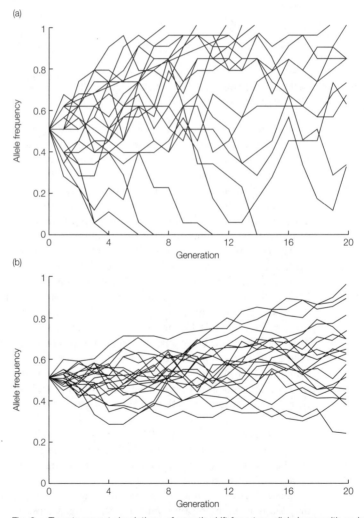

Fig. 2.   Twenty repeat simulations of genetic drift for a two-allele locus with an initial allele frequency of 0.5. In the small population (a) the large fluctuations in allele frequency result in fixation of the allele (when its frequency becomes 1.0) and homozygosity. The fluctuations are damped in the large population (b) and fixation and the loss of genetic variation is less likely to occur. Reprinted from Evolution, M. Ridley, 1993 *with permission from Blackwell Science.*

Random mortality will have the same effect as the above; the genotypes of the survivors may not be representative of the population before mortality occurred and unrepresentative genes will be passed to the next generation. This effect of mortality explains why random genetic drift is also apparent in asexual populations.

Genetic bottleneck

If a population undergoes an abrupt contraction of its numbers, say as a result of environmental catastrophe or overexploitation, it is said to have gone through a bottleneck. This is accompanied by a change in gene frequencies and decline in total genetic variation (*Fig. 3*). *Figure 3a* shows the change in population size over time for a population undergoing a bottleneck and then recovery. Shown like this it is easy to see why it is called a bottleneck.

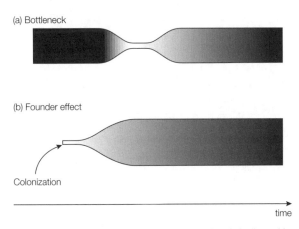

(a) Bottleneck

(b) Founder effect

Colonization

time

Fig. 3. The change in population size over time is indicated by the change in the width of the shaded band. (a) Genetic bottlenecks occur where the population size declines sharply, indicated by the narrowing of the shaded band (the intensity of the shading indicates the decline in genetic variation and the subsequent gradual increase through mutation); (b) populations founded by a few individuals display a similar initial lack of genetic variation.

If the population remains small, gene frequencies will be subject to genetic drift in successive generations, genetic variation will be lost and the population may become extinct (see Topic N2). Alternatively, population numbers could recover after the crash (as in *Fig. 3a*). This has happened in the northern elephant seal, *Mirounga angustirostris*, whose numbers were reduced to about 20 in the 1890s due to overhunting. There are now approximately 30 000 northern elephant seals and studies have demonstrated an extremely low level of genetic variation in this species with fixation for one allele at all of the 24 loci examined. Once genetic variation has been lost from a population it can only be restored through mutation (which will take many generations) or through mixing with another, genetically different population. The intensity of shading in *Fig. 3* indicates the accumulation of genetic variation through mutation.

Founder effect

The establishment of a new population in vacant habitat may involve only one or a few individuals (*Fig. 3b*). The level of genetic variation and the presence of particular genes in the new colony will be entirely dependent on the genotype of the colonists. This can cause **founder populations** to have low levels of genetic variation and alleles that are unrepresentative of the parent population. In the extreme case, a new population may be founded by a single pregnant female or a single seed of a self-compatible plant. If a colonist is carrying alleles that are rare in the parent population these may become common in the founder population.

The clearest example comes from humans. The Afrikaner population of South Africa is mainly descended from one shipload of 20 immigrants who landed in 1652. One of the original colonists was a Dutch male carrying the gene for Huntington's disease; the high incidence of this gene in the present day population is attributed to this founder effect.

02 SPECIATION

Key Notes

Reproductive species concept	The reproductive species concept focuses on the idea that species exist and maintain their genetic integrity and distinctness because they do not interbreed. The biological species concept and the recognition species concept both define species in terms of interbreeding. This view contrasts with the practical definition of species on the basis of distinguishing morphological characters.
Gene flow	Gene flow describes the movement of genes among populations through interbreeding, dispersal and migration. High gene flow causes populations to become genetically similar to one another. Restricted gene flow allows differentiation to occur among populations.
Biological species concept	This concept views species as a group of interbreeding natural populations that are reproductively isolated from other such groups. It allows populations that are indistinguishable morphologically to be classified as separate, sibling species because they do not interchange genes. Biological species are maintained by reproductive isolating mechanisms.
Recognition species concept	The recognition concept views a species as a group of individuals with a common Specific Mate Recognition System (SMRS). The SMRS consists of all aspects of mating such as compatibility of reproductive organs and gametes, courtship song and behavior and pheromones. This species definition emphasizes the factors that keep species together.
Allopatric speciation	Allopatric speciation occurs when the new species evolves in geographic isolation from the parent species. It may take place most readily in peripheral isolates, small populations at the extreme edge of a species range. The combined effect of a small atypical population and extreme environmental conditions can cause rapid and extensive genetic reorganization (a genetic revolution) leading to speciation .
Parapatric speciation	This form of speciation occurs where the speciating populations are contiguous but subject to different environmental (e.g. climatic) conditions in different parts of the range. Intermediate hybrids are found, as in ring species, but the large distances involved prevent the two types from merging completely.
Sympatric speciation	Sympatric speciation may occur where there is no geographical separation between the speciating populations but where they differ in host preference, food preference or habitat preference. Whether sympatric speciation happens at all is a contentious issue, although rapid changes in the host preference of phytophagous insects has been observed. Plants can undergo sympatric speciation through polyploidy.

Related topics	Adaptation (B1)	Genetic variation (O1)

Reproductive species concept

This concept focuses on the idea that species exist and maintain their genetic integrity and distinctness because they do not interbreed. Both the biological species concept and the recognition species concept (below) stress the importance of interbreeding in defining a species. As such, both are useful concepts when considering the history of a species, how it evolved and how speciation occurs in general. This contrasts with the more practical view, historically held by naturalists, whereby species are formally defined and named on the basis of distinguishing morphological characters. Although of undoubtable practical value when identifying specimens, this approach to species definition raises the question of how different populations have to be before they are classified as separate species. The tropical snail (*Partula* sp.) has two forms which are identical except in the direction in which the shell coils. Snails with opposite shell coils cannot mate, therefore they are true biological species even though they differ by only one character, which in turn is controlled by a single gene!

Gene flow

Gene flow describes the movement of genes among populations through interbreeding, dispersal and migration. The greater the level of gene flow among populations, the more **homogeneous**, or genetically similar, the populations will be. Restricted gene flow allows **differentiation** to occur among populations as adaptation and genetic drift can act more or less independently in each. Gene flow does not occur between species because they are reproductively isolated from one another and they do not interbreed.

Biological species concept

According to this widely held concept, a species is a group of interbreeding natural populations that are reproductively isolated from other such groups. This idea was first developed by Ernst Mayr in 1969. Species consist of individuals which share a common gene pool (see Topic O1), different and separated from other distinct individuals belonging to another species and gene pool.

The biological species concept allows populations that are indistinguishable morphologically to be classified as separate species because they do not interchange genes. The classic example of these **sibling species** is *Drosophila pseudoobscura* and *D. persimilis*. They are almost identical in appearance but do not interbreed and hence are reproductively isolated.

Biological species are maintained by **reproductive isolating mechanisms**. A reproductive isolating mechanism is any property that prevents gene flow between species such that species living in close proximity remain distinct. These are tabulated below (*Table 1*).

Isolating mechanisms are not necessarily absolute and biological species do not have to be wholly isolated in order to remain as species. For example, although matings between *Gryllus pennsylvannicus* males and *G. firmus* females (*Table 1*) are unsuccessful, matings with *G. pennsylvanicus* females produce viable hybrid offspring. Ecological separation with respect to soil type helps to keep these species distinct. In the north American frogs *Rana pipiens* and *R. palustris*, a combination of ecological, seasonal and geographic isolation contribute to the reproductive isolation between species.

Recognition species concept

The recognition concept views a species as a group of individuals with a common method of recognizing and responding to mates, i.e. a common **Specific Mate Recognition System** (SMRS). The SMRS consists of all aspects of mating such as compatibility of reproductive organs and gametes, courtship song and behavior and pheromones. This species definition emphasizes the

Table 1. *Dobzhansky's classification of reproductive isolation mechanisms. Modified from Dobzhansky (1970)*

1 *Premating or prezygotic* mechanisms prevent the formation of hybrid zygotes
   (a) *Ecological* or *habitat isolation.* The populations concerned occur in different habitats in the same general region, e.g. European mosquito, *Anopheles labrancuiae,* is found in brackish water, while *A. maculipennis* is found in running freshwater.
   (b) *Seasonal* or *temporal isolation.* Mating or flowering times occur at different seasons, e.g. *Pinus radiata* and *P. muricata* are found in close proximity in California but shed their pollen at different times.
   (c) *Sexual isolation.* Mutual attraction between the sexes of different species is weak or absent, e.g. in the European bush cricket, *Ephippiger,* females show a strong preference for the chirp pattern of the courtship song produced by males of the same species.
   (d) *Mechanical isolation.* Physical noncorrespondence of the genitalia or the flower parts prevents copulation or the transfer of pollen, e.g. some damselflies have very complex genitalia preventing intermating.
   (e) *Isolation by different pollinators.* In flowering plants, related species may be specialized to attract different pollinators, e.g. male bees pollinate bee orchids by 'copulating' with bee-mimicking flowers. Different species of bee orchid mimic different bee species making cross-fertilization impossible.
   (f) *Gametic isolation.* In organisms with external fertilization, female and male gametes may not be attracted to each other. In organisms with internal fertilization, the gametes or embryos of one species may be inviable in the physical environment of other species.

2 *Postmating* or *zygotic* isolating mechanisms reduce the viability or fertility of hybrid zygotes
   (g) *Hybrid inviability.* Hybrid zygotes have reduced viability or are invariable. In the north American crickets, matings between *Gryllus pennsylvannicus* males and *G. firmus* females fail to produce any offspring.
   (h) *Hybrid sterility.* The F_1 hybrids of one sex or of both sexes fail to produce functional gametes.
   (i) *Hybrid breakdown.* The F_2 or backcross hybrids have reduced viability or fertility.

factors that keep species together. The recognition concept views reproductive isolating mechanisms, the focus of the rival biological species concept (above), as an incidental by-product of divergence among SMRSs and not as an active part of the speciation process.

Allopatric speciation

Allopatric speciation occurs when the new species evolves in geographic isolation from the parent species (*Fig. 1a*). The species range, indicated by the circle in *Fig. 1a*, becomes subdivided by a barrier such as a new mountain range or the change in the course of a river. Gene flow between the two subpopulations becomes impossible allowing evolution to proceed independently in each. Natural selection may favor different genotypes on either side of the barrier and random genetic drift and mutations could contribute to divergence. Over time, divergence may proceed to the point that, were the two populations to meet again, they would not be able to interbreed and speciation would be complete.

This form of speciation may take place most readily in small populations at the extreme edge of a species range (note, in *Fig. 1a* the isolated population is smaller than the parent population). The peripheral population could become

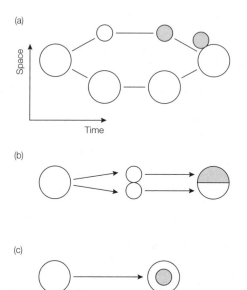

Fig. 1.   The degree and type of isolation required by different models of speciation.
(a) Allopatric speciation: the new species forms geographically separated from its parent
species; (b) parapatric speciation: the new species forms in an adjacent population;
(c) sympatric speciation: a new species emerges from within the range of the parent species.

isolated, for example, during contraction of the main species range in response
to changing climate. The isolated population would be subject to the founder
effect (see Topic O1) and could be genetically different from the parent population. The combined effect of a small atypical population and extreme environmental conditions can cause rapid and extensive genetic reorganization
through random genetic drift and strong natural selection, or, in other words,
a **genetic revolution**.

Distinct **peripheral isolates** have been observed on islands. For example, the
Papuan kingfisher of the *Tanysiptera* genus has three almost indistinguishable
forms on mainland New Guinea and strikingly different island species derived
from isolates. Speciation in the cichlid fishes (*Haplochromis*) in Lake Victoria is
thought to have occurred through the isolation of small founder populations
caused by changes in the drainage pattern of the rivers feeding the lake. Small
changes in, for example, male coloration and courtship behavior have resulted
in the formation of as many as 170 reproductively isolated species.

**Parapatric
speciation**

This form of speciation occurs where the speciating populations are contiguous
(*Fig. 1b*) and hence only partially geographically isolated. They are able to meet
across a common boundary during the speciation process. Where a species occupies a large geographical range it may become adapted to different environmental (e.g. climatic) conditions in different parts of that range. Intermediates,
or hybrids, will be found but the large distances involved prevent the two types
from merging completely.

For example, the herring gull *Larus argentatus* is a **ring species** whose distribution covers a large geographical area. Westwards from Britain towards North
America its appearance changes gradually, but it is still recognizable as the

herring gull. Further west in Siberia it begins to look more like the lesser black-backed gull *Larus fuscus*. From Siberia to Russia and into northern Europe it becomes progressively more like the lesser black-backed gull. The ends of the ring meet in Europe and the two geographical extremes appear to be two good biological species.

Sympatric speciation

Sympatric speciation describes a situation where there is no geographical separation between the speciating populations. All individuals are, in theory, able to meet each other during the speciation process (*Fig. 1c*). This model usually requires a change in host preference, food preference or habitat preference in order to prevent the new species being swamped by gene flow.

Whether sympatric speciation happens at all is a contentious issue. In theory it can occur where there is a polymorphism (see Topic O1) in the population conferring adaptation to two different habitats or niches. Reproductive isolation could then arise if the two morphs had a preference for 'their' habitat. There is some evidence for this in natural populations. For example, caterpillars of the ermine moth, *Yponomeuta padellus*, feed on apple and hawthorn trees. Females prefer to lay their eggs on the species on which they were raised. Caterpillars also prefer to feed on the plant on which their mothers were raised and adult moths prefer to mate with individuals from the same plant. The apple and hawthorn types are not completely isolated, but may represent an intermediate point in ongoing sympatric speciation.

A change in host–plant preference can happen very rapidly. The apple maggot, *Rhagoletis pomonella*, was first found on apples in 1864 having moved from its previous host, hawthorn (coincidentally, both hosts used by the ermine moth). Like the ermine moth, mating, feeding and egg-laying takes place on the host plant so the two forms no longer meet in the wild. Although only 100 generations old, they have also evolved extensive enzyme differences. Again, reproductive isolation is not complete, and may never become so as the apple and hawthorn forms are able to mate. Some evolutionary biologists would wish to see them classed as separate species.

An uncontentious example of sympatric speciation occurs in plants through **polyploidy**. Polyploidy is the spontaneous duplication of the entire genome resulting in an individual with a multiple of the original chromosome number. Polyploidy is common in plants, where it often results in larger, more vigorous forms. It is usually fatal in animals, although some amphibians are polyploids. The polyploid plant is no longer sexually compatible with the parent population but is able to establish a distinct population which may occupy a different habitat. The sand dune grass, *Spartina townsendii*, is a polyploid derived from the original *S. anglica*. It is more vigorous than the parent and has colonized large areas of sand dune in Britain.

P1 COMPONENTS AND PROCESSES

Key Notes

The concept of the ecosystem

The ecosystem concept was proposed by Tansley in 1935, and was originally defined to include all the animals, plants and physical interactions of a defined space. Modern ecologists tend to think of ecosystems in terms of energy flow, carbon flow or nutrient cycles.

Ecosystem components

The bodies of living organisms within a unit area constitute a standing crop of biomass: the mass of organisms per unit area of ground (or water), usually expressed in units of energy or dry organic matter (e.g. tons ha^{-1}). The great majority of biomass in a terrestrial community is vegetation. The primary productivity of a community is the rate at which biomass is produced per unit area by plants, the primary producers. The total fixation of energy by photosynthesis is referred to as gross primary productivity (GPP) of which a proportion (R) is lost from the community as respiration. The difference between GPP and R is known as net primary productivity (NPP) and represents the rate of production of new biomass that is available for consumption by heterotrophic organisms (bacteria, fungi and animals). The production of biomass by heterotrophs is called secondary production.

Ecosystems and the laws of thermodynamics

The first law of thermodynamics states that energy can neither be created nor destroyed. The second law of thermodynamics states that every transformation results in a reduction of the free energy of the system. Because energy transformation cannot be 100% efficient (from the second law), heterotrophs must have less energy, and must therefore be rarer than the plants they feed on. The complexity of ecological interactions means that it is not possible to construct predictive mathematical models of living systems based on these laws of thermodynamics.

Transfer efficiencies

The proportion of net primary production that flows through trophic levels depends on transfer efficiencies in the way energy is used and passed from one step to the next. A knowledge of just three categories of transfer efficiency is all that is required to understand the pattern of energy flow. These are consumption efficiency (CE), assimilation efficiency (AE) and production efficiency (PE). Consumption efficiency is the percentage of total productivity available at one trophic level (P_{n-1}) that is actually consumed (ingested) by a trophic compartment one level up (I_n). Assimilation efficiency is the percentage of food energy taken into the guts of consumers in a trophic compartment (I_n) which is assimilated across the gut wall (A_n) and becomes available for incorporation into growth or used to do work. Production efficiency is the percentage of assimilated energy (A_n) which is incorporated into new biomass (P_n). The remainder is entirely lost to the community as respiratory heat.

Energy flow through a community	Given that specified values can be obtained for net primary production and CE, AE and PE, it is possible to predict, using models, the pathway of energy flow at different trophic levels for different communities. From such modelling studies, which are supported by field data, the most significant finding is the overwhelming importance of the decomposer system. Overall, in a steady state community, losses through animal respiration balance NPP so that standing crop biomass remains the same.
Related topics	Primary and secondary production (P2) The community, structure and stability (Q1) Food chains (P3) Community patterns, competition and predation (Q3)

The concept of the ecosystem

The ecosystem concept has deep roots in ecology. The first statements of the idea of an ecosystem date back to 1877 with the writings of Forbes and Mobius. They stated that the unit of study of ecology must include the whole tangled mixture of plants and animals and their physical surroundings. From these ideas Tansley (1935) proposed the term ecosystem. Tansley's **ecosystem** would be considered to include all the animals, plants and physical interactions of a defined space. An ecosystem could be of any size depending on the communities to be studied. Modern ecologists tend to think of ecosystems in terms of energy flow, carbon flow or nutrient cycles.

Ecosystem components

Biomass is the mass of organisms per unit area of ground (or water). This is usually expressed in units of energy (e.g. $J\,m^{-2}$) or dry organic matter (e.g. tons ha^{-1}). The great bulk of biomass in communities is formed by plants. Biomass includes the whole bodies of the organisms even though parts of them may be dead. We may need to distinguish between the mass of dead material, the necromass from the living, active fraction. The **primary productivity** of a community is the rate at which biomass is produced per unit area by plants, the primary producers. It can be expressed either in units of energy (e.g. $J\,m^{-2}\,day^{-1}$) or of dry organic matter (e.g. $kg\,ha^{-1}\,year^{-1}$). The total fixation of energy by photosynthesis is referred to as **gross primary productivity** (GPP). A proportion of this is respired by the plant and lost from the community as **respiratory heat** (R). The difference between GPP and R is known as **net primary productivity** (NPP) and represents the actual rate of production of new biomass that is available for consumption by heterotrophic organisms (bacteria, fungi and animals). The rate of production of biomass by heterotrophs is called **secondary production**. The trophic structure of a community consists of the grazer system alongside a decomposer system. In addition, subcompartments are recognized within each trophic level so that distinctions are made between invertebrate and vertebrate categories and between microbes and detritivores. A joule of energy may be consumed and assimilated but will eventually be lost from the system as respiratory heat during work. The possible pathways in the grazer and decomposer systems are the same except that feces and dead bodies are lost to the grazer system (and enter the decomposer system), but feces and dead bodies from the decomposer system are simply sent back to the dead organic matter pool. This is highly significant as the energy available as dead organic matter may finally be completely metabolized with all the energy lost as heat. The exception to this situation is where matter is exported from a local environment (e.g. in streamflow) or where local abiotic conditions are unfavorable to decomposition processes, leaving incompletely metabolized high-energy matter, such as oil, coal and peat.

Ecosystems and the laws of thermodynamics

The first law of thermodynamics states that when energy is converted from one form into another, energy is neither gained nor lost. The first law is also called the law of conservation of energy. **The second law of thermodynamics** states that every transformation results in a reduction of the free energy of the system. Energy to fuel herbivore populations is limited to that flowing in the plant trophic level through photosynthesis. However, for herbivores to have use of this energy requires an energy transformation from plant carbohydrate to animal carbohydrate. Because this energy transformation cannot be 100% efficient (second law), the animals must have less energy, and must therefore be rarer than the plants they feed on. This trend continues through the successive energy transformations of each trophic level. Energy gets less and less and animals get rarer and rarer.

However, it is not possible to make purely physical and mathematical models of living systems based on these laws of thermodynamics. This is because the second law is far too simplistic a model of what is actually happening in living communities. The wastage of energy between trophic levels is far greater than could be accounted for by the inefficiency of physical transformation alone. Animals and plants use energy as well as just transforming it.

Transfer efficiencies

The proportion of net primary production that flows along these pathways depends on **transfer efficiencies** in the way energy is used and passed from one step to the next. A knowledge of just three categories of transfer efficiency is all that is required to predict the pattern of energy flow. These are **consumption efficiency** (CE), **assimilation efficiency** (AE) and **production efficiency** (PE).

Consumption efficiency is the percentage of total productivity available at one trophic level (P_{n-1}) that is actually consumed (ingested) by a trophic compartment one level up (I_n).

$$\text{Consumption efficiency (CE)} = \frac{I_n}{P_{n-1}} \times 100$$

In the case of secondary consumers it is the percentage of herbivore productivity eaten by carnivores. The remainder dies without being eaten and enters the decomposer chain. Consumption efficiencies of herbivores are very low, reflecting either the difficulty of utilizing plant material or the low herbivore densities. Average figures for consumption efficiencies are 5% in forests, 25% in grasslands and 50% in phytoplankton-dominated communities. Much less is known about the consumption efficiencies of carnivores. Vertebrate predators may consume 50–100% of production from vertebrate prey but only 5% from invertebrate prey. Invertebrate predators may consume up to around 25% of invertebrate prey.

Assimilation efficiency (AE) is the percentage of food energy taken into the guts of consumers in a trophic compartment (I_n) which is assimilated across the gut wall (A_n) and becomes available for incorporation into growth or used to do work. The remainder is lost as feces and enters the base of the decomposer system.

$$\text{Assimilation efficiency (AE)} = \frac{A_n}{I_n} \times 100$$

Assimilation efficiency cannot properly be ascribed to microorganisms as much of the digestion of dead organic matter is extracellular, and as no feces are produced assimilation efficiency is 100%. For herbivores, detritivores and microbivores, assimilation efficiencies are low (20–50%) while high values are achieved for carnivores (80%). Generally, animals are not well equipped to deal with living and dead plant material, mainly as a result of the polymeric composition (lignin, cellulose) of plant matter. However, plant seeds may be assimilated with an efficiency as high as 70%, and leaves up to 50%. Organic matter originating from animals poses much less of a problem for digestion and assimilation.

Production efficiency (PE) is the percentage of assimilated energy (A_n) which is incorporated into new biomass (P_n). The remainder is entirely lost to the community as respiratory heat.

$$\text{Production efficiency (PE)} = \frac{P_n}{A_n} \times 100$$

Production efficiency varies according to the taxonomic class of the organisms concerned. Invertebrates in general have high efficiencies (30–40%), losing little energy in respiratory heat. Among the vertebrates, ectotherms have intermediate values for PE (around 10%), whilst endotherms, with high energy expenditure associated with maintaining a constant temperature, convert only 1–2% of assimilated energy into production. Small-bodied endotherms such as shrews have the lowest PE values. Microorganisms have high production efficiencies, with short lives and rapid population turnover. In general, PE increases in size in endotherms and decreases with size in ectotherms.

Energy flow through a community

Given that specified values can be obtained for net primary production and CE, AE and PE, it is possible to predict the pathway of energy flow at different trophic levels for different communities. From such modelling studies, the most significant finding is the overwhelming importance of the decomposer system. *Figure 1* shows a generalized model of trophic structure and energy flow for a terrestrial system. In a model grassland system, although 29% of NPP was consumed by herbivores, only 2% of secondary production was based on this. Of every 100 J of NPP, 55 J find their way into the decomposer system per year, while less than 1 J finds its way into grazer production. Overall, in a steady state community, losses through animal respiration balance NPP so that standing crop biomass remains the same. These model predictions agree well with observations in real communities.

There have been only a few studies in which all the community compartments have been studied together. These studies show that the decomposer system is responsible for the majority of secondary production, and therefore respiratory heat loss in every community. The grazer system has its greatest role in plankton communities, where a large proportion of NPP is consumed alive and assimilated to a high efficiency. The grazer system is less important in terrestrial communities due to low herbivore consumption and assimilation efficiencies. Grazer systems are almost nonexistent in small streams and ponds where primary productivity is very low. These systems are dependent on terrestrial systems for a source of energy. Deep ocean benthic communities are similar

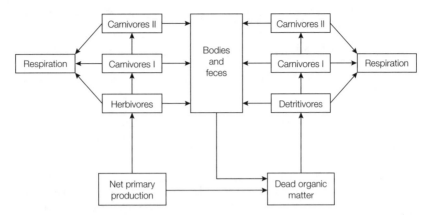

Fig. 1.   A generalized model of trophic structure and energy flow for a terrestrial community.

to streams and ponds as the water is too deep for appreciable photosynthetic activity to occur, and energy is derived from dead phytoplankton, bacteria, animals and feces which sink from the autotrophic community in the euphotic zone.

P2 PRIMARY AND SECONDARY PRODUCTION

Key Notes

Primary production	Global net primary productivity is approximately 120×10^9 tons dry weight per year on land, and 50×10^9 tons per year in the sea. This productivity is very unevenly distributed across the Earth. The most productive systems are found amongst swamp and marshland, estuaries, reefs and cultivated land. Productivity decreases moving away from the equator, indicating the importance of temperature and radiation.
Relationship of productivity to biomass	Community productivity can be related to the standing crop that produces it by comparing the ratio of productivity (P) to standing crop biomass (B). The resulting P:B ratios (i.e. kg produced year^{-1} kg^{-1} standing crop) average 0.042 for forests, 0.29 for other terrestrial systems, and 17.0 for aquatic communities. An alternative way of looking at P:B ratios would be to define biomass in terms of weight of living tissue, which would reduce these large differences between communities. However, accurate measurement of the proportion of biomass alive is difficult.
Secondary production	Secondary production is defined as the rate of production of new biomass by heterotrophic organisms. Heterotrophs are organisms, such as animals and fungi, with a requirement for energy-rich organic molecules. Secondary production by heterotrophs is inevitably dependent on primary productivity. Generally, in grazer systems, that part of the trophic structure of a community which depends on the consumption of living plant biomass, secondary productivity is an order of magnitude less than primary production resulting in a pyramidal structure. However, there are exceptions to this.
The relationship between matter and energy	Once energy is transformed into heat, it can no longer be used by living organisms to do work or to fuel the synthesis of biomass. The heat is lost to the atmosphere and can never be recycled. Life on Earth is possible because a fresh supply of solar energy is made available every day. In contrast, nutrients such as carbon can be reused. Chemical nutrients, the building blocks of biomass, can be used again, and recycling is a critical feature. Unlike the energy in solar radiation, nutrients are not in unalterable supply. If plants, and their consumers were not eventually decomposed, the supply of nutrients would become exhausted and life on Earth would cease.
Related topics	Solar radiation and plants (F1) Components and processes (P1) Plants and consumers (G2) Food chains (P3)

Primary production

The best current estimate of **global terrestrial net primary productivity** is 120×10^9 tons dry weight per year, and in the sea $50–60 \times 10^9$ tons per year.

However, this productivity is not evenly distributed across the Earth. World productivity maps provide estimates of annual net primary productivity and plant standing crop biomass. By comparing these maps with the distribution of incident solar radiation it can be seen that solar radiation alone does not determine primary productivity. The discrepancies between the maps occur because incident radiation can only be efficiently captured when water and nutrients are available and when temperatures are suitable for plant growth. Many areas of land receive adequate radiation but lack water, while most areas of the oceans are deficient in mineral nutrients.

A large proportion of the globe produces less than 400 g m^{-2} year^{-1}. This includes over 30% of land area and 90% of the ocean. The most productive systems are found amongst swamp and marshland, estuaries, reefs and cultivated land. There is a general trend of increasing productivity with latitude, with the lower latitudes being most productive, indicating that temperature and radiation are important factors in primary productivity. However, in the oceans, nutrient limitation tends to be the most important factor controlling primary productivity, with highest productivity occurring where there is an upwelling of nutrient-rich waters irrespective of the latitude.

Small differences in topography can result in large differences in community productivity. In tundra for example, a distance of several metres, from beach ridge to meadow with impeded drainage can change primary production ten fold, from <10 to 100 g m^{-2} year^{-1}. Therefore, although there is a general latitudinal trend, there is also a broad spectrum of variation at a given latitude resulting from different microclimates.

All biotic communities depend on a supply of energy for their activities. In most terrestrial systems this is supplied by the photosynthesis of green plants. Organic matter and energy generated within the community are called **autochthonous**. In aquatic communities, the autochthonous input is through the photosynthesis of large plants and attached algae in shallow waters, and by plankton in open oceans. However a significant proportion of the organic matter in such communities often comes into the community as dead organic material that was formed from outside. This is termed **allochthonous** material and arrives in rivers or may have been blown by the wind. The relative importance of these two sources of material and energy is dependent on the size of the water body and the types of terrestrial communities which deposit organic material into it. A small lake is likely to derive much of its energy from a terrestrial source because its circumference, across which terrestrial litter passes, is large in relation to lake area. In contrast, a large, deep lake will only derive limited organic matter from the outside (smaller circumferences : surface area ratio) and therefore phytoplankton production will be dominant. In oceans organic input from terrestrial sources is negligible and the phytoplankton are all-important. Estuaries are often areas of high productivity where phytoplankton dominate in large estuary basins and seaweeds dominate in small open basins. Continental shelf communities derive a proportion of their energy from terrestrial sources and their shallowness often results in significant littoral production by seaweed communities.

Relationship of productivity to biomass

We can relate community productivity to the standing crop that produces it by comparing average values of productivity (P) and standing crop biomass (B) for all community types. The resulting P:B ratios (i.e. kg produced year^{-1} kg^{-1} standing crop) average 0.042 for forests, 0.29 for other terrestrial systems, and

17.0 for aquatic communities. The main reason for these differences is that the large proportion of forest biomass is dead and much of the living tissue is not photosynthetic. In grassland, more of the biomass is living and involved in photosynthesis. In aquatic communities, particularly when dominated by phytoplankton production, dead cells do not accumulate and photosynthetic output is high. In addition, there is a rapid turnover of phytoplankton biomass. Perhaps a more realistic way of looking at P:B ratios is to define biomass in terms of weight of living tissue. This would then reduce the large differences in ratios described. However, it is difficult to assess what biomass is living and which is dead.

Secondary production

Secondary production is defined as the rate of production of new biomass by heterotrophic organisms. Heterotrophs may be defined as organisms with a requirement for energy-rich organic molecules. Unlike plants, heterotrophs such as animals, fungi and most bacteria cannot manufacture complex, energy-rich compounds from simple molecules. They derive their matter and energy either directly, by consuming plant material, or indirectly from plants by eating other heterotrophs. Plants, the primary producers, comprise the first trophic level in a community; primary consumers occur at the second trophic level, secondary consumers (carnivores) at the third, etc. Since secondary production is dependent on primary productivity, a positive relationship exists between the two variables in communities. This general relationship can be seen in both terrestrial and aquatic communities. Generally, in **grazer systems** (that part of the trophic structure of a community which depends on the consumption of living plant biomass) secondary productivity is an order of magnitude less than primary production. This results in a pyramidal structure in which plant productivity provides a broad base upon which a smaller productivity of primary consumers depends, with a still smaller productivity of secondary consumers after that. Trophic levels may have a **pyramidal biomass structure** (*Fig. 1a*). However, there are many exceptions to this structure. Food chains based on trees will have larger numbers (but not a larger biomass) of herbivores per unit area than tree, while chains dependent on phytoplankton production may give **inverted pyramids** of biomass, with a highly productive but small biomass of short-lived algal cells maintaining a larger biomass of longer-lived zooplankton (*Fig. 1b*).

(a)

11	Carnivores
132	Herbivores
703	Producers

(b)

| 21 | Zooplankton |
| 4 | Phytoplankton |

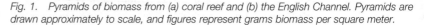

Fig. 1. Pyramids of biomass from (a) coral reef and (b) the English Channel. Pyramids are drawn approximately to scale, and figures represent grams biomass per square meter.

The production of herbivores is invariably less than that of the plants on which they feed. This loss of energy occurs for a number of reasons. Firstly, not all the plants are grazed. Much dies and supports the decomposer community (bacteria, fungi and detritivores). Secondly, not all of the consumed plant biomass is assimilated and available for incorporation into consumer biomass. Some is lost in feces, again passing to the decomposers. Thirdly, not all energy which has been assimilated is actually converted to biomass. A proportion is lost as respiratory heat. This occurs because no energy conversion process is ever 100% efficient (some is lost as unusable random heat, consistent with the second law of thermodynamics) and also because animals do work which requires energy, again released as heat. These three energy pathways occur at all trophic levels (*Fig. 2*).

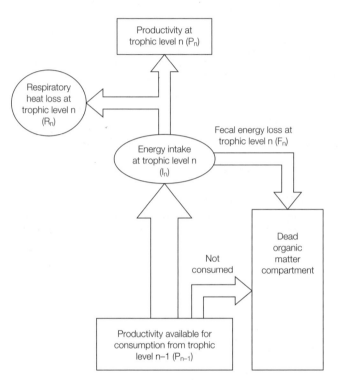

Fig. 2.   The pattern of energy flow through a trophic compartment.

The relationship between matter and energy

Once energy is transformed into heat, it can no longer be used by living organisms to do work or to fuel the synthesis of biomass. The heat is lost to the atmosphere and can never be recycled. Although energy can pass back and forth between the dead organic matter compartment and the decomposer system, this should not be described as cycling. Life on Earth is possible because a fresh supply of solar energy is made available every day. In contrast, nutrients such as carbon can be reused. The relationship between energy flow and nutrient cycling is shown in *Fig. 3*. Chemical nutrients, the building blocks of biomass can be used again, and recycling is a critical feature. Unlike the energy in solar radiation, nutrients are not in unalterable supply, and the process of locking some up into living biomass reduces the supply remaining to the rest of the community. If plants and their consumers were not eventually decomposed, the supply of nutrients would be exhausted and life on Earth would cease. The decomposer system plays the major role in nutrient recycling.

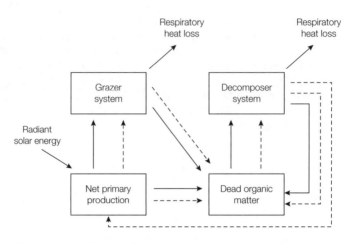

Fig. 3. Diagram to show the relationship between energy flow (———) and nutrient cycling (– – – –).

The picture described in *Fig. 3* is an oversimplification as not all of the nutrients released during decomposition are necessarily taken up again by plants. Nutrient recycling is never perfect and some nutrients are lost from land in aqueous or gaseous forms. In addition, a community receives additional supplies of nutrients from the weathering of rock and from rainfall.

P3 FOOD CHAINS

Key Notes

Pathways of nutrient flow	Autotrophic organisms assimilate inorganic resources into packages of organic molecules. These become the resources for heterotrophs which then become a resource for another consumer. At each link in this food chain we can recognize three pathways to the next trophic level: decomposition, parasitism and predation. Consumers may be generalists (polyphagous), taking a wide variety of prey species or may specialize on single species or a range of closely related species (monophagous).
Interactions between trophic levels	A characteristic of an ecosystem is the number and nature of the species that occupy its various trophic levels. The relationship between constituents of one trophic level and constituents of adjacent trophic levels may be described by a food chain. This is a chain of eating and being eaten that connects, for example, carnivorous animals to their ultimate plant food. Many food chains exist in any given ecosystem and can be combined into food webs. Ecosystems vary considerably in the pattern of their energy-nutrient webs.
Top-down or bottom-up?	It has been argued that the earth is green and vegetated because herbivore numbers are regulated by their predators (top-down control), whilst all other trophic levels are regulated by competition for resources (bottom-up control). This simple model is attractive but of doubtful value. Herbivore species are highly constrained in the range of plant tissue they can eat, due to plant defences, and hence may be limited by competition even if the world is green. Further, at a gross level, plants are not energy-limited but space limited, so any space cleared by herbivory opens opportunities for more plants.

Related topics	The nature of predation (J1)	Primary and secondary production
	Predator behavior and prey response (J2)	(P2)
	The nature of parasitism (K1)	The community, structure and stability (Q1)
	Components and processes (P1)	Community patterns, competition and predation (Q3)

Pathways of nutrient flow

As we have seen, autotrophic organisms (e.g. green plants) assimilate inorganic resources into packages of organic molecules (proteins, carbohydrates, etc.). These become the resources for heterotrophs (organisms that require resources in an organic form) which then take part in a chain of events in which each consumer of a resource becomes, in turn, a resource for another consumer. At each link in this **food chain** we can recognize three pathways to the next trophic level.

- **Decomposition,** in which the bodies (or parts of bodies) of organisms die and, together with waste and secretory products, become a food resource for decomposers (bacteria, fungi and detritivorous animals).

- **Parasitism,** in which the living organism is used as a resource while it is still alive. A parasite is a consumer that lives in close association with its host, feeding on its tissue, such as aphids and tapeworms (see Topic K1).
- **Predation,** in which the food organism, or part of it, is eaten and killed (see Topic J1).

Interactions between trophic levels

One of the characteristics of an ecosystem is the number and nature of the species that occupy its various trophic levels. It is generally found that although many species can occupy each trophic level in any ecosystem (for example, buffalo, grasshoppers and field mice may all occupy the same trophic level in a prairie ecosystem), certain species may predominate at each level. A few species of trees may dominate in a forest ecosystem, or a certain type of grass may predominate in a field ecosystem. There are also complex relationships between the constituents of one trophic level and constituents of adjacent trophic levels. Ecologists sometimes refer to this relationship between trophic levels as a **food chain**: a chain of eating and being eaten that connects large and carnivorous animals to their ultimate plant food. An example from the sea would be:

phytoplankton ⇒ copepod ⇒ adult herring ⇒ seal

Food chains are written with the arrows pointing to show the direction of food passing from animal to animal up the chain to the top of the system. Many food chains exist in any given ecosystem and can be combined; they are somewhat more accurately described as **food webs**. The web is a better analogy to the actual nature of the interactions between trophic level components. *Figure 1* shows an example of a terrestrial food web. The occupants of a primary consumer level may eat (draw energy and material from) a number of producer species. Similarly, carnivores draw their food from a number of individual species at the primary consumer level. Ecosystems vary considerably in the design of their energy-nutrient webs.

Top-down or bottom-up?

Hairston *et al.* (1968) suggested that the fact that 'the Earth was green' and the terrestrial environment largely vegetated was a paradox, given that herbivores could apparently eat this extensive resource. They argued that this could be explained because herbivore numbers are regulated by their predators (top-down control), whilst all other trophic levels are regulated by competition for resources (bottom-up control). This simple model is attractive, but of doubtful value. Plants have a variety of effective means of resisting herbivore attack, being physically and chemically protected (see Topic J2). Herbivore species are therefore highly constrained in the range of plant tissue they can eat, specializing on a narrow range of species and commonly feeding on a restricted range of individuals within that species and on restricted parts of those individuals. Herbivores may thus be limited by competition even if the world is green. Demonstrations that predator species may be limited by a shortage of herbivore prey do not indicate that herbivore numbers are limited by predation.

Further, plants are not energy-limited, in the sense that they are usually exposed to more solar radiation than can be used in photosynthesis, but are space-limited and any unused land can be viewed as an unused area of photosynthetic potential. This means that when a plant is destroyed or grazed by a herbivore, a space is cleared for another plant to grow. So as fast as a gap is torn in the vegetation, the gap is filled by another plant. The energy available to the herbivore trophic level is necessarily a small fraction of the energy

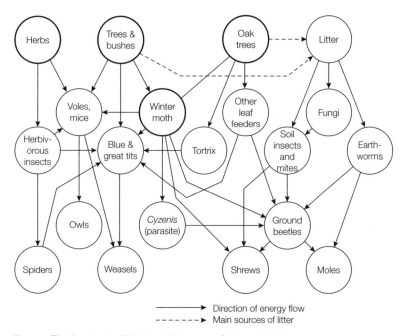

Fig. 1.   The food web of Wytham Wood near Oxford, UK. Re-drawn from Varley, 1970, Symposium of the British Ecological Society with permission from Blackwell Science.

available to the plants. So the green parts of the Earth stay because herbivores as a whole trophic level cannot command an energy flux sufficient to keep land free of plants.

Q1 THE COMMUNITY, STRUCTURE AND STABILITY

Key Notes

The community	The community is an assemblage of species populations that occur together in the same place at the same time. It has properties determined by the interactions among individuals such as competition and parasitism. The community can also be viewed from the broader perspective in terms of species diversity, food-webs, energy flow and the interactions among guilds of species.
Community structure	The species diversity of a community depends on the number of different species it contains (the species richness) and the evenness of species abundance across species. Diversity indices can be calculated to take into account both of these factors. Another way of representing species richness and evenness together is to plot the relative abundance of species against rank order of species abundance. Diversity can be measured over the three different spatial scales of the local community, the region and the broadest geographic scale (e.g. the continent). This yields alpha (α)-, beta (β)- and gamma (γ)-diversity.
Community boundaries	Communities do not end abruptly but grade into one another as individual species encounter the limits of their environmental tolerance. This can be demonstrated through gradient analysis, which focuses on the overlapping distributions of individual species. The statistical methods of classification and ordination allow community boundaries to be superimposed on this continuum. Classification techniques separate ecologically different species or sites yielding objective classes. Ordination does not attempt to draw boundaries but groups species or sites according to how similar they are in their distributions.
Guilds	Guilds are groups of species that occupy similar niches; for example, insects feeding on broad-leaved trees form one guild. Some studies have found constancy in the proportion of total species in certain guilds within a community. This indicates that there may be certain common 'rules' governing community structure.
Community complexity, diversity and stability	There are two components to stability – resilience and resistance, which describe the community's ability to recover from disturbance and to resist change. Complexity is thought to be important in determining resilience and resistance. However, more complex communities are not necessarily the most stable; increased complexity has been shown to lead to instability. In addition, different components of the community (e.g. species richness and biomass) may respond differently to disturbance. Communities with a low productivity (e.g. tundra) tend to be the least resilient. In contrast, weak competition permits coexistence among species and reduces community instability.

<table>
<tr><td>Trophic complexity and stability</td><td>Food chain length may influence the resilience of the community. Models of communities with different levels of trophic connectance show that complexity reduces resilience and stability. However, such studies should be interpreted with caution, as real communities may possess important attributes not found in the communities of null models. Stability also depends on environmental conditions – a fragile (complex or diverse) community may persist in a stable and predictable environment, while in a variable and unpredictable environment only simple and robust communities will survive.</td></tr>
<tr><td>Related topics</td><td>Resource partitioning (I3)   Food chains (P3)</td></tr>
</table>

The community

The community is an assemblage of species populations that occur together in the same place at the same time. It is the biological part of an **ecosystem**, as distinct from its physical environment. The community has interesting and complex properties arising from the interaction between species. Competition, predation, parasitism, and mutualism occurring among individuals appear to lie behind many patterns in **community organization**. At the other extreme, the community can be viewed from the wider perspective in terms of species diversity and distribution, food-webs, energy flow and the interactions among **guilds** of similar species.

Early this century the community was perceived as a **'superorganism'** whose member species were somehow bound together such that the community was born, lived, died and evolved as a whole. This view has been largely rejected in favor of the **'individualistic'** concept which focuses on the community as a collection of individual species where community patterns can be explained by processes at the level of the individual.

Community structure

A community can be characterized by its diversity which is a function of the number of different species it contains and their abundance. Diversity depends on **species richness** (the number of species) and on the **evenness** or equability of species abundance. **Diversity indices** such as the **Shannon Index** take into account both of these components. Two hypothetical communities comprising the same species can differ greatly in structure and in their diversity, depending on their relative abundance distributions (*Fig. 1*). Communities in which the

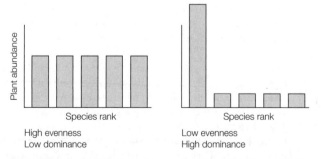

Fig. 1. Two hypothetical dominance-diversity distributions with the same species richness (5), but different levels of species evenness. The right-hand community is dominated by one species and has a lower overall diversity than the species-even community (left).

species are all more or less equal in abundance exhibit evenness, whereas communities with one or a few abundant species and many rare ones show **dominance**.

One way of demonstrating species richness and evenness is to plot the relative abundance of species against their **rank order**; that is, to plot the largest value first followed by the next largest and so on. This yields either a straight line relationship or a characteristically-shaped curve. *Figure 2* shows five of these relationships plotted on the same axes for the relative abundance of plant species in old fields in five stages of abandonment in southern Illinois, USA. For each line, the most abundant species appears on the left, the least abundant on the right. It is apparent that the species richness and the species evenness increase over time as more species colonize the site and a more complex community is established.

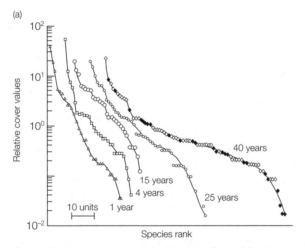

Fig. 2.   *Relative abundance data (in terms of cover) for plant species in old fields in five stages of abandonment in southern Illinois, USA.*

It is useful to distinguish between diversity measurements over a range of **spatial scales** – a common convention is to divide these into three categories:

(i)   **alpha (α)-diversity** is the diversity of species within a habitat or community, as described above;

(ii) **beta (β)-diversity** measures the rate of change in species composition along a gradient from one community to another on a regional scale;

(iii) **gamma (γ)-diversity** reflects the broadest geographical scale, representing species diversity across a range of communities in a region or a number of regions.

Community boundaries

The vegetation of one area differs in characteristic ways from that of another area allowing communities to be distinguished subjectively, for example on the basis of dominance by particular tree species. Early plant ecologists sought to classify plant communities and draw boundaries around vegetation types. However, these boundaries are artificial as communities do not end abruptly but grade into each another. This can be explained by focusing on individual species distributions through **gradient analysis**. Each species has a

(a)

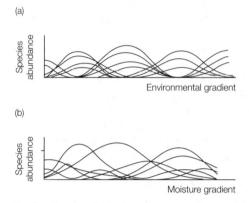

Environmental gradient

(b)

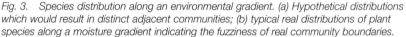

Moisture gradient

Fig. 3.   Species distribution along an environmental gradient. (a) Hypothetical distributions which would result in distinct adjacent communities; (b) typical real distributions of plant species along a moisture gradient indicating the fuzziness of real community boundaries.

unique distribution, determined by an environmental gradient such as latitude, which will overlap with those of other species (*Fig. 3*). Since the limits of individual species are not sharp, but tail off gradually, the same must be true of community boundaries.

An objective method for classifying communities and superimposing boundaries on this continuum is to employ statistical methods which look for similarity between species distributions or the species composition of sites. **Classification** techniques separate different species or sites to yield distinct classes which can be taken as representative of different communities (*Fig. 4*). In Britain, the National Vegetation Classification (NVC) has characterized and described vegetation types using a classification technique called TWINSPAN (two-way indicator species analysis). Species data from any community can be compared to the NVC database of community types and the community named and classified accordingly.

Ordination does not attempt to draw boundaries but groups species or sites according to how different their distributions are. For example, species ordination uses a number of theoretical axes along which species are assigned positions; plots of these axes reveal clumps of species that have similar

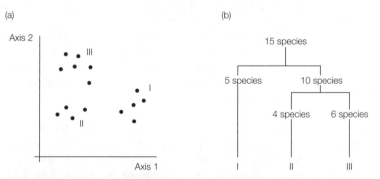

Fig. 4.   Ordination and classification of species abundance data. (a) Ordination places species with similar distributions together to give three loose groups; (b) classification separates out the most different species (group I) first and then divides those remaining along the same lines as the ordination.

requirements, occur together and belong to the same community (*Fig. 4*). Questions can then be asked as to the underlying environmental factors determining the patterns revealed by ordination. Ordination and classification are often employed together as complementary techniques.

Guilds

Species can be categorized into **guilds**, which are groupings of species that occupy similar niches; for example, insects feeding on broad-leaved trees form one guild. Ecological guilds derive their subsistence from common pools of resources. Where interspecific competition has been shown to occur it is usually among members of the same guild. Some studies have found constancy in the proportion of total species in certain guilds within a community, especially the ratio of predator species to prey species, which indicates that there may be certain common 'rules' governing community structure.

Community complexity, diversity and stability

Complexity is a function of the number of interconnections among community elements. Complexity increases as the number of interacting species in the community increases. These interactions can be **horizontal** such as competitive interactions which tend to occur among species at the same trophic level, or **vertical**. **Trophic interactions,** such as plant–herbivore, prey–predator, host–parasite, are vertical interactions as they involve species at different trophic levels within the community.

Stability involves two components: **resilience**, the ability of a community to return to its original state quickly following displacement, and **resistance**, the ability to avoid displacement. One question that has exercised ecologists is does complexity enhance stability? Intuitively, complex communities are considered stable, because the impact of sudden population change in one species will be cushioned by the large number of interacting species and will not produce drastic effects in the community as a whole. For example, it has been suggested that such buffer mechanisms operate in tropical rain forest communities where insect outbreaks are unknown. This contrasts with simple cultivated communities where pest outbreaks are common.

The analysis of complexity has usually emphasized the trophic structure of communities, although in some studies competitive interactions have been considered also. Some studies involving theoretical communities have concluded that increased complexity leads to **instability**. Although these models have been criticized for being unrealistic, the consensus is that complex communities are not necessarily stable.

Experiments on natural communities yield a variety of results. Species-poor (simple) and species-rich (complex) Serengeti plant communities were perturbed by increasing buffalo grazing. This caused a *reduction* in diversity in the species-rich communities only. In this case, the more complex communities were less resistant. In contrast, species diverse grassland communities in Yellowstone National Park were more resistant to summer drought than species-poor communities.

The Serengeti study demonstrates a decline in stability with increasing complexity. However, it also showed that **biomass stability** was greatest in the more complex communities. Another study in Minnesota, USA, found that species-poor grasslands took longer to recover their biomass following drought than diverse grasslands. Thus, in terms of energy, complex communities may be the more stable. As biomass stability suggests, there may be different kinds of stability for different properties of the community.

The flux of energy through a community also appears to have an important influence on its resilience. Models of contrasting communities show that the higher the energy input and turnover of the system, the less time the community takes to return to equilibrium after disturbance. Tundra communities (see Topic S3) with very low turnover appear to be the least resilient.

There is little theoretical support for the notion that a complex species network brings about stability. However, there is mounting evidence that competition and diversity are related. Intense competition leads to low diversity, while weak competition permits coexistence and hence higher diversity. Theories of stability have tended to treat horizontal (competitive) and vertical (trophic) interactions alike. Thus, weak competition reduces instability in theoretical communities.

Trophic complexity and stability

Trophic complexity increases with the number of trophic levels and vertical connections in the food web (see Topic Q3). A community can be considered to be only as resilient as its least resilient species. Consider a lake ecosystem where the density of phytoplankton has been reduced by manipulation or disturbance. This will reduce the density of species at higher trophic levels. The phytoplankton may be able to recover quickly, but the fish will recover more slowly. Zooplankton will be abundant and will keep the density of phytoplankton down. The resilience of this system is limited by that of the predatory fish. This indicates that food chain length may influence the overall resilience of the community.

Food-web connectance is a function of the number of vertical links in the food-web and is considered to influence community stability. In general, theoretical work has demonstrated that if food webs are randomly constructed, increasing the number of species in the food web, increasing the connectance and increasing the intensity of trophic interactions all result in a **decrease** in stability. This is in direct opposition to the holistic view of communities where a dynamic balance is thought to be maintained by the network of species present in the community. These findings should be interpreted with caution however, as some models incorporating more realistic community attributes have shown the reverse. Real communities are different from the randomly constructed communities of null models; it is conceivable that they are complex in a particular way which enhances stability.

In natural communities, stability appears to depend on the balance between the stability of the community and the variability of the environment. In a stable environment, a community that is complex and dynamically fragile may persist (e.g. tropical rain forests). In a variable environment, only dynamically robust, and simple communities will persist.

Q2 ISLAND COMMUNITIES AND COLONIZATION

Key Notes

The species–area relationship

The number of species on an island (or in any area) will increase with the size of the island. The increase is initially rapid, tailing off at the maximum number of species for a given habitat. A plot of log species number against log area gives a linear relationship. For oceanic islands or islands of habitat, the slopes of these log–log plots mostly fall within the range 0.24–0.34. For subareas within continuous habitat, the slope is around 0.1. The effect of increasing species diversity with increasing area is more pronounced on islands than within continuous habitat.

Island biogeography

MacArthur and Wilson's theory of island biogeography states that the number of species found on an island is determined by a dynamic equilibrium between the immigration of new colonizing species and the extinction of previously established ones. As the number of colonizing species increases, the number of immigrants arriving on the island decreases over time. In contrast, as competition among species becomes more intense, the extinction rate increases. The point at which extinction and colonization rates are equal gives the number of species at equilibrium. The model also accounts for the increase in species number with increasing island size and decreasing distance from a source of colonists. Evidence in favor of the model is given by observations of recolonization of defaunated islands and from the loss of species from recently isolated islands. Extinction and colonization account for the depauperate flora and fauna of islands when compared to the adjacent mainland.

Islands and metapopulations

Metapopulation theory has superseded island biogeography in explaining the behavior of populations in 'islands' of fragmented habitat. A metapopulation consists of a number of populations that exchange individuals through immigration and emigration. Unlike islands, habitat patches are embedded in a landscape mosaic that can influence the quality of the patch and the species it will contain.

Related topics

Intraspecific competition (I2)
Rare species, habitat loss and
  extinction (V1)

Conservation strategies (V2)

The species–area relationship

The number of species in an area will increase with increasing area. This can be demonstrated by sampling increasingly larger areas of habitat and recording the number of new species found. Generally, in a uniform habitat, a plot of the total number of species (S) against area sampled (A) produces a characteristic curve (*Fig. 1*). The increase in S is initially rapid as the most abundant species

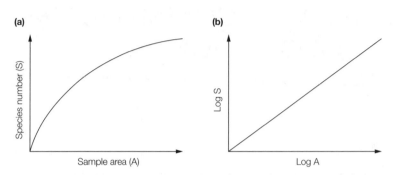

Fig. 1.   (a) The rise in the number of species with increasing sample area; (b) plotted as logarithms, this produces a straight line.

are recorded, but the rate of increase declines as the sample area gets larger and only the rarities remain undetected. This tailing off occurs as the total number of species present in the given area is reached. When both axes are plotted on a log scale the relationship is linear.

The species–area relationship applies to mainland habitats and to islands. 'Islands' are not necessarily oceanic islands surrounded by sea, but could be **habitat islands**; fragments surrounded by less hospitable habitat (e.g. a woodland surrounded by farmland).

The species–area exercise can be undertaken for different groups of organisms, such as birds or flowering plants. For some groups, an increase in area will result in many more species being found, while few species of another group may be found. In other words, the steepness of the line would vary between groups. A typical relationship between the number of species living on an island and the island's area is illustrated in *Fig. 2*. For organisms ranging from birds to ants to land plants, in both oceanic and habitat islands, the slopes of these log–log plots mostly fall within the range 0.24–0.34.

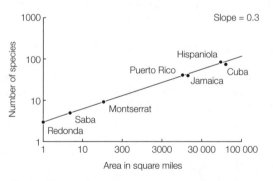

Fig. 2.   The number of amphibian and reptile species on oceanic islands in the West Indies, showing the increase in species number with island size. Redrawn from MacArthur and Wilson, The Theory of Island Biogeography, *1967, Princeton University Press.*

The same analysis for increasingly large samples within a habitat yields a slope of 0.1 (*Fig. 3*). The fact that the type of relationship is the same for continuous habitats and for islands reflects the increase in **habitat diversity** with area. The difference in the magnitude of the slope illustrates that the effect of increasing area on species number is more pronounced on islands.

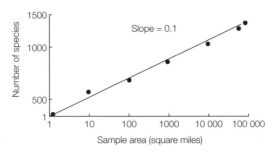

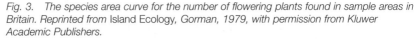

Fig. 3. The species area curve for the number of flowering plants found in sample areas in Britain. Reprinted from Island Ecology, *Gorman, 1979, with permission from Kluwer Academic Publishers.*

Island biogeography

MacArthur and Wilson (1963) developed the theory of island biogeography. Briefly, the theory states that the number of species of a given taxon found on an island results from a **dynamic equilibrium** between the **immigration** of new colonizing species and the **extinction** of previously established ones. Consider an empty island some distance from a mainland source of potential colonists. Some species are better dispersers than others, and these will arrive first, followed eventually by those with poorer dispersal. As the number of colonizing species increases, the numbers of new immigrant species arriving on the island decreases, because fewer new immigrants are available from the mainland source (*Fig. 4a*).

At the same time as colonization is occurring, some island species will become extinct. As more species arrive and begin to compete for resources, the extinction rate increases. The rate at which one species is lost and a replacement is gained is the **turnover rate**. In *Fig. 4a*, the point at which extinction and colonization rates are equal (*S*) gives the number of species at equilibrium.

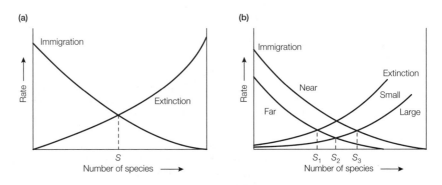

Fig. 4. The MacArthur and Wilson model for the equilibrium number of species on a single island. (a) The point at which the rate of immigration intersects with the rate of extinction determines the equilibrium number of species (S) for a given taxon on the island; (b) the extension of the model to include both island area and distance from a source of immigrants. Small, distant islands will have the fewest species (S_1) while large, near islands will attain the most species (S_3) at equilibrium.

The effect of differing **island size** and **proximity** to a source of colonists is also shown in *Fig. 4b*. Small islands lose species more quickly than large ones and islands close to a source of colonists gain species more quickly than remote ones. Large islands are also greater targets for colonists. The time that a new island has had to acquire species is also important.

Testing the model is difficult; it is rarely possible to observe the colonization of a new island. Small islets off the Florida Keys were fumigated to remove all insects and recolonization allowed to occur. At the end of the experiment, those islands closest to the main stands of mangroves had the greatest number of species, and the islands richest and poorest in species before the experiment were also those richest and poorest in species after recolonization. These findings are broadly in support of the MacArthur and Wilson model. However, the actual species present were not necessarily the same as before the experiment, and species turnover was high.

A useful example is provided by studies of native freshwater birds on New Zealand islands (*Fig. 5*). The general linear relationship between log S and log A holds, but there is considerable scatter and some notable exceptions. White Island lies below the line, which indicates that it has fewer species than would be expected. It is an active volcano! The open squares are islands more than 100 km away from the mainland, illustrating the effect of distance in depressing species number. The islands of the Chatham group (open circles) are extraordinary in that they are about 650 km from the mainland yet are very species rich. This is thought to be because of the unusually high habitat diversity of these islands. Finally, the New Zealand source pool itself is depauperate, a consequence of its own relative isolation. Many islands have been recently colonized by exotics, indicating that the equilibrium number of species may not have yet been reached, in spite of the age of these islands.

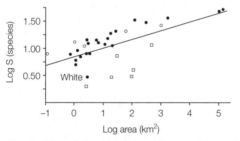

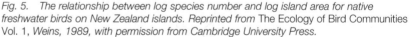

Fig. 5. *The relationship between log species number and log island area for native freshwater birds on New Zealand islands. Reprinted from* The Ecology of Bird Communities Vol. 1, *Weins, 1989, with permission from Cambridge University Press.*

Evidence in favor of the model is also provided by islands that have been recently created when parts of a previous land-bridge to the mainland have been submerged. If an equilibrium number of species is determined partly by a relationship between extinction rate and island area, these islands should *lose* species as they reach a new equilibrium determined by their size. This process is referred to as **relaxation**. Comparison among lizard communities on islands in the Gulf of California reveal that, after correction for area, the number of species present has fallen from 50–75 species to about 25 during a period of 4000 years. This lends weight to the 'equilibrium' interpretation of island diversity.

We can expect smaller islands (or smaller habitats) to support fewer species on two counts: extinction and immigration. These factors explain why islands generally support fewer species than a nearby mainland. For example, Britain has 44 species of indigenous terrestrial mammals, but Ireland, just 20 miles further into the Atlantic, has only 22. Both the flora and fauna of Ireland are said to be **depauperate**. This discrepancy might arise because of the difficulty mammals experience crossing water; however, it affects bats equally and Ireland is also depauperate in birds. There are no woodpeckers, no little or tawny owls and no marsh or willow tits. The relative sizes of Britain and Ireland, and their effect on colonization and extinction account for this pattern.

Islands and metapopulations

Natural habitats are becoming increasingly fragmented as a result of human activity. Habitat fragments have been considered as islands by conservation biologists interested in the number of species they will support.

Metapopulation theory is now more commonly employed to explain population dynamics in fragmented habitats. A **metapopulation** consists of a number of populations that exchange individuals through immigration and emigration. This approach is more realistic than the extreme isolation of the island model, because maintenance of populations depends on movement among patches and not just colonization from a large single source. Habitat patches, unlike islands, are embedded in a landscape mosaic. The surrounding landscape can influence the quality of the patch and block movement from and to it. As the landscape between patches becomes increasingly inhospitable and fragmentation increases, the number of **edge species** increases at the expense of **interior species**. Interior species may persist in fragments if there is a large source of colonists nearby to 'top up' the population, and/or if the landscape incorporates **corridors** or **greenways**, facilitating movement among metapopulations (see also Topic V2).

Q3 COMMUNITY PATTERNS, COMPETITION AND PREDATION

Key Notes

Community assembly	The concept of assembly rules attempts to explain how natural communities vary from random assemblages derived from the range of available species (the species pool). Community assembly may be influenced by habitat type, species colonization and establishment, and/or by interspecific relationships such as predation, parasitism and competition. Assembly rule studies investigate patterns in community make-up, comparing real community patterns with those generated by 'null' community models.
Competition	Competition can be an important force shaping community structure but is not necessarily significant at the time of investigation. The ghost of competition past can leave a strong imprint on a community (e.g. as niche differentiation). Some studies of competition have shown that only one member of a guild of ecologically similar species tends to be present in the community, suggestive of competitive exclusion of other similar species. The distribution of supertramp bird species on islands also supports the theory that communities are structured by competition.
Grazers	Grazing animals have two effects on plant communities: (i) their selective feeding affects species abundance in the community, and (ii) grazing suppresses the growth of competitive species thus enhancing and maintaining the diversity of less competitive species. When grazing intensity is very high, diversity can be reduced as species are forced to local extinction.
Carnivores	Selective predation and prey switching can leave rarer species unpredated. This behavior can lead to the coexistence of a large number of relatively rare species in the same community.
Keystone species	A keystone species has a significant and disproportionate effect on the community. Keystone species can be top predators such as the northern sea otter; however, the term can be usefully applied to any species whose removal would have a significant effect on community structure.
Related topics	Intraspecific competition (I2) Community responses to Resource partitioning (I3) disturbance (R2) Succession (R1)

Community assembly

Ecologists have long been interested in how the observed make-up of the community differs from assemblages of species drawn at random from the **species pool**. This question can be investigated by removing species in order to study how the community restructures itself – which species become the new dominants, which increase, and which decrease. Another approach is to attempt to reconstruct the sequence in which species were added to the community when it was first formed.

Several factors act to determine which subset of species from the species pool will occur in the local community (*Fig. 1*). Firstly, not all species will be able to colonize a location. Secondly, even if colonization does occur, a species cannot be considered a member of a community until it has established a population. Establishment depends on the suitability of the habitat, the area size, the size of the colonizing population and the other species that exist in the area when the colonists arrive. Predators may reduce the number of colonists, preventing establishment of a persistent population. Competitive interactions with ecologically similar species that are already members of the community may prevent a species from joining that community.

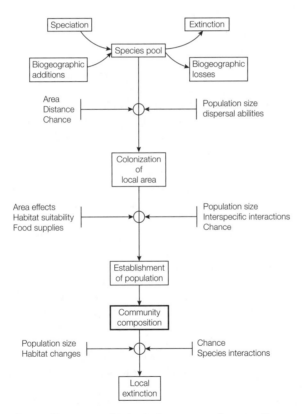

Fig. 1. *Factors contributing to the process of community assembly. Reprinted from* The Ecology of Bird Communities vol. 1, *Weins, 1989, with permission from Cambridge University Press.*

Community assembly is ongoing as communities are subject to a continual stream of colonists and also to the subtraction of species through local extinction. Habitat change may render the area unsuitable for some species, while others may suffer the effects of interactions with predators and competitors that develop after their initial establishment.

Different aspects of community assembly have been emphasized by different studies and may apply in different types of community. The **'first in wins'** idea favors the importance of random colonization, while **'dominance control'** stresses the importance of interspecific competition in determining species persistence. The 'first in wins' hypothesis applies to transient early-successional communities, while dominance control is apparent in late succession stages (see Topic R1). Historic factors, including the ghost of competition past (see Topic I3) the age of the community and the timing and sequence of previous colonization are also potential determinants of community assembly.

An alternative approach to community assembly is that of the 'null' model which assumes that community assemblages are based on chance alone and are made up of species drawn at random from the available species pool. Patterns in real communities can be compared to those in null communities to establish whether they could have arisen by chance alone or whether other factors, such as competition, are responsible.

Competition

There is no doubt that competition *can* be an important force shaping community structure. However, there are many reasons why current competition may not be strong.

(i)   In a patchy environment, strongly competing species may be able to coexist because they do not encounter the same resources.

(ii) Natural selection may have favored the avoidance of competition through niche differentiation (the ghost of competition past).

(iii) Species may only compete during periods of population outbreak or resource scarcity.

Competition among bumble bee (*Bombus*) species appears to determine which species are present in the community. In Colorado, USA, there are four guilds of bumble bees. Long proboscis species feed on flowers with long corollas, (petal tubes), short proboscis species feed on flowers with short corollas and medium proboscis species feed across the range of flower lengths. There is also a nectar-robbing species with a short proboscis that is able to bite through the base of the corolla tubes to access the nectar. This species is placed in a guild on its own. As would be predicted for communities molded by competition, each guild has only one representative per community. In addition, there is evidence that current competition is strong between the long proboscis species, *Bombus oppositus,* and the medium proboscis species, *B. flavifrons*. Removal of one of these species results in increased utilization of flowers formerly exploited by the other species. Interspecific competition of this type demonstrates that competition between, as well as within, guilds can be important in shaping community structure.

In the islands of the southwest Pacific, some bird species are absent from species-rich (large) islands but are found in small, species-poor communities. These **'supertramps'** have high dispersal abilities, high reproductive potential and unspecialized habitat affinities. In contrast, some species are only found in species-rich communities and tend to have specialized habitat requirements. It was inferred that supertramps are outcompeted and excluded from species-rich islands by guilds of *K*-selected specialists which are better adapted to some aspect of the supertramp's niche. Rather than these communities being a random collection of species, it appears that competition plays a large part in their assembly.

Two of the island species, the cuckoo doves *Macropygia mackinlayi* and *M. migrirostris*, are very similar in their ecology. They exhibit a classic **'checkerboard'** pattern, having mutually exclusive distributions in the archipelago where each island in the archipelago supports only one of the two species (*Fig. 2*). Checkerboard distributions have been produced in null communities and are not common enough to provide proof of the importance of competition. However, the observations that some species combinations never occur and that species-poor and species-rich island communities consist of different species lend weight to the competition hypothesis.

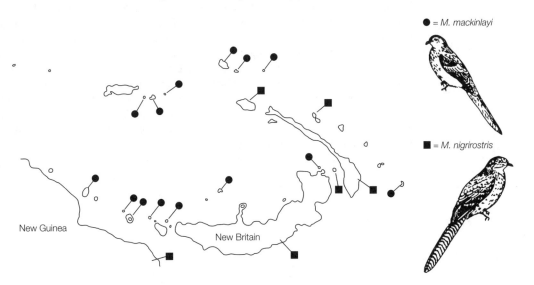

Fig. 2.   'Checkerboard' distribution of Macropygia *cuckoo doves in the Bismark region. Islands have either one species, or neither (no symbol). Reproduced from Pietrewicz and Kamil,* Foraging Behavior: Ecological, Ethological and Physiological Approaches, *1981, pp. 311–332, with permission from Garland Press.*

Niche differentiation is sometimes apparent as morphological differentiation, (e.g. in body size.) Hutchinson noted that potential competitors drawn from communities have weight ratios of approximately 2.0 (or length ratios of the cube root of 2.0 which is 1.3). This spacing demands that species utilize different resources. The same pattern emerges from disparate taxa such as raptors, finches and desert rodents. The canine diameter of felids and mustelids exhibit very even spacing while spacing in other body traits is not clear. Canines are used to seize and kill live prey so are likely to be associated with specialization on certain prey types and the partitioning of prey resources. In the bumble bee example, coexisting species have a proboscis length ratio of 1.32. Desert rodents which differ in body size by a ratio of less than 1.5 coexisted less frequently than would be expected by chance alone.

In general, there is strong evidence that competition is an important driver of community composition However, the importance of competition will vary from community to community and may affect only a small proportion of the interactions between species.

Grazers   Grazing animals have a dual influence on the structure of the plant communities on which they feed. Firstly, many grazers are **selective** and avoid unacceptable food plants while actively seeking out the more palatable ones, thus having an affect on **species abundance** in the community. Secondly, grazing suppresses the growth of aggressive, dominant species (e.g. grasses) that would otherwise outcompete less competitive species, such as annual and perennial forbs. In this way **grazing influences community diversity**; low or no grazing pressure results in a sward dominated by a few species, while moderate grazing causes an increase in the number of species because less competitive species are able to persist (e.g. orchids). When rabbit grazing intensity on grassland is very high, diversity can be reduced as plant species are forced into local extinction. The effect of grazing is a specific application of Connell's intermediate disturbance hypothesis (see Topic R2). This pattern is apparent in other systems. *Figure 3* shows the effect of grazing by the periwinkle, *Littorina littorea*, on algal-species-richness in intertidal pools. When *L. littorea* are rare the pools are dominated by *Enteromorpha*. Intermediate grazing pressure prevents competitive exclusion of other algae by *Enteromorpha* and other species are able to coexist. At very high densities, *L. littorea* consumes all the palatable species leaving only a few inedible ones. This correlation between grazing intensity and species richness provides circumstantial evidence in favor of the intermediate disturbance model.

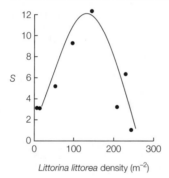

Fig. 3. The effect of Littorina littorea *density on the species richness (S) of algae on a rocky shore. Redrawn from Leverich* et al., Phlox drummondi, American Naturalist **113**, *881–903, 1979, University of Chicago Press.*

Carnivores   Carnivorous predators tend to show a preference for particular types of prey which may not reflect the abundance of those species in the community. **Prey switching**, that is switching from one prey type to another, taking disproportionately more of the most common prey type (Topic J2), leaves the rarer species (relatively) unpredated. This behavior may lead to the coexistence of a large number of relatively rare species in the community.

Keystone species   A keystone species has a significant and disproportionate effect on the community. Like the keystone in the center of an arch, its removal leads to the collapse of the structure, in this case the extinction or large changes in abundance of several other species. The northern sea otter (*Enhydra lutis*) used to occur through most of the subtidal northern Pacific ocean but is now much restricted in its distribution. Isolated populations in the Aleutian Islands have provided ideal conditions for elucidating the role of the sea otter as a keystone predator. Communities with sea otters have small urchin populations as a result of the otters' feeding activities. Healthy beds of macroalgae result with few mussels, limpets and chitons. The rich detritus produced by the algae supports the high productivity of nearshore fish populations and their predators. In contrast, where sea otters are only present as occasional vagrants, large populations of sea urchins flourish, subjecting the algae to massive grazing pressure and creating space for bottom-dwelling invertebrates. The reduced primary productivity has a 'knock-on' effect on fish populations.

Another classic example of a keystone predator is the starfish, *Pisaster ochraceus* which feeds primarily on barnacles and mussels. Removal of the predator has dramatic consequences for the intertidal community, leading to dominance by the mussel population (*Mylilis californicus*), and the exclusion of algae and browsing species such as whelks and limpets.

Keystone species do not have to be top predators; pollinating insects play a crucial role in maintaining community structure and could be considered to be keystone species. The term can be usefully applied to any species whose removal would have a significant effect on community structure.

R1 SUCCESSION

Key Notes

Succession – the classical model	Ecological succession is defined as a continuous, unidirectional, sequential change in the species composition of a natural community. This sequence of community is termed a sere, and culminates in the climax community. Early successional stages are characterized by pioneer species, low biomass and often low nutrient levels. Community complexity increases as succession progresses, often peaking in the mid-successional stage. A mid-successional community is characterized by high biomass, high levels of organic nutrients and high species diversity.
Autogenic succession	Autogenic succession is self-driven, resulting from the interaction between organisms and their environment. Primary succession occurs on a newly formed substrate such as glacial till. Nutrient enhancement and litter accumulation by pioneer species allow new species to colonize. Secondary succession follows disturbance, for example by flooding, fire or human activity. In both types of autogenic succession pioneer species colonize quickly, making opportunistic use of resources before the invasion of more competitive species. Shading leads to dominance by shade-tolerant species which tend to be slow colonizers.
Degradative succession	Degradative succession is a type of autogenic succession involving colonization and subsequent decomposition of dead organic matter. Different species invade and disappear in turn, as the degradation of the organic matter uses up some resources and makes others available. This process leads to the production of humus and is important in soil formation.
Allogenic succession	Allogenic succession results from external environmental factors, such as long term climatic change (e.g. ice ages) or environmental change over a short time (e.g. sediment accretion). Changes in community structure over time are apparent from pollen analysis of sediment cores. Successional progress depends on species tolerance to environmental conditions such as salinity.
Successional processes	Succession is strongly influenced by three processes. (i) Facilitation: changes in the abiotic environment that are imposed by the developing community and allow other species to invade. (ii) Inhibition: species of one stage resist invasion by later successional species such that invasion is only possible following disturbance or death. (iii) Tolerance: late successional species invade because they are able to tolerate lower resource levels and can outcompete early successional species. Highly competitive species, which are tolerant of low resource levels will replace opportunistic good colonizers and come to dominate the climax community.

Fluctuations and the climax community	There is no single climax community for a geographic area, but a continuum of climax types varying along environmental gradients. Climax communities are not stable, but are in a state of continual flux. Unidirectional succession to the climax does not always happen as succession can be arrested at an earlier stage. In general, biomass and species diversity increase with succession but often peak at an intermediate stage and not at the climax community.
Related topic	Community response to disturbance (R2)

Succession – the classical model

Ecological succession is the change in species composition and community structure and function over time. Succession is usually defined as a continuous, unidirectional, sequential change in the species composition of a natural community. The term succession was first used to describe the transition in abandoned old fields in eastern North America. After abandonment there appears to be a predictable sequence from grass and weedy herbaceous plants to shrubs such as sumac and hawthorn, eventually developing into a forest of maple, oak, cherry or pine:

Annual $\Rightarrow$ Herbaceous $\Rightarrow$ Shrubs $\Rightarrow$ Early $\Rightarrow$ Late
weeds   perennials   successional   successional
                                                                 trees   trees

Such a sequence of communities is termed a **sere** and each of the distinct and recognizable successional stages is a seral stage. Each seral stage is a snapshot of a continuum that is changing over time, but each has its characteristic species composition. The sere is a generalization – some seral stages may be missed completely. Seral stages may persist for a few years or a few decades depending on the type of stage and the environmental conditions.

Eventually succession slows as the community reaches a steady equilibrium with the environment. This final seral stage is termed the **climax community**. In theory, at this point the community is stable and self-replicating. In its extreme form, the climax concept predicts that there is only one final community for a geographical region and that all succession will progress towards this **monoclimax**.

According to the classical model, structural complexity and organization of an ecosystem increase and mature over time as succession proceeds. Early successional stages are characterized by a few species known as **pioneer species**, low biomass and often low nutrient levels. Net community production is greater then respiration, resulting in an increase in biomass over time. Food chains are short and species diversity is low. The mature stage in succession is characterized by high biomass, high levels of organic nutrients, and gross production that about equals respiration. Food chains are complex and levels of competition are high. Accompanying these changes is an increase in species diversity. Species diversity often declines as the climax community is approached and the community becomes dominated by the most competetive species. *Figure 1* shows how environmental conditions can influence the accumulation of species. On mesic sites, forest diversity peaks at an intermediate stage in the succession, but continues to increase on xeric (dry) and intermediate sites.

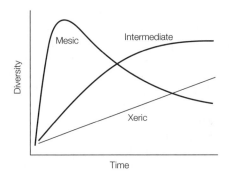

Fig. 1.   Species diversity of forest sites over time for xeric, intermediate and mesic moisture conditions. Diversity increases over time as the climax community is approached on xeric and intermediate sites. On mesic sites diversity peaks in mid-succession. Redrawn from Ecology and Field Biology, 5th edn, R.L. Smith, 1996, Addison Wesley Longman.

Autogenic succession

Succession can be thought of as autogenic or self-driven. Changes in the environment occur as a result of the interaction between the organisms and that environment. Succession that begins on newly formed substrates not occupied by any organisms and where there is no organic material present, such as lava flows, newly exposed rock faces, alluvial deposits and glacial moraines, is termed **primary succession**. Succession where the vegetation cover has been disturbed by humans, animals or by fire, wind or floods, is called **secondary succession**.

In primary succession initial conditions are often severe. At first, the only organic matter will be wind-blown and no soil is present. On exposed boulder clay till deposited by retreating glaciers in Glacier Bay, Alaska, the nutrient poor clay is first colonized by mosses and shallow-rooted herbaceous species such as mountain avens (*Dryas* sp.). These early colonists alter the soil conditions allowing new species to colonize the habitat. Mountain avens has nitrogen fixing symbionts that increase the nutrient content of the substrate. Litter accumulation also leads to soil development, allowing colonization by shrub and eventually tree species such as cottonwood and hemlock. Alder (also a nitrogen fixer) has a strong acidifying effect, reducing the pH of the surface soil and creating conditions suitable for invasion by sitka spruce. The alder is eventually displaced by the spruce leading to a mature sitka spruce–hemlock forest.

Primary succession to woodland is a slow process, taking in the region of a hundred years or more. More rapid primary succession can be observed on intertidal boulders overturned by wave action. Bare substrate is rapidly colonized by the pioneer green alga *Ulva*, followed by several species of red algae. The red algae are slower to establish and can only invade where the *Ulva* has died, been grazed off or removed by some other disturbance. Selective grazing by a species of crab particularly partial to *Ulva* accelerates succession to the tougher, longer-lived red or brown algae.

Old field succession is an example of secondary succession. Here the pioneer species are opportunistic annuals that can respond quickly to the appearance of a new habitat. The germination of *Ambrosia artemisiifolia* is triggered by disturbance, unfiltered light, fluctuating temperature and reduced CO_2 concentration, all of which are associated with a newly cleared site. Summer annuals such as *Ambrosia* are replaced by winter annuals which establish earlier in the season, giving them a headstart in the competition for space, light and other resources. Late successional species are shade-tolerant and slow colonizers.

Degradative succession

The term degradative succession describes a particular type of autogenic primary succession: the colonization and subsequent decomposition of dead organic matter. Different species invade and disappear in turn, as the degradation of the organic matter uses up some resources and makes others available.

For example, pine needles fall in August and are first colonized by fungi which digest and soften them allowing other fungal species and mites to penetrate. After about 2 years in the A_0 layer (see Topic G3) the tightly compressed needle fragments are invaded by other soil microfauna which feed on the fungus as well as the needles. Bascidiomycetes attack the needle fragments, digesting cellulose and lignin. After about 7 years the needles have been completely decomposed, forming an acidic humus with almost no biological activity. All degradative successions terminate when the organic substrate is completely metabolized.

Allogenic succession

Serial replacement of species can result from external environmental factors, such as geophysico-chemical changes. This is in contrast to autogenic succession (above) which is dependent on the biological action of the organisms on their environment.

Allogenic succession has occurred over most of North America and northern Europe in response to climate warming, following the retreat of the last Pleistocene ice sheet about 10 000 years ago. The ice sheet advanced and retreated three times in Europe causing a similar advance and retreat in the biota, each advance consisting of a somewhat different mix of species. As the climate ameliorated, easily dispersed, light-demanding trees such as pine and birch advanced northward to be replaced by the slower dispersing, shade-tolerant species such as oak and ash. The next glacial period favored spruce, fir and eventually treeless tundra vegetation.

These changes have taken place over thousands of years and are evidenced by pollen cores taken from lake deposits. An example of a pollen map from Pennsylvania is given in *Fig. 2*, showing the transition over time (depth) from fir to pine and then to oak, hickory, beech and hemlock (with grass in forest openings) which occurred as the climate warmed.

Allogenic transition over shorter time scales occurs where sediment is accreting (e.g. on sand dunes or estuaries). Silt is accreting in the Fal estuary in Cornwall, England causing the salt marsh to extend seawards and the valley woodland to invade the landward limits of the marshland. The recent spread of the salt marsh grass, *Spartina anglica*, is also contributing to the extension of salt marsh. Species are able to colonize at a particular height above sea level according to their tolerance of inundation by brackish water at high tides. Pioneer grasses and sedges are the only higher plant species tolerant of the brackish mudflats, while slightly higher up tidal scrub and then tidal woodland have developed. This geographic transition mirrors the changes apparent from the woodland soil profile, and hence the historic origins of the woodland community.

Successional processes

The examples above and studies involving models of succession indicate that it is strongly influenced by three processes.

(i) Facilitation: changes in the abiotic environment that are imposed by the developing community. Pioneer species 'prepare the ground' allowing other species to invade, e.g. mountain avens and alder on boulder clay till.

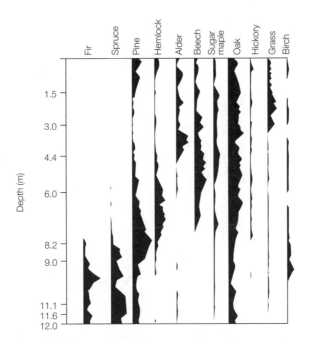

Fig. 2.   Pollen diagram for Crystal Lake, Harstown Bog, Crawford County, Pennsylvania. *From bottom to top: initially, spruce-fir forest with some oak invaded as the ice sheet retreated. This was replaced by pine, oak and birch as the climate warmed. The slower colonizing species, such as beech, hickory and hemlock were next to dominate.*

(ii) Inhibition: species of one seral stage resist invasion by later successional species such that invasion is only possible following disturbance or by replacing individuals as they die, e.g. green algae (*Ulva*) on intertidal boulders.

(iii) Tolerance: late successional species invade because they are able to tolerate lower resource levels and can outcompete early successional species for limited light and space, e.g. old field succession.

   The tolerance model suggests a predictable sequence of species replacement based on species' strategies for exploiting resources. Tilman, in his **resource ratio hypothesis**, argues that species dominance is determined by the relative availability of two resources, light and nutrients. During succession, nutrient availability increases with litter accumulation and soil formation, while light levels decrease as a result of shading. *Figure 3* shows how species A, with the highest requirement for light and the lowest requirement for nutrients is the first to colonize, followed by a predictable sequence of invaders until eventually the community is dominated by nutrient hungry, shade-tolerant species E.

   Nobel and Slatyer also stress the importance of individual species properties in determining their place in succession. These properties are their method of recovery after disturbance and their ability to reproduce in the face of competition. Species unable to reproduce in competitive environments and able to recover rapidly after disturbance (e.g. by means of a seedling pulse from a seedbank) occur early in the succession. Competitive species only able to invade slowly through seed dispersal into the habitat dominate later.

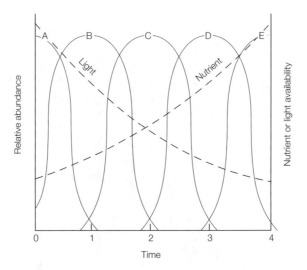

Fig. 3.   The resource ratio hypothesis of succession. Reprinted from Ecology, 3rd edn, Begon, Harper and Townsend, 1996, with permission from Blackwell Science.

Typically, early successional species are *r*-selected, having developed efficient colonization mechanisms to escape from competition. Late successional species have evolved characteristics enabling them to persist longer in competitive situations and are thus responding to *K*-selection. In general, good colonizers are poor competitors and *vice versa*, i.e. there is a trade-off between competitive strength and colonization ability.

The three primary strategies proposed by Grime – competitive (C), stress-tolerant (S), and ruderal R (see Topic M1) – allow any species of plant to adopt a combination of strategies, trading off performance in one type of environment with performance in another. Using this model, succession in a temperate forest clearing involves a transition from ruderal (pioneer) species to fast growing trees and shrubs which have a competitive strategy (*Fig. 4*). At later stages in the course of succession the proportion of stress-tolerant species increases. Secondary succession on sites with low soil fertility has a more shallow parabola, as productivity is constrained.

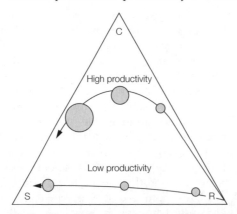

Fig. 4.   Successional trajectories in relation to Grime's triangle of competitive, stress-tolerant and ruderal (CSR) strategies. Reprinted from Plants in a Changing Environment, Bazzaz, 1996, with permission from Cambridge University Press.

Fluctuations and the climax community

Studies have shown that the model of the **single** climax community is inaccurate. The climax community will vary depending on local variations in climatic and edaphic conditions and in response to other factors such as the disturbance regime and the influence of animals, particularly herbivores. There is no single climax but a continuum of climax types varying along environmental gradients.

The **stable** climax community also proves elusive in practice. No community is stable for long because of natural disturbances such as storms and fire. A forest climax may take 300 years to develop but the probability of disturbance may be so high that succession never reaches this stage. Nonsuccessional, short term, reversible changes in the floristic composition of a community, or **fluctuations**, are also common. These changes are the result of environmental stochasticity and range from changes in species dominance to more subtle variations such as changes in the age structure of a species.

Succession was considered a predictable unidirectional process, but it is now known that this is not always the case. For example, **cyclic succession** occurs on a small scale in most communities. Where a forest tree dies and falls, pioneer species germinate in the gap to be replaced by mid- and late-successional species until the gap is again filled by trees. Thus, even old, seemingly stable communities are in a state of flux and are a shifting mosaic rather than a steady state. In disturbed habitats, cyclic succession will prevent the development of a climax community and arrest seral development at an early stage. Although succession is undoubtedly an important and widespread process, the physical and biological structure of many communities may be dominated by other influences.

R2 COMMUNITY RESPONSES TO DISTURBANCE

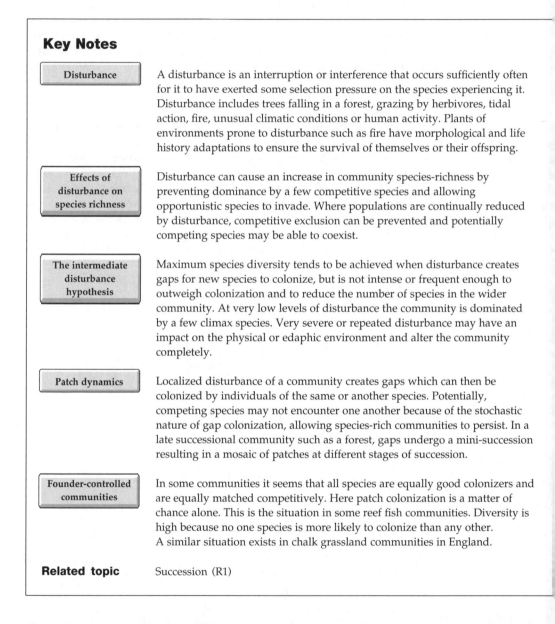

Key Notes

Disturbance	A disturbance is an interruption or interference that occurs sufficiently often for it to have exerted some selection pressure on the species experiencing it. Disturbance includes trees falling in a forest, grazing by herbivores, tidal action, fire, unusual climatic conditions or human activity. Plants of environments prone to disturbance such as fire have morphological and life history adaptations to ensure the survival of themselves or their offspring.
Effects of disturbance on species richness	Disturbance can cause an increase in community species-richness by preventing dominance by a few competitive species and allowing opportunistic species to invade. Where populations are continually reduced by disturbance, competitive exclusion can be prevented and potentially competing species may be able to coexist.
The intermediate disturbance hypothesis	Maximum species diversity tends to be achieved when disturbance creates gaps for new species to colonize, but is not intense or frequent enough to outweigh colonization and to reduce the number of species in the wider community. At very low levels of disturbance the community is dominated by a few climax species. Very severe or repeated disturbance may have an impact on the physical or edaphic environment and alter the community completely.
Patch dynamics	Localized disturbance of a community creates gaps which can then be colonized by individuals of the same or another species. Potentially, competing species may not encounter one another because of the stochastic nature of gap colonization, allowing species-rich communities to persist. In a late successional community such as a forest, gaps undergo a mini-succession resulting in a mosaic of patches at different stages of succession.
Founder-controlled communities	In some communities it seems that all species are equally good colonizers and are equally matched competitively. Here patch colonization is a matter of chance alone. This is the situation in some reef fish communities. Diversity is high because no one species is more likely to colonize than any other. A similar situation exists in chalk grassland communities in England.
Related topic	Succession (R1)

Disturbance

The term 'disturbance' refers to interruption or interference that occurs sufficiently often for it to have exerted some selection pressure on the species experiencing it. This is distinct from catastrophes, which are more devastating in effect, but occur too infrequently to allow species to evolve a response.

Disturbance includes trees falling in a forest, grazing by herbivores, tidal action or human activity. Disturbance can also include severe climatic conditions such as a harsh winter or a storm.

The way in which species can evolve responses to disturbance is demonstrated by plants of fire-prone ecosystems, such as North American chaparral. Some species have thick insulating bark to protect the cambium from heat damage. Another defence is the accumulation of a mat of needles that encourages frequent, low intensity fires which prevent the build-up of flammable debris that would burn less frequently but more intensely. In some species, mature individuals are totally destroyed by fire on a regular basis, but the heat of the fire triggers seed germination. The seedlings develop in a newly created seedbed fertilized with ash. This occurs in jack pine and lodgepole pine, resulting in even-aged stands.

Effects of disturbance on species richness

Disturbance can cause an increase in community species richness. This is because disturbance prevents dominance by a few competitive species and allows opportunistic species to invade. During the development of a community, the competitive interactions between species lead to the replacement of opportunistic species by more persistent competitive species (see Topic R1). This provides fewer opportunities for new colonizing species and, as a result of competitive exclusions, species richness may be expected to decline over time. However, if populations are continually reduced by disturbance, competition will be, to some extent, prevented and species will be able to coexist. *Figure 1* shows the outcome of a Lotka–Volterra simulation of this situation. Undisturbed, species 1 rapidly becomes extinct. However, with the populations of both species periodically reduced by half, competitive exclusion is prevented and the species are able to coexist. This situation is analogous to a community experiencing repeated disturbance or a fluctuating environment.

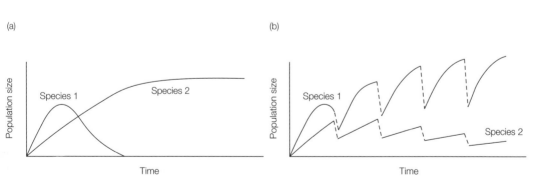

(a)   (b)

Fig. 1.   The effects of nonequilibruim conditions on the outcome of competition between an r-selected and a K-selected species. (a) with no disturbance, competitive exclusion is reached; b) both species are able to coexist where periodic, density-independent reductions in the population are imposed. Species 1: high r, low K. Species 2: low r, high K.

The intermediate disturbance hypothesis

Maximum species diversity will be achieved when the rate of disturbance is sufficient to create gaps for colonization, but does not exceed the rate at which new species colonize. High rates of disturbance tend to destroy the larger part of the community. This idea is called **Connell's intermediate disturbance hypothesis**, which predicts that the highest species richness will be maintained by an intermediate level of disturbance. This intermediate level of disturbance occurs where its scale and frequency allows fugitive or supertramp species to survive, without reducing species richness in the rest of the community. For example, if grazing pressure is very high or if fire is very frequent, disturbed areas will not progress beyond the pioneer stage and community diversity will be low. As the intensity or frequency of disturbance decreases, more time or space is allowed for the invasion of species and diversity increases. At very low levels of disturbance the community is dominated by climax species and diversity is reduced through competitive exclusion (*Fig. 2*).

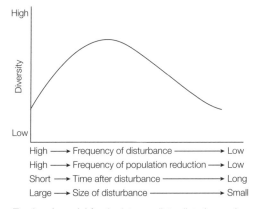

Fig. 2. A model for the intermediate disturbance hypothesis.

Intertidal boulders in California vary in their frequency of disturbance. Small boulders are overturned more often and have a greater percentage of bare, algae-free areas. The green alga, *Ulva* was the dominant species on these small boulders. The largest, least disturbed boulders were colonized by the long-lived, slow-growing red alga, *Gigartina*. As predicted by the intermediate disturbance hypothesis, intermediate boulders were the most diverse, supporting a mixture of pioneer and climax species.

Intense disturbance may totally alter a community. Clear felled areas of white fir in northern California were sprayed with herbicides which prevented stump regrowth and successional recovery from disturbance. This had a knock-on effect on the soil microbial community because the soil was no longer shaded and was exposed to extremes of temperature. Soil structure and its water holding capacity were lost. The original white fir forest was replaced by a totally different community capable of surviving without mycorrhizae, dominated by brome grass, bracken and manzanita.

Patch dynamics   Localized disturbance of a community creates a gap. For example, in grassland gaps may be created by burrowing animals, heavy grazing or severe frost. Disturbance of this kind has the effect of removing organisms and creating space which can then be colonized by individuals of the same or another species. Patch dynamics have been simulated by models which view communities as consisting of cells that are colonized at random by individuals of a number of species. In these models, because potentially competing species do not necessarily encounter one another (due to the stochastic nature of colonization), species-rich communities can persist indefinitely. A species may be competitively excluded locally but persist in the patchy environment as a whole.

In a late-successional community such as a forest, dominated by K-selected, competitive species, disturbance opens up gaps that are colonized by opportunistic, pioneer species. These are later replaced by other species that are slower to colonize and mature. The gap undergoes a mini-succession. Diversity in the gap is initially low, reaching a peak in mid-succession and falling as competitive exclusion by climax species takes place.

In fresh gaps, regrowth occurs from three sources: seeds, plants established prior to gap formation and the growth of branches of trees on the gap periphery. Shade-intolerant saplings are 'released' and race to fill the gap. Larger gaps remain open sufficiently long for buried seeds to germinate and contribute to the gap community. In tropical forests, some species are adapted to the filtered light regime of gaps and are found nowhere else. Nectar- and fruit-eating birds are more abundant in neotropical forest gaps than under the closed canopy because understorey plants in gaps flower and fruit more abundantly where light levels are less limiting.

Localized, unsynchronized disturbance leads to a mosaic of patches at different stages of succession. This mosaic has a higher species diversity than a community subject to a uniform disturbance regime.

Founder–controlled communities

What happens if all species are equally good colonizers and are equally matched competitively? In this case, sequential replacement of species as a newly opened patch undergoes succession will not occur. After disturbance or death of the occupant(s) of a patch, which species colonizes that patch is a lottery. Colonization is determined by **priority effects** – the first species to arrive will be the one that comes to occupy the site. Because any species may become the replacement, community species-richness is maintained at a high level.

This model also describes the patch dynamics of some coral reef communities such as Heron Reef, part of the Great Barrier Reef. Space on the reef is at a premium and all species breed frequently, flooding the reef with larvae in order to maximize their chance of colonization. There is no evidence of succession – once a territory is won it is held for life. The high diversity of this community is promoted by the unpredictability of territory aquisition.

A similar situation exists in chalk grassland communities in England. Any small gap is rapidly exploited by a seedling. Which seeds develop is largely a matter of chance as many species have overlapping germination requirements. However, in this case gaps are not identical and not all seeds present will be able to germinate. Species colonization also depends on the season – a gap created in late summer will become occupied by a different species than one created in spring. For example, grasses will germinate in late summer gaps, while annuals such as flax and violets tend to colonize spring gaps.

S1 ECOSYSTEM PATTERNS

Key Notes

Vegetation and climate	The vegetation of the earth is divided into distinct blocks or formations which broadly reflect climatic conditions. This pattern arises as a result of the adaptation of plant form to temperature and water availabilty. The great plant formations first mapped by Koppen are what we now call biomes. Biomes represent the divisions of the major community types of the world, consisting of distinctive combinations of plant and animal species and characterized by an approximately uniform life form of vegetation, such as grass or deciduous trees.
Ecotones	On the local and regional scale communities vary as the individual species respond to gradients in environmental conditions. The boundaries of individual species, communities and biomes are not distinct and abrupt, but blurred and gradual. Biomes merge into one another along ecotones. A vegetation map superimposes boundaries on this continuum, indicating approximately where one biome ends and another begins, and reflects the limit of tolerance of the dominant life form.
Temperature and distribution	Damage by frost is probably the single most important factor limiting plant distribution. In some cases it is possible to relate the distributional limits of a species to a lethal temperature. It is common to find a close correspondence between the distributional limits of a species and an isotherm. Raunkiaer's classification of life forms, based on the vulnerability of plant meristems to frost, is a fair predictor of plant distribution.

Related topics	Adaptation (B1)	Community. The structure and
	Coping with environmental variation (B2)	stability (Q1)
		Grasslands (S2)
	Solar radiation and climate (C1)	Tundra (S3)
	Microclimate (C2)	Forests (S4)
	Temperature and species distribution (E3)	Deserts, semi-deserts and shrubland (S5)
	Solar radiation and plants (F1)	

Vegetation and climate

Plant geography is determined by the response of plant species to different energy regimes. The climate of the earth occurs in distinct blocks which can also be distinguished according to their vegetation. This correspondence arises because plants are selected by their environment to balance heat budgets and manage water in ways that maximize survival.

In desert environments, which lie in the southwestern areas of continents and behind mountain ranges, characteristic drought-tolerant vegetation occurs. If leaves are present at all they are small, needle-shaped structures that do not retain heat but dissipate it through convection thus never rising above the temperature of the ambient air. Many desert cacti lack leaves entirely and are shaped like pillars; they avoid over-heating by minimizing the amount of heat absorbed.

In the warm and wet tropics plants dissipate heat by evaporating water. The leaves in a tropical rain forest are flat, numerous, and stacked layer on layer to diffuse incident light. In temperate latitudes, summer cooling is effected through evaporation, but in winter many plants drop their leaves entirely in order to conserve water at a time when ground water tends to be frozen and unavailable. Further north, the optimal solution to maximizing productivity appears to be an evergreen needle-leaf form which is resistant to freezing. The temperature of needles remains close to that of ambient air in both summer and winter.

In addition to temperature, the pattern of water availability is fundamental to plant distribution. Small, needle-shaped leaves are also a means by which plants can minimize water loss and these leaves usually have thick cuticles. Grasses, which can die down completely above ground are able to survive periods of cold or drought, emerging from underground shoots when conditions improve.

The first **climate map** and the standard **Koppen classification** of climate were based on **vegetation maps**. Koppen worked on the hypothesis that the distribution of vegetation reflected the underlying pattern of climate. He mapped the plants of the world in order to produce a map of climate. These great plant formations are what we now call **biomes**. Biomes represent the divisions of the major community types of the world, and are characterized by specific climate conditions.

Ecotones

The presence and abundance of individual species will depend on ambient conditions which will vary along environmental gradients. These individual species responses underlie the variation in communities on the local and regional scale. The physical and biological structure of a community will change in response to conditions such as moisture, altitude and soil type. This type of variation is known as **zonation**. Changes in soil type can result in a clear boundary between species distributions and hence between communities. However, often the boundaries of individual species, communities and biomes are not distinct and abrupt but blurred and gradual because of the nature of the underlying environmental variation (see also Topic Q1). Biomes consist of large areas of approximately uniform habitat that merge into each other along **ecotones**. For example, in the Northern Hemisphere there is a north–south gradient in temperature. The tropical forest of low latitudes is replaced by subtropical forest further north, followed by temperate deciduous forest, which grades into coniferous forest and is in turn replaced by tundra. Between the boreal forest of Canada and the eastern deciduous forest of the United States lies a broad transitional band which contains assorted broad-leaved and coniferous trees in mixed communities and where some conifers are deciduous.

A **vegetation map** is a simplification of the pattern of vegetation on the ground and represents an attempt to superimpose boundaries on a continuum, indicating approximately where one biome ends and another begins (*Fig. 1*). The boundaries of biomes reflect the limits of tolerance of the dominant plant life form (e.g. deciduous, broad-leafed trees to temperature and water availability). *Figure 2* shows the distribution of eight major terrestrial biomes with respect to mean annual temperature and mean annual precipitation.

Temperature and distribution

It is more difficult to attribute a precise role to temperature when single species are examined. In some cases we can relate the distributional limits of a species

Tundra

Coniferous forest

Temperate deciduous forest

Montane forest, alpine tundra complex

Mixed hardwood-conifers

Mediterranean chaparral

Tropical forests

Semidesert, arid grassland, tree savannah

Tropical savannah, thorn forest

Desert

Grassland, Steppe

Fig. 1. Major biomes of the world.

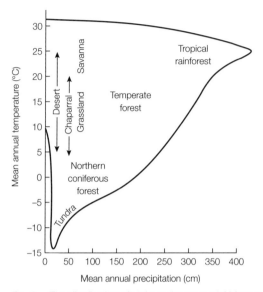

Fig. 2. *The distribution of eight major terrestrial biomes with respect to mean annual temperature and mean annual precipitation.*

to a **lethal temperature** which precludes survival. Damage by frost is probably the single most important factor limiting plant distribution. For example, the saquaro cactus is liable to be killed when temperatures remain below freezing for 36 hours, but if there is a daily thaw it is not under threat. In Arizona, the northern and eastern edges of its distribution correspond to a line joining places where on occasional days it fails to thaw. It is common to find a close correspondence between the distributional limits of a species and an **isotherm**, or a line on a map joining locations having the same mean temperature at a particular time of year. The wild madder (*Rubia peregrina*) for example produces new shoots in January for the following spring, and shoot production is inhibited by low temperatures. Its northern distributional limit is correlated with the 4.5°C January isotherm (*Fig. 3*). This is not a threshold temperature below which production ceases; rather a January average of less than 4.5°C is indicative of a location where temperatures are frequently too low for shoot production and a population of wild madder cannot be sustained.

Christen Raunkiaer classified plant life forms according to the position of meristem tissue that is present throughout the year (*Fig. 4*). Plant meristems produce new shoots and are particularly vulnerable to damage by frost. In cold climates, these perennating meristems are found close to, or under the ground as tubers and roots. Trees and shrubs have persistent leaf-producing buds held above ground. In contrast, the above-ground tissues of grasses die back during winter while their perennating buds remain protected deep in the ground. A community with a high proportion of perennating tissue well above ground is characteristic of warmer, wetter climates. A community where most of the species have meristems on the surface or underground is found in colder, drier climates.

Fig. 3.   The northern limit of the distribution of the wild madder (Rubia peregrina) is closely
correlated with the position of the January 4.5°C isotherm. Reprinted from Ecology, 2nd edn,
Begon et al., 1990, with permission from Blackwell Science.

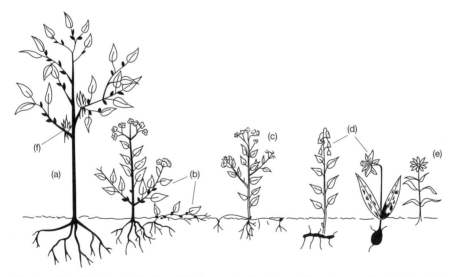

Fig. 4.   Raunkiaer's classification of life forms showing the position of perennating buds and shoots:
(a) exposed and high above the ground – warm, moist climates; (b) on or a short distance above the
ground – cool, dry climates; (c) at the surface, protected by soil and litter – cold, moist climates; (d) buried
deep in the ground on a bulb or rhizome, protected from freezing and desiccation – cold moist climates;
(e) annuals able to survive unfavorable conditions as seed – deserts and grasslands; (f) epiphytes – warm,
humid climates. Redrawn from Ecology and Field Biology, 5th edn, R.L. Smith, 1996, Addison Wesley
Longman.

S2 GRASSLANDS

Key Notes

Primary regions

Grasslands occur where rainfall is intermediate between that of deserts and forests. There are two major types of grassland depending on the temperature: tropical grassland (savannah), and temperate grassland (steppe, prairie and pampas). Grasslands occupy vast areas in North America, northern Europe and Africa, blending into desert or forest where the climate is unsuitable.

Climate and soils

Tropical grasslands receive abundant rain (1200 mm or 60 inches max.) in the wet season, but usually suffer a prolonged dry season during which fires are common. Low soil moisture impedes nutrient cycling and reduces nutrient availability. Temperate grasslands have between 250 and 600 mm (10–30 inches) of rainfall per annum. Grassland soils receive a large amount of organic matter and are very rich, making them well suited to the growing of arable crops such as corn and wheat.

Major vegetation

Grasslands have a high primary productivity and relatively low biomass. Managed grasslands are used for crops and for rangeland. Grassland communities are dominated by grass species but frequently include trees, such as the acacia. Grazing is a significant influence on the structure and composition of grassland communities and tends to maintain high production and species diversity. Temperate grasslands include broad-leaved perennials which either flower early in the season or after the grasses have died down.

Grassland animals

The African savannah supports large populations of grazing and browsing animals. These herbivores in turn support large numbers of mammalian carnivores. The uniformity of the vegetation structure, the scarcity of trees and the short growing season limit the diversity of birds and amphibians.

Environmental concerns

Intensive grazing of managed grasslands can lead to the destruction of grassland communities, soil erosion and desertification. The original temperate grassland fauna has been almost extinguished by hunting and the conversion of grasslands to arable cultivation and rangelands. Large herbivores, burrowing animals and predators are now either extinct in the wild or rare. In order to accommodate the migratory behavior of herds, national parks in tropical grasslands must be very large, or be connected by corridors of suitable habitat.

Related topics

Soil formation, properties and classification (G3)
Primary and secondary production (P2)

Ecosystem patterns (S1)
Conservation strategies (V2)

Primary regions Grassland regions occupy a broad band between forests and deserts. Unlike most trees, which require higher levels of rainfall, the characteristics of grasses enable them to survive periodic drought and also adapt them to the stresses of fire and grazing. Tropical grassland, or **savannah**, often with scattered trees, is most extensive in Africa, but is also found in Australia, South America and southern Asia. Temperate grassland occurs across large areas of eastern Europe and Asia (**steppe**), central North America (**prairie**), and South America (**pampas**). Grasslands comprise vast areas of uniform vegetation, yet like other biomes, the transition between grasslands and forest or desert is not abrupt. For example, temperate prairie grassland in central USA blends into warm desert along its southern margin as it becomes too dry for grasses, and into temperate deciduous forest to the north where water availability allows trees to dominate *(Fig. 1)*.

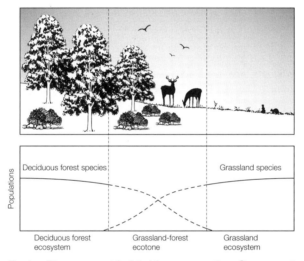

Fig. 1. Biomes are not isolated from one another. One ecosystem blends into the next through a transitional region, an ecotone, that contains many species common to the two adjacent systems. From Environmental Science, *5th edn, Nebel and Wright, 1996. Reprinted with permission from Pearson Education Inc., Upper Saddle River, NJ.*

Climate and soils *Tropical grasslands*
Tropical grasslands may receive up to 1200 mm (60 inches) of rain in the wet season, but none during the prolonged dry season. During the dry season there is a high risk of fire and some savannah areas are burnt every year. Low soil moisture for most of the year may limit microbial activity and restrict the recycling of nutrients.

Temperate grasslands
Temperate grasslands are moderately dry, having 250–600 mm (10–30 inches) of rainfall per annum, but the amount of available water will depend on the temperature, seasonal distribution of rainfall and the water-holding capacity of the soil. The climate is continental with hot summers and cold winters.

Commonly, because grassland species are short lived, grassland soils receive a large amount of organic matter and can contain as much as 5–10 times as much humus as forest soils. These rich soils (e.g. the chernozems of the Eurasian steppes) are well suited to the growing of arable crops such as corn and wheat, and the major grain-producing belts in North America and Russia are located in grassland regions.

Major vegetation Unsurprisingly, the dominant plant life forms are perennial grasses, ranging from tall species (5–8 feet) to short (6 inches or less) species which may form clumps or turfs. The root systems of grasses can be divided into three general types, according to depth and coarseness, although their development is greatly affected by soil conditions. In short-grass species most of the roots are found within 30–90 cm of the surface, whereas tall and mid-grasses often penetrate to 1.5–1.8 m. The bunch grasses have dense, fibrous rooting systems that radiate in all directions from the base of the stem. In the drier temperate grasslands, bunch grasses such as blue grama (*Bouteloua gracilis*) develop a network of fine roots concentrated in the upper 15 cm of the soil, so that moisture from sporadic precipitation events is absorbed efficiently during the growing season. Grassland communities are often quite complex, containing a significant number of nongrassy herbaceous species and frequently trees. Net primary productivity of grasslands is usually high, but biomass is relatively low because of the absence of persistent, woody tissue.

Grasses are highly adapted to survive grazing and flourish best where grazing animals are present. If grazers are excluded they may be replaced by dicotyledons. Leaf production is able to continue after grazing because the growth meristems are situated near the soil surface beyond the reach of grazing animals. The structure and composition of grasslands depend on the grazing intensity. Moderate grazing has been shown to increase the palatability of the sward, by stimulating the growth of new tillers (leaves) and delaying or preventing flowering, while maintaining high production and ground cover. Grazing animals also cause nutrient enhancement of patches through urine and feces deposition. Grazing can increase species diversity by creating gaps in the sward, or vacant niches, for invasion by other species.

Tropical grasslands
Acacia is characteristic of the African savannah, providing shade, food for browsing animals and supporting many invertebrate species. Savannah trees have thick bark which insulates the living cambium against the heat of grassland fires. Fire is an important factor in North America and in Africa in preventing the invasion of shrubby species. Perennial grasses are able to recover rapidly from light burning and this may occur regularly with the accumulation of dry, flammable leaf litter (*Fig. 2a*). Fierce burning is more likely to damage or kill the vegetation and recovery to the original biomass will be slow.

Temperate grassland
Associated with the grasses of temperate grasslands are broad-leaved perennials. These flower early in the growing season, before the grasses reach their maximum height; the larger broad-leaved perennials flower towards the end of the season after the grasses have flowered and begun to die down.

Grassland animals *Tropical grasslands*
The African savannah still supports large populations of grazing animals, such as wildebeest (*Connochaetes taurinus*), herds of which migrate across the Serengeti following the rain in search of the best grazing. Other abundant grazers include zebra *(Equus burchelli)*, buffalo *(Syncercus caffer)* and Thompson's gazelle *(Gazella thompsoni)*. Browsers include impala *(Aepyceros melampus)*, giraffe *(Giraffa camelopardalis)* and black rhino *(Diceros bicornis)*. These herbivores support large numbers of mammalian carnivores such as spotted hyena (*Crocuta crocuta*), lion (*Panthera leo*), leopard (*P. pardus*), cheetah (*Acinonyx jubatus*) and

African wild dog (*Lycaon pitus*). These animals and much of their habitat are protected, and they provide an important source of income through the tourist revenue they attract.

Temperate grasslands

The original temperate grassland fauna consists of migratory herds of grazing animals, burrowing mammals and associated predators. There are not many species of birds in temperate grasslands, probably because of the uniformity of the vegetation structure and the absence of trees. The short growing season gives little time for amphibians and reptiles to develop from egg to adulthood. Those species that are present tend to give birth to live young, the eggs hatching while they are still inside the mother so that they are able to look for food immediately.

Environmental concerns

The less productive grasslands have been utilized as rangelands for raising beef cattle, sheep and goats. Heavy grazing can alter the species composition of the grassland by replacing palatable species with unpalatable ones. Intensive use, especially by large herbivores, can cause damage by trampling which alters the ratio between water infiltration and runoff and may result in soil degradation and erosion (*Fig. 2b*). This can proceed to **desertification** where regeneration is not possible because of the loss of topsoil and continued grazing pressure.

In order to accommodate the migratory behavior of herds, national parks must be very large, sometimes crossing national boundaries, or be connected by corridors of suitable habitat. The demands of expanding local human populations cause conflict, resulting in overgrazing by domestic animals and

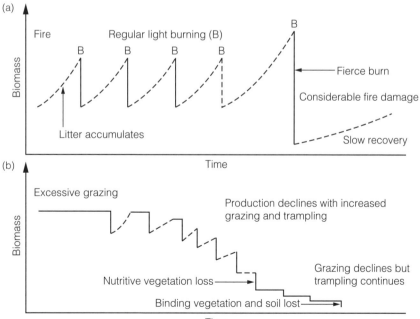

Fig. 2.   The response of Savannah grassland to burning and excessive grazing. (a) Light burning at short intervals allows the biomass to return to its previous level. Recovery after prolonged, fierce burning results in slower recovery. (b) Excessive grazing results in removal of vegetation cover and breakdown in soil texture and the eventual degradation of the grassland.

poaching. The near extinction of elephants in some areas may lead to an increase in woodland and a decrease in grassland, with obvious consequences for grazers and browsers. Much of the original temperate grassland fauna has been almost extinguished by hunting and the conversion of grasslands to arable cultivation and rangelands. Once the huge areas of steppe supported great herds of Przewalski's horse (*Equus przewalskii*). This species is now extinct in the wild, surviving only in zoos where it breeds well – a reintroduction programme is underway. Burrowing animals, such as the ground squirrels and gophers of the prairies have suffered decline, and with them predatory species such as wolves and coyotes.

S3 TUNDRA

Key Notes

Primary regions	The **arctic tundra** forms a circumpolar band between the Arctic Ocean and the polar ice caps to the north and the coniferous forests to the south. Smaller, but ecologically similar regions found above the tree line on high mountains are called **alpine tundra**.
Climate and soils	For most of the year temperatures are below that required for plant growth. Precipitation is low (usually less than 250 mm) and occurs mainly as snow. Below a certain depth the ground remains permanently frozen forming permafrost. The low productivity and limited microbial activity result in thin soils.
Major vegetation	Tundra has a low productivity, but a high species richness. The vegetation consists of low-growing mat- and hummock-forming plants. The long day length during the summer, combined with higher temperatures, allows primary productivity at this time of the year to be an order of magnitude higher than in the winter.
Tundra animals	Some tundra animals are only present as summer migrants. Permanent residents such as reindeer are migratory, ranging over vast areas in order to find enough food.
Environmental concerns	Tundra vegetation and soils are very slow to recover from disturbance. Since the discovery of oil in the tundra, vegetation has been lost and soils eroded as disturbance causes the permafrost to melt. The control of development and protection of the tundra is an international problem.
Related topics	Primary and secondary production (P2) Ecosystem patterns (S1) Conservation strategies (V2)

Primary regions

The coniferous boreal forests of the Northern hemisphere give way to the north to the **arctic tundra** which forms a circumpolar band which in turn gives way to either the Arctic Ocean or the polar ice caps. **Alpine tundra** consists of ecologically similar regions found in smaller patches above the tree line on high mountains. These are superficially barren, treeless regions where extreme environmental conditions severely limit plant growth. Tundra occurs where the low temperature and short growing season prevent the development of forest.

Climate and soils

The temperature falls below that required for plant growth for most of the year. There is an 8–10 week long growing season when temperatures are moderate and, at high latitudes, the day length is long. Precipitation is low (usually less than 250 mm) and occurs mainly as snow but, because of the low evaporation rate, water is not a limiting factor. Below a certain depth the ground remains permanently frozen forming a feature known as **permafrost**. The low productivity and limited microbial activity result in thin soils which are frozen in the winter and water-logged and marshy in the summer. The annual freezing and thawing gradually turns the soil over, sorting out the soil particles by size. On flat ground, this forcing of the larger stones to the surface results in a polygonal pattern. In alpine regions, temperatures vary markedly with slope angle and exposure. Precipitation increases with altitude and a pronounced rain shadow often occurs on the lee-side of the range. The air temperature in tundra regions tends to be highest near to the ground, an effect particularly pronounced in the spring when the air within 2 cm of the surface can be 15°C warmer than that higher up (*Fig. 1*).

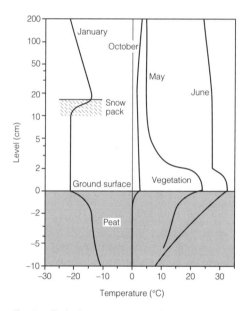

Fig. 1.   Typical temperature gradients in air, snow vegetation and soil in arctic tundra, showing the increase in temperature at the surface.

Major vegetation   The productivity of tundra regions is low, but a large number of species is found in this extreme habitat. The vegetation consists of low-growing mat- and hummock-forming plants such as sedges (*Carex* spp.), lichens and dwarf willows (*Salix* spp.) that are adapted to survival in conditions of extreme cold. The stature of these plants enables them to take advantage of warmer temperatures close to the ground. The perennating buds which produce new growth in the spring are buried or are held very near the ground where they are somewhat sheltered (see Topic S1). During much of the year plants are protected from exposure to high wind speeds by a covering of snow. Lichens, mosses, grasses and sedges form the basis of a species-rich community. In alpine tundra, a characteristic pattern of vegetation distribution is apparent, associated with the dominant environmental variables of snow cover, wind exposure and topography (*Fig. 2*). Snow cover and altitude are the major influences on the vegetation pattern shown in *Fig. 2*, with the more tolerant cushion plants and *Geum* turf occurring on exposed sites.

The long day-length during the summer, combined with higher temperatures allows primary productivity at this time of the year to be an order of magnitude higher than in the winter. Hollows in the ground formed by the action of frost fill up with water to form shallow ponds which are also highly productive. The combined aquatic and terrestrial net production of the tundra in the summer supports resident and migratory animal populations.

Fig. 2.   *Alpine vegetation patterns associated with environmental gradients in the Rocky Mountains, indicating the strong influence of snow cover and topographic site.*

Tundra animals

The extreme seasonality of the tundra means that some animals are only found during the summer. For example, migratory birds such as geese, sandpipers, ducks and other waterfowl breed on the tundra in summer, feeding on the vegetation and on emergent insects. Permanent resident mammals remain active all year, such as musk oxen and reindeer, which dig through snow to feed on lichen during the winter. These animals are also migratory, ranging over vast areas in order to find enough food. Lemmings, arctic hares, lynx, bears, arctic fox, ptarmigan and snowy owl are also part of the resident tundra community.

Environmental concerns

The low productivity and thin soils characteristic of tundra mean vegetation recovery following disturbance is very slow. Degradation of sewage and other pollutants is also inhibited because of the low temperature. Opening up the Arctic for oil and mineral exploration in recent years has proved to be the biggest threat to this relatively pristine ecosystem, although now such activities are preceded by comprehensive environmental impact studies. Disturbance of the vegetation cover by vehicle damage results in greater absorption of solar energy by the dark surface soils causing the ground ice to melt. Even a single

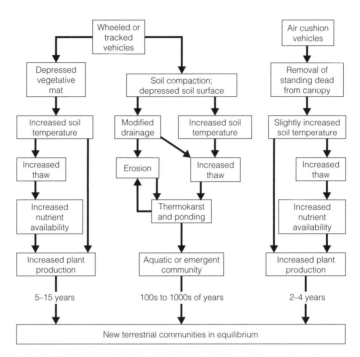

Fig. 3. Flow chart showing the impact of air-cushion vehicles and wheeled or tracked vehicles upon the Arctic tundra at Barrow, Alaska.

pass of a vehicle can result in water-filled ruts which continue to deepen because of the increased thermal conductivity of the saturated peat. The result can be deep ponds and a change in the composition of the plant cover which may take hundreds of years to restore (*Fig. 3*). Air-cushion vehicles have minimal impact when used atop snow in winter, but in summer the compression of the vegetation mat reduces its insulating properties and allows soil temperatures to rise. This in turn accelerates decomposition, releasing nutrients and stimulating plant growth. Another concern is the spillage of crude oil, which kills plants and leaves on contact, damages roots and reduces nutrient availability. The control of development and the protection of the arctic tundra is the subject of international conservation strategy (see Topic V2).

S4 FORESTS

Key Notes

Primary regions
Forest type is dependent on rainfall and temperature (latitude or altitude). Coniferous boreal forest is characteristic of cold climates and high elevation. Temperate forests occur at lower latitudes where there is sufficient rainfall, giving way to tropical rain forest toward the equator.

Climate and soils
Boreal forests have long, cold winters with light precipitation in the winter and more rain in the summer. The soils are podzols. The climate of temperate forests is also seasonal, with higher rainfall. Soils are well developed and rich. In the tropics, the forest climate is nonseasonal and warm with frequent heavy rainfall. Soils are leached, acidified and nutrient poor.

Major vegetation
Forests tend to have a high net primary productivity and also a high biomass. Boreal forests are dominated by coniferous tree species which are adapted to minimize evapo-transpiration and tissue damage from freezing. Temperate forests consist mainly of broad-leaved, deciduous species and the complex layered structure of the vegetation leads to high primary production. The high biodiversity of tropical rainforests results from their great age and complex physical environment.

Forest animals
Temperate and boreal forests support herbivorous mammals (e.g. deer), and predatory species such as wolves. Forests are important habitats for birds and, in temperate forests, small mammals which inhabit the dense understorey. Forest ecosystems also contain a wide range of specialist and generalist insect herbivores. Tropical rain forests support tremendous animal diversity, particularly of insects, amphibians, reptiles, birds and small mammals. Many species are tree dwelling, feeding on fruit and/or seeds.

Environmental concerns
Tropical rain forests are the most diverse communities on earth. The clearing of tropical rain forests results in biodiversity loss, depletes the soil and may lead to erosion. Forest biodiversity represents a valuable global resource as many plant species have unique chemical properties that can be beneficial to mankind. Forest clearance contributes to global warming, and may cause flooding. Many temperate and boreal forests are suffering damage from acid rain.

Related topics
Ecosystem patterns (S1)
Biological resources and
  gene banks (V3)

Greenhouse gases and global
  warming (W2)

Primary regions
The range of environmental conditions that will allow forest development is very wide (see Topic S1). However, different temperature and rainfall regimes support different forest communities and a gradation of forest types is found along a north–south gradient. **Coniferous boreal forest**, also known as taiga, is characteristic of cold climates and found in the northern regions of North

America, Europe and Asia, extending southward at high elevation. **Temperate forests** occur further south, in western and central Europe, in eastern Asia and eastern North America. Where temperature and rainfall are higher, **tropical rain forest** has developed, occurring in an equatorial belt in northern South America, Central America, western and central equatorial Africa, southeast Asia and on various islands in the Indian and Pacific oceans.

Climate and soils   Because forest biomes cover a wide geographical area, they also encompass a wide range of climatic and edaphic conditions.

Boreal forest
The coniferous forests of the north have long, cold winters with light precipitation in the winter and more rain in the summer. Water is often limiting because winter precipitation falls mainly as snow and permafrost is widespread. The soils are podzols, acidic and humus rich, with a deep litter layer that accumulates because microbial activity is slow at low temperatures.

Temperate forests
The climate of temperate forests is also seasonal, with temperatures falling below freezing in winter and warm humid summers. Rainfall ranges from 700 mm to 190 mm per year. Soils are well developed and rich.

Tropical forests
In the tropics, the forest climate is nonseasonal and the mean annual temperature is 28°C. Rainfall is frequent and heavy with an annual average greater than 220 mm. The heavy rainfall causes the soils to become leached and acidified. Microbial activity and nutrient cycling is very rapid with little nutrient storage in the soil, and consequently the soils are nutrient poor.

Major vegetation   In general, forests tend to have a high net primary productivity and also a high biomass due to the large quantity of persistent woody material present.

Boreal forests
Boreal forests are dominated by coniferous tree species such as spruce (*Picea* spp.), fir (*Abies* spp.), larch (*Larix* spp.) and pine (*Pinus* spp.) with a smaller number of broad-leaved species such as birch (*Betula* spp.) and poplar (*Populus* spp.) also occurring. The needle-like leaves of conifers minimize transpiration and reduce water loss. The conical growth form of many conifers is an adaptation against snow damage, preventing snow from settling on the crown of the tree. Their low evapo-transpiration rates and their ability to produce a form of antifreeze in their leaves allows them to remain in leaf throughout the winter.

Temperate forests
Temperate forests consist of broad-leaved, deciduous trees (oak, hickory, maple, ash, beech) with some coniferous species also present (pine, hemlock). There is a dense shrubby understorey and, where light can penetrate, a rich ground flora consisting of herbaceous plants, ferns, lichens and bryophytes (*Fig. 1a*). Many understorey and ground flora species such as bluebells (*Scilla nonscripta*) and wood anemones (*Anemone nemorosa*), which occur in English oak forests, flower early in the summer before the dominant trees close canopy and light becomes limiting. The complex layered structure of the vegetation and the resulting large

(a)   (b)

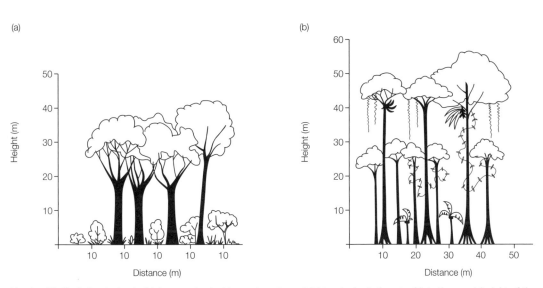

Fig. 1.   Vertical structuring in (a) temperate deciduous forests and (b) tropical rain forests. Note the great height of the tropical forest canopy and the presence of epiphytes.

biomass of autotrophs mean the community is able to fix a large amount of energy during the spring, summer and early autumn.

Managed woodlands that have been clear-felled and replanted do not have the same ground flora diversity as primary woodlands. Plants such as dog's mercury (*Mercurialis perennis*) and herb Paris (*Paris quadrifolia*) are characteristic indicators of ancient woodland in Britain.

Tropical forests

Tropical rainforests are characterized by a great diversity of towering trees, the tallest being 60 m or so. The high temperature and rainfall provide optimal conditions for plant growth and net productivity is the highest of any terrestrial biome. Their high biodiversity is thought to be a result of their long history; tropical rain forests have never been glaciated, therefore species have been able to accumulate, developing complex communities over time. In addition, the complex physical environment provides many niches for specialist species.

The dense canopy prevents light from reaching the forest floor precluding the development of any ground flora. However, the vertical structuring of rain forests is complex with shade-tolerant trees growing beneath the canopy of dominant emergent species overlain with epiphytes and lianas (*Fig. 1b*). Nutrient cycling within the canopy is probably important as many species produce aerial roots which absorb nutrients in the same way as roots anchored in the soil.

Forest gaps form when a tree falls, creating a new habitat with higher light intensity, greater temperature variation and more available soil moisture. These gaps are initially colonized by fast-growing pioneer species, whose germination from the seed bank is stimulated by light. Gaps are eventually dominated and closed by slower growing, canopy-forming trees. The vertical stratification of tropical rain forests and the dynamics of gap formation result in a very spatially heterogeneous habitat. This heterogeneity is often cited as an explanation for the high species diversity of tropical rain forests. As rain forest regions have never been glaciated, another factor in their diversity may be their great age. The long period free from severe climatic disruption may have allowed the community more time for the evolution of more species and more complex species interactions.

Forest animals   *Boreal forest*
Coniferous forests support a broad range of herbivorous mammals, from moose to mice, hares and red squirrels. Predatory species such as wolves and European bears have been hunted almost to extinction, although a few isolated populations remain. Coniferous forests are important nesting grounds for many migratory birds such as warblers and thrushes. They support a diverse community of seed-eating birds including crossbills, which are uniquely able to extract seeds from unopened cones.

Temperate forests
Animal diversity in temperate forests intensively managed for timber tends to be low. Non-native coniferous species are often planted and, although these grow quickly and are in high demand for paper pulp and softwood, they do not provide food and habitat for native woodland fauna. Less disturbed temperate forests in North America will support chipmunks, racoons, deer, coyotes and bears as well as numerous species of song bird, woodpeckers, owls and hawks. In English oak woods, birds such as tits, warblers, finches, woodpeckers and jays occur. Small mammals include bank vole (*Clethryonomys glareolus*), wood mouse (*Apodemus sylvaticus*), shrews (*Sorex* spp.) and hedgehogs (*Erinaceus europeus*). All of these are ground-dwelling mammals, living under the cover of the ground flora.

The high primary productivity of forests allows them to support a wide range of specialist and generalist insect herbivores. Insect diversity tends to reflect plant, and particularly tree, diversity and increases with decreasing latitude.

Tropical rain forests
Tropical rain forests are renowned for their enormous plant and animal biodiversity. Insects, amphibians, reptiles and bird species are particularly abundant. Monkeys and other small mammals are the dominant noninsect herbivores. There are a few large predatory species, such as tigers and jaguars. Much of the animal life is confined to the tree canopy, feeding on the year-round supply of fruit. Mammals, other than bats and primates, are poorly represented in rain forests and are primarily arboreal.

Like the trees, the animal communities are stratified: the emergents are inhabited mostly by birds and insects which live out their whole lives without ever leaving the canopy. The abundance of fruits, seeds, buds, leaves, and insects has led to the evolution of many specialist-feeding bird species. Investigations into how so many different species coexist have centered on the partitioning of resources in time and space. The forest layers are occupied by different species and species are also separated by their diurnal activity patterns. Many ground-, tree- and canopy-dwelling species are only active during either the day or night (*Fig. 2*).

Because individual trees and different species bear fruit at different times, many primate species have to defend large home ranges in order to meet their requirements. Species such as chimpanzees are able to achieve this because they live in communal groups and can share the territorial role.

Other species concentrate on leaves. Leaves are a nutritionally poor food

Fig. 2.   Space–time partitioning of resource use by non-flying mammals of the lowland rainforest in Sabah, Indonesia.

source and are very difficult to digest because of their cellulose content. Sloths have multicompartmented stomachs which hold cellulose-digesting bacteria. Even with this assistance, full digestion takes about a month. This low energy diet is the reason why sloths are slothful – they must conserve energy by remaining inactive. They are unable to escape from predators but are afforded camouflage protection by blue-green algae that live on their fur, turning it green!

Environmental concerns

Tropical rain forests cover only 7% of the land surface of the earth, but probably contain at least half the world's species. The clearing of tropical rain forests to provide farmland and grazing land for cattle results in biodiversity loss and also depletes the soil and may lead to erosion. The high plant biodiversity contained within these forests represents a valuable global resource. Many plant species have unique chemical properties that can be beneficial to mankind. For example, the rosy periwinkle produces a chemical, vincristine, which is effective in the treatment of leukaemia. Pharmaceutical companies are focusing research and development on potential natural cures. However, rainforest destruction is continuing in many areas unabated.

The island of Madagascar is one of the biologically richest areas on earth, with over 8000 endemic species of flowering plant. Many of the plants and animal species living in the rain forests along the east coast are endangered through habitat loss and degradation. Some notable species, including the dodo (*Raphus cucullatus*) have become extinct. Before human colonization 1500 to 2000 years ago, forest is thought to have covered most of eastern Madagascar

(11.2 million ha); by 1950 only 7.6 million ha remained, and this had reduced to 3.8 million ha or to 34% of the original area by 1985 (*Fig. 3*). Deforestation has been greatest in areas of highest population density in the south of the island. Forest still remains in the north and on steep slopes in the south. A similar pattern of forest loss is seen in Costa Rica, where deforestation has followed road development. Deforestation results in loss of habitat and habitat fragmentation, the isolation of remaining patches of habitat.

There are a number of reasons why rain forest destruction is occurring so rapidly. Population pressures and the transmigration of landless people and the urban poor into sparsely populated areas is encouraged in some areas, e.g. the Amazon Basin and western New Guinea. Rain forest trees are also being clear-felled or selectively logged for their valuable timber. In the Amazonian region of Brazil many areas of forest are being replaced, first with crops of rice, maize or cassava for a few years, and then with cattle pasture. These pastures degrade as soil fertility is lost through erosion and leaching and as phosphate is fixed in forms unavailable to plants.

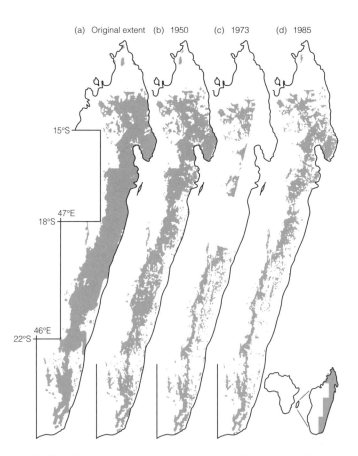

Fig. 3. Distribution maps of rain forest in eastern Madagascar: (a) before human colonization; (b) 1950, based on vegetation maps; (c) 1973, based on incomplete aerial photography; and (d) 1985, based on Landsat images.

Shifting agriculture (or slash and burn) is widespread within tropical rain forests and a major cause of deforestation. This type of farming is practiced by small farmers and large machinery is not used, consequently soil damage is slight and little erosion occurs. Once abandoned, small clearings surrounded by primary forest will undergo succession and may return to forest. However the sustainability of shifting agriculture depends on the length of the fallow period between crops, the size of the plots, and their proximity to primary forest. At least 20 years fallow is required for the forest to reestablish, which is often more than human population pressure will allow. Short fallow periods may reduce the regenerative capacity of the forest and lead to the establishment of grassland. Rainforest clearance by burning also affects the global carbon cycle and can contribute to global warming through the release of carbon dioxide into the atmosphere (see Topic W2).

Temperate forests across northern Europe are suffering damage from acid precipitation derived from industrial pollution. Forestry practices such as clear felling leaves the soil exposed and can lead to erosion and may have consequences for drainage. Clear felling has been implicated as a causal factor in flooding downstream of deforested areas.

S5 DESERTS, SEMI-DESERTS AND SHRUBLAND

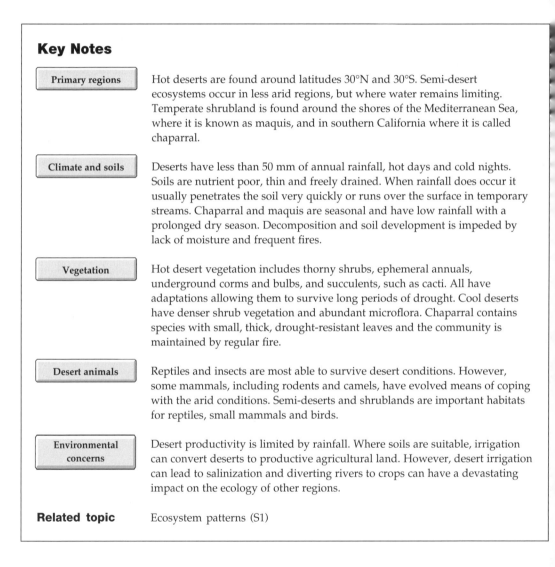

Key Notes

Primary regions

Hot deserts are found around latitudes 30°N and 30°S. Semi-desert ecosystems occur in less arid regions, but where water remains limiting. Temperate shrubland is found around the shores of the Mediterranean Sea, where it is known as maquis, and in southern California where it is called chaparral.

Climate and soils

Deserts have less than 50 mm of annual rainfall, hot days and cold nights. Soils are nutrient poor, thin and freely drained. When rainfall does occur it usually penetrates the soil very quickly or runs over the surface in temporary streams. Chaparral and maquis are seasonal and have low rainfall with a prolonged dry season. Decomposition and soil development is impeded by lack of moisture and frequent fires.

Vegetation

Hot desert vegetation includes thorny shrubs, ephemeral annuals, underground corms and bulbs, and succulents, such as cacti. All have adaptations allowing them to survive long periods of drought. Cool deserts have denser shrub vegetation and abundant microflora. Chaparral contains species with small, thick, drought-resistant leaves and the community is maintained by regular fire.

Desert animals

Reptiles and insects are most able to survive desert conditions. However, some mammals, including rodents and camels, have evolved means of coping with the arid conditions. Semi-deserts and shrublands are important habitats for reptiles, small mammals and birds.

Environmental concerns

Desert productivity is limited by rainfall. Where soils are suitable, irrigation can convert deserts to productive agricultural land. However, desert irrigation can lead to salinization and diverting rivers to crops can have a devastating impact on the ecology of other regions.

Related topic

Ecosystem patterns (S1)

Primary regions

Hot deserts are found around latitudes 30°N and 30°S. The main desert regions are in northern and southwestern Africa (the Sahara and Namib deserts), parts of the Middle East and Asia (Gobi desert), Australia, the Great Basin and south-western United States, and northern Mexico. In southern Africa the trade winds from the Indian Ocean bring precipitation to the slopes of the east cost (*Fig. 1*). This early loss of moisture means that little rain falls in the region of the Kalahari and rainfall declines steadily from east to west. The winds blowing from the

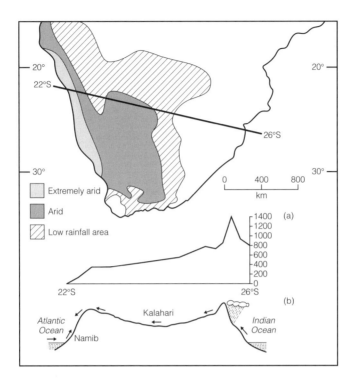

Fig. 1. Map of southern Africa showing desert and low rainfall areas. (a) Mean annual precipitation (mm) along the transect indicated on the map. (b) Cross section through southern Africa, indicating the Kalahari Basin and the prevailing wind direction.

Atlantic Ocean and to the Namibian Plateau in the west do not bring significant rain and the Namib is a hot, dry desert.

Semi-desert ecosystems occur in less arid regions. Warm semi-desert scrubs occur throughout the world in dry, warm and tropical climates. Cool semi-deserts occur in parts of North America, central Asia and mountainous regions where the climate is too dry for grassland.

Temperate shrubland is found around the shores of the Mediterranean Sea, where it is known as **maquis** and in southern California where it is called **chaparral**.

Climate and soils Deserts have less than 50 mm of annual rainfall and its timing is extremely unpredictable. Deserts usually have hot days and cold nights. Daytime temperature depends on latitude; for example, there are two types of desert in North America, a 'hot' desert in Arizona and a 'cool' desert in Washington State. These deserts differ in the vegetation communities they support. Temperate deserts are often dry because they lie in rain shadows where mountains intercept moisture from the sea. Moist air is forced to rise and release rain before it reaches the desert region. In extreme deserts, rainless periods can span many years and the only available water is found deep underground or as nighttime dew. Desert rainfall is usually erratic in its seasonal distribution and average rainfall figures are of little biological significance. The biological effectiveness of rainfall depends on its timing and the interval between rainfall events.

The sparse vegetation and low productivity mean that there is little accumulation of organic matter, resulting in soils that are nutrient poor, thin and freely drained. When rainfall does occur it usually penetrates the soil very quickly or runs over the surface in temporary streams.

Chaparral and maquis have rainfall in the region of 300 to 800 mm and a rain-free period of about 4 months. The mean temperature ranges from 10°C in the winter to 25°C in the summer. Decomposition and soil development is impeded by lack of moisture and frequent fires.

Vegetation

Hot desert vegetation is very sparse, and may include widely spaced thorny shrubs, which can shed their leaves and become dormant during dry periods. Desert annuals only germinate immediately following rain. They are opportunistic, **ephemeral** species able to grow and flower rapidly, carpeting the desert floor for short periods. **Geophytes** survive underground as corms or bulbs. **Succulents**, such as the cacti of America or the *Euphorbia* of Africa, have adaptations allowing them to survive long periods of drought. They have thick cuticles, sunken stomata and low surface-area-to-volume ratio, all of which reduce water loss.

It is clear that plant growth in desert environments is limited by water availability. When the limiting factor is removed plant biomass in the Namib dune system increases dramatically (*Fig. 2*). Plant nitrogen levels (a measure of protein content) also increase and are of great significance to herbivorous animals, whose distribution reflects the patchiness of the food supply. Some plant species are adapted to respond rapidly to the advent of rain, germinating, growing, flowering, and setting seed using the moisture from just one rain shower. For others the time interval between rain showers is very important. The availability of rainfall to plants depends on the soil-type. Rainfall will penetrate sandy soil before it evaporates, while a clay soil which drains less well is more susceptible to evaporation. Deep-rooting plants are able to access the water that has penetrated into sandy soils, and may not be able to survive in areas of a higher clay content.

Cool deserts have denser shrub vegetation, such as the North American sagebrush, which can remain green all summer. Extensive shallow root systems in combination with deep tap roots up to 30 m long provide access to scarce rainfall and groundwater. Microflora, including mosses, lichens and blue-green algae remain dormant in the soil but, like desert annuals, are able to respond quickly to cool, wet periods.

California chaparral contains species with small, thick, drought resistant leaves. Shrubs of *Quercus, Ceanothus,* and *Arctostaphylus*, among others, form a low aromatic scrub. This is very flammable in the summer and fire is an important aspect of the ecology of this system. If fires occur less frequently than every 12 years or so, fire intolerant species such as *Prunus ilicifolia* and *Rhamnus crocea* invade.

Desert animals

Reptiles and insects are able to survive desert conditions by virtue of their waterproof body coverings and dry excretions. Some mammals (e.g. a few species of nocturnal rodents) are adapted to overcome the shortage of water by excreting concentrated urine and have evolved means of keeping cool that do not use water. They can survive without ever having to drink water. Other animals, such as camels, must drink periodically but are physiologically adapted to withstand long periods of dehydration. Camels can tolerate a loss of 30% of their total water content and can drink 20% of their body weight in 10 minutes.

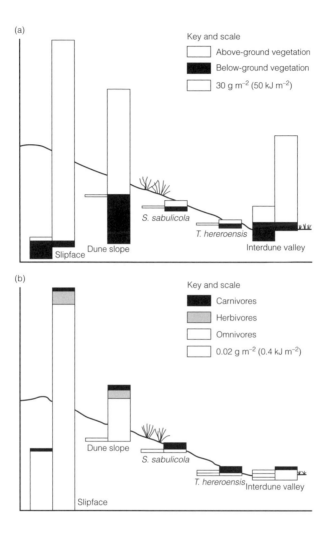

Fig. 2. The change in biomass of (a) plants and (b) animals from a dry period (left bar) to a wet period (right bar) measured on different parts of the Namib dune system.

Semi-deserts and shrublands are important habitats for reptiles, small mammals and birds.

Environmental concerns

Productivity depends on rainfall almost linearly, as rainfall is the dominant factor limiting plant growth (see Topic S1). In the Californian Mojave desert, a 100 mm annual rainfall results in 600 kg ha^{-1} of net production. An increase to 200 mm will cause a rise in net production to 1000 kg ha^{-1}. In the cooler deserts, where loss of water to evaporation is less, 200 mm of rainfall produces 1500–2000 kg ha^{-1}.

Such is the potential of desert regions that, where soils are suitable, irrigation can convert deserts to productive agricultural land. However, it is questionable whether desert irrigation is sustainable. The large quantities of water evaporating from the soil cause salts to be left behind where they may accumulate to toxic levels – a process known as **salinization**. Diverting rivers and draining lakes to meet the demands of agriculture can have a devastating impact on the ecology of other areas. For example, because water has been diverted for irrigation, the water level of the Aral Sea (in the former Soviet Union) has dropped by 9 m and is expected to drop a further 8–10 m. The surrounding shoreline and exposed lake bottom are nearly desert, and a thriving fishing industry has been destroyed.

S6 SALTWATER BIOMES

Key Notes

Primary saltwater regions	The saltwater biomes are open oceans, continental shelves, the intertidal zone and coral reefs, salt marsh, mudflats and mangroves. Oceans cover 70% of the world's surface. Physical factors, namely tides, currents, temperature, pressure (depth) and light intensity, determine the makeup of biological salt water communities.
Open oceans	The open ocean is the most extensive biome in area, but it tends to be nutrient poor and hence unproductive. The surface photic zone, where light can penetrate, contains phytoplankton and zooplankton. Below the photic zone carnivorous and detritivorous animals occur, feeding on material from the communities above. Light levels and productivity decline with depth. Bottom or benthic fauna is sparse except in regions of hydrothermal vents where communities based on chemotrophic bacteria thrive.
Continental shelves	Continental shelves support kelp forests and fisheries, which are among the most productive of marine ecosystems. Diverse coral reef communities occur in warm and very shallow water. Corals are colonial animals which produce structurally complex calcareous skeletons on which live algae, invertebrates and herbivorous and carnivorous fishes.
The intertidal zone	Intertidal rocky shores are dominated by algae. Zonation of algal and animal communities with exposure and distance from the sea occurs. Sandy beaches provide an unstable, abrasive and nutrient-poor substrate inhabited by filter-feeding burrowing animals which are themselves food for wading birds.
Salt marsh, mudflats and mangroves	Salt marsh occurs in sheltered areas protected from wave action and is dominated by salt-tolerant higher plants. Mudflats and estuarine silts are fine substrates rich in organic matter and low in oxygen. Their high invertebrate density supports fish and bird populations. Mangrove forests replace salt marsh in tropical regions and support a rich fauna. Mangroves exhibit physiological and morphological adaptations to the high salinity and anoxic conditions.
Environmental concerns	The open ocean is used as a dumping ground for numerous pollutants, including oil, sewage, hydrocarbons and metals. Some become magnified in the food web and can contaminate fish stocks. Recreational and commercial development of intertidal regions has led to habitat destruction and pollution. Dredging, sewage pollution and over-fishing have degraded coral reefs.
Related topic	Fishing and whaling (T2)

Primary saltwater regions Saltwater biomes include a number of categories with distinct physical conditions. **Open oceans** cover about 70% of the earth's surface, grading into areas of **continental shelf** occurring around coasts to a depth of about 200 m, and including **coral reefs**. The **intertidal zone** occurs at land margins and includes **sandy beaches** and **rocky shores**. Where there is less tidal influence **salt marsh, mudflats** and **mangroves** can develop.

The major characteristics of saltwater regions are shown in *Table 1*. Physical factors, namely tides, currents, temperature, pressure (depth) and light intensity, determine the makeup of biological salt water communities. These communities in turn have a considerable influence on the composition of bottom sediments and gases in solution in seawater and also in the atmosphere.

Table 1. Physical characteristics of saltwater ecosystems

Ecosystem	Physical characteristics
Mud flats and salt marsh	Slow-moving water, e.g. in estuaries, results in the deposition of fine silts. High organic matter content and small air spaces result in anaerobic conditions. Salt marsh occurs in sheltered intertidal areas which are inundated with salt water at every high tide.
Intertidal	Dominated by tidal cycle. Distance from the sea determines duration of exposure/submergence and the structure of rocky shore communities. Sandy coasts occur in areas of wave deposition.
Continental shelf	Shallow water (130 m on average) allows light penetration to the benthos. Latitude and ocean currents determine temperature and nutrient status.
Deep ocean	Surface dominated by wave action. Deep ocean environment is stable with constant temperature, little or no light penetration and high pressure.

Open oceans

Ocean waters can be up to nearly 10 km deep. The vast majority of this water receives little light, is nutrient poor and is unproductive. Oceans are inhabited by **pelagic** communities. Light penetration typically occurs to about 150 m producing a surface zone where photosynthesis is possible, known as the **photic** zone. Marine **phytoplankton** occur including microscopic algae, flagellates, diatoms and bacteria ranging in size from 2–200 μm. These phytoplankton support large numbers of **zooplankton** consisting largely of the larval stages of marine and intertidal invertebrates. Planktonic organisms are unable to control their movements and drift with the wind and ocean currents. Those organisms that are able to swim, such as larger larvae, fish, turtles, seals, sharks and penguins are part of the **nekton**.

Below the photic zone carnivorous and detritivorous animals occur, feeding on the gravitational shower from the communities above. These organisms are either blind or able to produce their own light for the detection of food, confusion of predators or the attraction of mates. As productivity declines with depth, population densities become very low and prey scarce. Deep ocean fish are adapted to infrequent feeding opportunities interspersed with long periods of starvation; they are able to consume very large prey and can conserve energy by lowering their metabolic rate when food is not available. Bottom or **benthic** fauna is sparse except in regions of hydrothermal vents where communities based on chemotrophic bacteria thrive, deriving energy from the heat and sulfur spewed from these volcanic vents. Three hundred thermal vent species have been described, from sulfur bacteria to limpets, tubeworms and fish, and are unique to this habitat.

Continental shelves

Continental shelves support some of the most productive of marine ecosystems (see Topic T2), particularly in areas of upwelling where currents bring nutrients to the surface. Here kelp forests are found, formed from brown algae such as *Laminaria* spp. which anchors itself to the substrate, growing to a length of 50 m or more. The continental shelf benthos supports a diverse fauna including polychaete worms, molluscs, sea squirts, bryozoans, sponges, sea spiders, crustaceans and echinoderms. Continental shelves can also support large populations of fish; for example, most anchovies come from fisheries along the South American continental shelf.

Reef-building corals grow in warm and very shallow water (above 20°C and less than 50 m deep) and provide the substrate for complex communities of extraordinary fish and invertebrate diversity. Parrot fish feed directly on the coral, crunching up their protective limestone skeleton to consume the polyps within. Some fish graze the algae which would otherwise coat and smother the coral, while others act as cleaners, consuming parasites, cleaning wounds and removing dead skin from other fish.

The intertidal zone

Intertidal **rocky shores** are dominated by algae which attach by specialized holdfasts. Higher plants are unable to colonize the rock substrate. Algae give way to lichens above high tide level where exposure and desiccation are greatest. **Zonation** of algal and animal communities with distance from the sea is characteristic of rocky shores (*Fig. 1*). Dramatic variation in temperature, salinity, illumination and wave exposure occurs with distance from the sea and the length of time any one part of the shore is submerged at high tide. The upper limits to species distribution are generally determined by the degree of

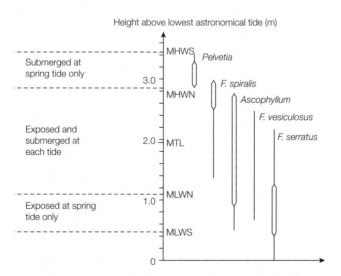

Fig. 1. Zonation of seaweeds on the Isle of Cumbrae, Scotland. MHWS, mean high water spring tide; MHWN, mean high water neap tide; MLWS, mean low water spring tide; MHWN, mean low water neap tide. Reprinted from Ecology, *Chapman and Reiss, 1992, with permission from Cambridge University Press.*

tolerance to these exposure factors. The lower limit is probably set by competition with other species. For instance, removal of the brown algae, *Fucus serratus* and *Ascophyllum nodosum* will cause the upper-shore species *Fucus spiralis* to move down-shore to occupy the vacant niche, eventually to be replaced again as the original species recolonize. Zonation of rocky shore animals is also apparent, with barnacles and limpets encrusting rocks closest to the sea while winkles and whelks occur further up the shore. Algal distribution is also determined by the degree of exposure to wave action. Species of kelp grow at the lowest tidal levels on moderately sheltered coasts but are replaced by *Alaria esculente* in more exposed areas (*Fig. 2*). Black lichen *Verrucaria maritima* occurs near the high-tide mark and is especially prominent on exposed, wave-swept headlands. *Fucus vesiculosis* is generally found on sheltered shores, while *Fucus serratus* is more tolerant of wave action.

Sandy beaches provide an unstable and abrasive substrate which is not readily colonized and is nutrient poor. Minute organisms inhabit the sand, such as sea cucumbers and molluscs less than 2 mm long. Cockles and elongated, flattened molluscs such as razor shells are able to bury themselves in the sand, emerging to feed on plankton and detritus at high tide. Worms, such as lugworms, can also be plentiful, producing characteristic casts at low tide. Beaches with rich invertebrate life support large numbers of wading birds including sanderlings, ringed plovers, oyster catchers and curlews.

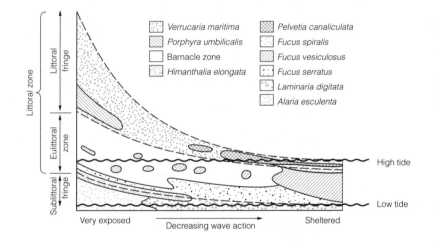

Fig. 2. The distribution of common European seaweeds in relation to tidal range and exposure to wave action.

Salt marsh, mudflats and mangroves

Salt marsh occurs in sheltered areas protected from wave action and provides a stable substrate for colonization by salt-tolerant higher plants. Salt marsh vegetation is dominated by grasses such as *Spartina*, with glasswort (*Salicornia europaea*) also common. Salt marshes provide important feeding and overwintering grounds for geese and wading birds.

Unlike salt marshes where most of the organic matter is washed out with the tide, **mudflats** tend to retain organic matter deposited by the tide because of their small particle size. Estuarine silts consist of river sediments which are very rich in organic matter. The absence of oxygen greatly restricts the organisms able to survive in these muds and silts, but those that can reach very high densities and provide a rich food source for other organisms, including birds. In estuaries, where the water is brackish, phytoplankton, benthic microflora and invertebrates are very abundant making estuaries important nursery grounds for shellfish and fish.

Mangrove forests replace salt marsh in warmer climates and develop on highly saline, anoxic muds. They cover 60–70% of the coastline of tropical regions. The dominant plants are mangroves which have shallow, widely spreading roots which emerge from the trunk above ground and act as props providing stability in the almost fluid mud. Fine nutritive roots are responsible for nutrient uptake and many species have root extensions, called pnue-matophores, that take in oxygen for the roots (*Fig. 3*). Some mangroves (e.g. *Bruguiera* spp.) produce roots with 'knees' which stick up at intervals above the sediment. These knees are covered in lenticels which draw in air. When they are covered by the incoming tide the pressure in the air spaces declines as oxygen is used in respiration; on re-exposure the negative pressure serves to draw in more air, replenishing the oxygen supply. Another challenge of living in a highly saline environment is obtaining sufficient fresh water to replace that lost through evapotranspiration. Mangrove leaves exhibit xeromorphic adaptations and have waxy cuticles or are covered in fine hairs. Leaves may also be succulent, acting as water storage organs. Mangroves can be classified according to three main strategies for maintaining water balance. Some are salt excluders and are able to avoid salt uptake (e.g. *Rhizophora*), others excrete salt through specialized glands in their leaves (e.g. *Avicennia*), and others show remarkable salt tolerance (e.g. *Excoecaria*).

The distribution of mangroves depends on the stability of the sediment and how fast it is accreting, salinity (a function of rainfall, inundation regime and freshwater input), substrate type and the dispersal of propagules by coastal currents. Mangrove forests are faunally rich with a unique mixture of terrrestrial and marine life. Fiddler crabs and tropical land crabs burrow into the mud during low tide and live on prop roots during high tide. In the well devlloped Indo-Malaysian mangrove forests mud skippers (*Periophthalmus*) are found. These are fish that live in burrows in the mud and are able to crawl about on top of it like amphibians.

Environmental concerns

The open ocean is used as a dumping ground for numerous pollutants, including oil and other hydrocarbons, sewage and metals. Light oil fractions either evaporate, dissolve in the water or are absorbed by particles and sink to the bottom. Bacteria can digest this light oil, but heavier oil persists as tarry deposits on the surface and the sea bed. Oil kills bottom life and the soluble fractions are highly toxic. Seabirds become covered in surface oil which can poison them as they clean their feathers, or cause hypothermia. The vast size and depth of the sea has made it a natural disposal route. The bottom 105 km^2 of the New York Blight is covered with a black toxic sludge of sewage and industrial waste. Some pollutants such as metals, found in sewage and industrial effluent, become magnified in the food web and can contaminate fish stocks.

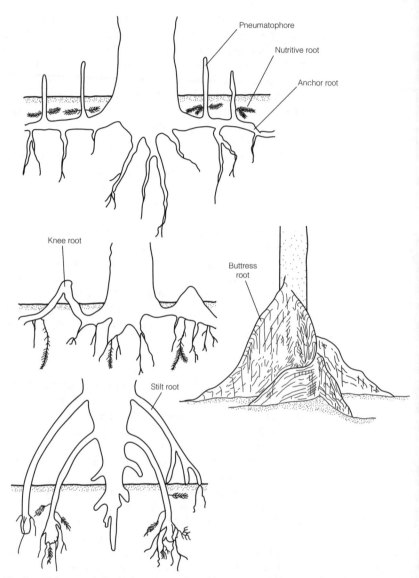

Fig. 3. Major morphological root types found in mangroves show modifications providing anchorage and increased oxygen uptake.

The proximity of coral reefs to the shore make these ecosystems particularly sensitive to disturbance and pollution arising from beach-based tourism. Construction on shore can have devastating effects for the coral as it is buried by sediment and sand produced during building and excavation. Coral reefs are also threatened by commercial fishing, pollution and global warming.

Recreational and commercial development of intertidal regions has led to habitat destruction and pollution of the intertidal zone.

S7 FRESHWATER BIOMES

Key Notes

Primary freshwater and wetland regions	Freshwater biomes include lakes, rivers, bogs, marshes and swamps. These systems are fed by water and nutrients leaching from the surrounding catchment area.
Streams and rivers	The physical characteristics of streams and rivers alter along their length; they change from being small and turbulent close to their source to wider and slower at their mouth. Plant and animal diversity and production tends to be highest in the middle regions where flow rates and substrate allow the growth of macrophytes. River sediments deposited in floodplains give rise to fertile soils and form the basis of many of the world's most productive agricultural regions.
Lakes and ponds	Lakes have very little or no current, allowing the water body to acquire vertical stratification with illuminated, warm water at the surface and dark, cold water below. Lakes can be nutrient rich (eutrophic) or nutrient poor (oligotrophic).
Environmental concerns	Canalization, commercial development and pollution have resulted in a loss of aquatic biota and bankside vegetation in the lower reaches of many rivers. Wetlands have been lost to agriculture for grain production and grazing. Eutrophication can occur through organic and inorganic pollution, resulting in a loss of plant diversity and algal blooms. This is a particular problem in small lakes and ponds and in semiclosed systems such as the Norfolk Broads.
Related topic	Air, water and soil pollutants (S7)

Primary freshwater and wetland regions

Freshwater biomes include open water systems such as lakes and rivers as well as water-logged regions known as bogs, marshes and swamps. Bogs occur on impervious substrates where rainfall is high. They are dominated by low-growing plants able to tolerate waterlogged and nutrient-poor conditions, such as *Sphagnum* moss and insectivorous sundews. Swamps, such as the Florida Everglades, are tree-dominated wetlands occurring in tropical and warmer temperate regions.

Freshwater contains dissolved gases, nutrients, trace metals and organic compounds as well as organic and inorganic particles. These chemicals originate from rainwater which washes substances out of the atmosphere, direct deposition and from the leaching of soils and rocks from the surrounding **catchment area**.

Streams and rivers

Streams and rivers differ greatly, depending on their size. They also vary along their length from their source in upland areas to their mouth where the river meets the sea. In general as the mouth of a river is approached:

- the speed of water flow decreases, the water becomes less turbulent and oxygen levels fall;
- the volume of water increases having accumulated as the river passes through its catchment;
- the energy of the river decreases, suspended material is deposited and the river bed becomes composed of finer particles and eventually silt;
- the river bed becomes less steep because the larger volume of water erodes a broader channel;
- human influences increase; many rivers flow through farmland and urban or industrial areas and receive agricultural run-off, treated sewage and other effluent which may raise the organic content of the river leading to eutrophication (see below).

Streams high in the catchment that are unpolluted will support caddis fly (*Trichoptera*) and blackfly (*Simulium* spp.) larvae feeding on fine organic particles. The water will be too turbulent and nutrient poor for all but aquatic mosses, liverworts and algae. Plankton communities, consisting of algae, photosynthetic bacteria, crustaceans and rotifers, can develop further downstream where the volume of moving water is increased and the current is reduced. Fish, reptiles, birds and mammals may be present.

As water flow continues to decrease, particularly at the edges of a growing channel, plankton communities become more complex and sediment is deposited, providing a rooting medium for larger aquatic plants (**macrophytes**) and a habitat for benthic organisms such as oligochaete worms, chironomid larvae and molluscs. Emergent plants, which grow up beyond the water's surface provide physical habitat for invertebrates, fish and epiphytic algae, which in turn provide food for other organisms. River sediments, deposited in the lower reaches, tend to be highly fertile and form the basis for human population settlement and productive agriculture. In less developed countries, e.g. Bangladesh, the benefits of high agricultural yield are precariously balanced against the risk of flooding.

The Amazon is the largest river in the world, with a catchment area of 6.15 km^2, and discharges 200 000 m^3 s^{-1} into the sea, 10 times that of the Mississippi. The Amazonian headwaters flow through deep, eroded valleys and small forest streams. In the lowlands, the Amazonian tributaries meander across the floodplain, forming ox-bow lakes as they become cut off from the main stream. The flat topography of the Amazonian basin and the large seasonal fluctuations in discharge result in an extensive floodplain which is alternately flooded and drained with the annual rise and fall in river level. The floodplain forms an extensive mosaic of habitats differentiated into those that have an annual dry phase and those that are permanently flooded (*Fig. 1*). Many of the plant species rooted into the fertile floodplain sediment are capable of growing extremely rapidly (20 cm per day) and are able to cope with rising water levels. Fast vegetative reproduction allows rapid colonization of exposed habitats as water levels rise and fall. The species diversity is very high, a characteristic attributed to the great habitat diversity and the regular flood pulse. The alternation of flood and drying phases does not allow time for interspecific competition to develop, and the flood pulse constrains communities at the early seral stages (see Topic R1). Elimination of the flood pulse leads to dry areas becoming colonized by forest and permanently wet areas dominated by a few species of aquatic plants.

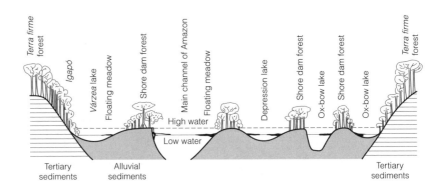

Fig. 1.   Cross section through the middle Amazonian floodplain.

Lakes and ponds   A large, slow moving river is very similar to a shallow lake with plankton, submerged and emergent macrophytes and sediment deposits. Lakes have very little or no current, allowing the water body to separate out into layers depending on temperature and chemical composition – that is, to acquire **vertical stratification**. The illuminated, warm water at the surface is less dense than the dark, cold water below and this difference in density prevents the two from mixing. The layer of warm water is called the **epilimnion**. The cooler water below, the **metalimnion,** becomes colder with depth. For every 1 m depth, the temperature declines by 1°C, a drop called the **thermocline**. When the temperature of the water reaches 4°C and its greatest density, it lies as a layer at the bottom called the **hypolimnion**.

The thermocline acts as a barrier between the epilimnion and the hypolimnion. Very deep lakes, such as Lake Malawi in East Africa (706 m), are so deep that vertical stratification, established many thousands of years ago, is now disturbed only at the surface and is permanent below 100 m or so. In shallower lakes the stratification persists during the summer when surface waters are warm (*Fig. 2*). Nutrients present at the bottom of a lake are not available to phytoplankton in the upper layer and the surface may suffer nutrient depletion in late summer. By autumn, the surface waters start to cool and sink, displacing the warmer water beneath it so that it also cools. Nutrients are recharged and oxygen is incorporated as the water circulates and is stirred by wind. This circulation is called **overturn**. A slight **temperature inversion** develops in winter because water below 4°C is less dense than warmer water and floats on the surface. In the spring, another overturn occurs as the surface water heats up. The whole water body is now nutrient and oxygen rich, becoming more stratified as the season progresses.

Depending on the substrate and the geology of the surrounding catchment, lakes can be nutrient rich (**eutrophic**) or nutrient poor (**oligotrophic**). **Eutrophication** can also occur through organic and inorganic pollution.

Environmental concerns   The biodiversity of river flood plains tends to decline in the lower reaches because of pollution and agricultural or urban development. The channel profile is often altered as rivers are straightened or canalized, reducing the potential

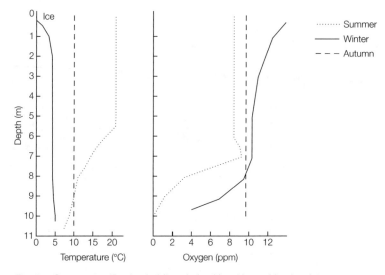

Fig. 2.   Oxygen stratification in Mirror Lake, New Hampshire, in winter, summer and late fall.

for bankside vegetation and also the diversity of niches available for aquatic organisms. Wetlands have been lost to drainage for agriculture and many of those remaining are contaminated by pesticides. Britain has only about 20% of its original wetlands remaining. Lakes can also suffer from industrial pollution; heavy metal contamination of fish stocks in the American Great Lakes has made them unfit for human consumption.

An example of a system suffering from human-induced eutrophication is the British Norfolk Broads. These are shallow manmade lakes derived from flooded peat cuttings. The increase in the human population of the catchment and the dramatic increase in the discharge of sewage effluent containing phosphate (a pollutant found in detergent) which occurred in the 1960s altered these lakes from being nutrient poor (20 μg phosphate l^{-1}) to nutrient rich (360–1000 μg phosphate l^{-1}). Algal growth increased, changing the crystal clear waters into, in the extreme, a murky soup, shading out and smothering macrophytes. Once-common species such as stoneworts are now uncommon, and the holly-leafed niad is now found in only one broad. These species are replaced by robust, tall plants tolerant of epiphytic algae, such as horned pondweed and water crow foot. Extreme eutrophication results in the loss of all macrophtyes, domination by algae, and an increase in phytoplankton (*Fig. 3*).

Phosphate removal has been attempted through mud suction. Another solution is to treat the symptom rather than the cause by boosting the zooplankton population and allowing it to control the algae and phytoplankton, thus 'cleaning up' the water and allowing macrophytes to recover. **Biomanipulation** through fish removal and the creation of artificial habitats and shelter for zooplankton have produced dramatic results, improving both the ecological and amenity value of the Broads.

Phase 1

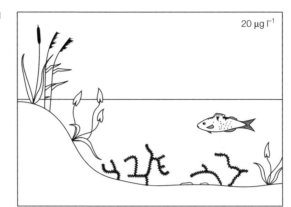

Phase 2

Phase 3

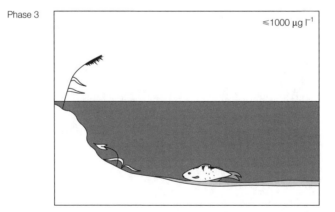

Fig. 3.   Phases of water quality for the Norfolk Broads, UK. Phase 1: pre 1900, clear water and a covering of low growing species. Phase 2: mid twentieth century, clear water with higher nutrient levels and luxuriant growth of taller plant species. Phase 3: high nutrient levels, abundant algal growth has shaded out macrophytes. Phosphate levels are indicated.

T1 HARVESTING THEORY

Key Notes

The goals of harvesting	To maximize the returns gained by exploiting a population, management should aim to harvest the maximum yield that the population can produce sustainably over the long term.
Maximum sustainable yield	One approach to deduce the maximum long-term yield is known as the maximum sustainable yield, or MSY approach. As a population increases in density from very low numbers, the birth rate outstrips the death rate, so the net recruitment (births – deaths) into the population rises. As the population density approaches the maximum the environment can support, the death rate increases and the birth rate falls. The maximum net recruitment therefore occurs at an intermediate density, when intraspecific competition is relatively low yet there are many reproductive individuals in the population. This maximum net recruitment is the maximum number that can be harvested from the population sustainably – the MSY.
Quota limitation	Harvesting may be limited by controlling the quota (the biomass or number of individuals culled) in a given period. Quota controls thus allow the harvesters to remove a fixed number of the prey species every season or year. Quota limitation is commonly applied in marine fisheries to achieve a MSY with limited success. The theory demonstrates that quota-limitation is intrinsically risky and may cause overexploitation and even extinction of the harvested populations.
Effort limitation	The risk inherent in quota-limitation can be alleviated by adopting an approach of limiting effort. This has a clear advantage – as a prey species becomes rarer, more effort will be required to seek out and catch the diminishing numbers.
Environmental fluctuation	The ability of models of population growth to predict the harvesting potential of a population is severely compromised by environmental variation which may affect recruitment or mortality.
Dynamic pool models	Simple models do not consider the age structure of populations, which limits their predictive power, as the mortality rates and reproductive output of individuals is age-specific. Usually, the individuals harvested are the larger and older members of the population, which have the highest reproductive potential. Dynamic pool models explicitly consider the recruitment, growth and mortality of different age classes, allowing a better model of the population to be derived. This approach means that, for example, the impact of varying mesh size in a marine fishery can be tested theoretically.

The goals of harvesting

Harvesting economics demands that the **maximum possible yield** is taken from the population. However, unlike a coal seam or a diamond mine, a biological population will replenish itself, provided it is not harvested to extinction. It is not good management practice to harvest a population almost to extinction, because this will demand a long recovery period until reharvesting can occur. Understanding what population size should be left (and which individuals), such that the **long-term yield is maximized,** is the concern of harvesting theory.

Maximum sustainable yield

A simple theoretical solution to predict the maximum long-term yield is known as the **maximum sustainable yield** or **MSY** approach. As a population increases in density the birth rate initially rises more rapidly than the death rate, but the death rate increases as the population density approaches the **carrying capacity, K** (the maximum density the environment can support) (*Fig. 1a*). At the carrying capacity, the birth rate and death rates are equal and the population is constant (see Topic H1). The difference between the birth rate and the death rate is known as the **net recruitment**. Maximum net recruitment, therefore, occurs at an **intermediate density**, when intraspecific competition is relatively low yet there are many reproductive individuals in the population (*Fig. 1a* and *b*). This **maximum net recruitment** occurs at population density N_m (*Fig. 1b*) and represents the maximum number that can be harvested from the population sustainably – **the MSY**. This approach to harvesting theory based on a simple **net recruitment** model is known as the **surplus yield** approach.

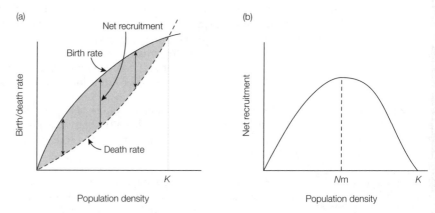

Fig. 1.   (a) The change of birth and death rates in relation to population density, illustrating that the net recruitment (the excess of births over deaths) is maximal at an intermediate population density; (b) the change of net recruitment with density – a recruitment curve. The maximal recruitment occurs at Nm.

The MSY concept has several limitations: (i) it assumes a **constant unvarying environment** and a constant recruitment curve; (ii) it ignores **the age structure** of the population, and hence does not take account of age-specific variation in survival and fecundity; (iii) the **population data** required to estimate the recruitment curve are often of poor quality. In spite of these serious shortcomings, MSY has been the dominant model in harvesting fisheries, whaling, wildlife and forestry.

Quota limitation

One way of limiting harvesting to achieve the **MSY** is to control the **quota**, the biomass or number of individuals culled in a given period. Quota control allows the harvesters to remove a fixed number of the prey species every season or year, and is politically popular, as it allows harvesters to estimate their income. The **MSY quota** is that which balances exactly with the net recruitment curve (*Fig. 2a*). If harvesting is maintained at this level, the population's **recruitment will exactly balance the number removed by harvesting** and the population will remain stable at density N_m. However, quota limitation is intrinsically risky as the equilibrium is **unstable**. If the population is knocked below N_m and harvesting continues at the MSY level, the number removed by harvesting will exceed the population's ability to replace individuals and **extinction** will result (*Fig. 2a*). If the MSY quota is only slightly overestimated, the population will proceed directly to extinction. Only if a quota significantly below the MSY quota is adopted will a **stable equilibrium** result.

Quota limitation has commonly been applied to marine fisheries to achieve a MSY, with limited success.

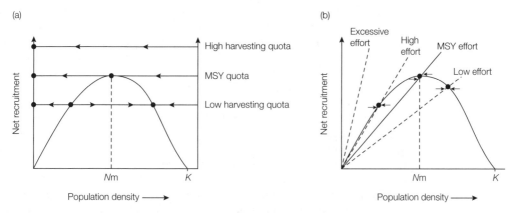

(a)

High harvesting quota

MSY quota

Low harvesting quota

Nm   *K*

Population density ⟶

(b)

Excessive effort   High effort   MSY effort

Low effort

Nm   *K*

Population density ⟶

Fig. 2.   (a) The effect of various levels of fixed-quota harvests on a population. Arrows indicate the trajectory of the population for a given harvesting regime and density. Closed circles represent equilibrium points. The only stable equilibria are (i) when the population is extinct and (ii) when the harvesting quota is low and the population density high; (b) the effect of various levels of fixed-effort harvests on a population. Closed circles represent equilibrium points. All equilibria are stable, and only excessive effort results in extinction, regardless of the starting population density.

Effort limitation   Regulation of **harvesting effort** can reduce the risk inherent in quota limitation. This approach has a clear intuitive advantage – as a prey species becomes rarer, more effort will be required to seek out and catch the diminishing numbers. The effects of four different effort regimes on a harvested population are illustrated in *Fig. 2b*. Because there is a fixed effort, the yield varies with population size, and can be expressed as a straight line through the origin of slope varying with intensity of effort. The **MSY effort** exactly balances the population's recruitment, but if the population falls below N_m, and harvesting continues at the MSY effort level, the result will be a lower yield, but the population will not be pushed to extinction . Similarly, if the effort applied is slightly too high, a stable equilibrium at a lower population density will be established. This is wasteful harvesting of resources as a higher long-term yield will be attained if less effort is applied. In fact, the **diminishing returns** of seeking to harvest at declining population density means that the optimal economic effort lies somewhat below the MSY effort, and is the effort which returns the **maximum economic yield** or **MEY**.

Effort controls are commonly applied to **fisheries** and **wildlife** harvests – game hunting of mammals may be controlled by limiting the number of **gun licences**, salmon river fishing by **rod licences** and European marine fish stocks by a limitation on boat numbers. However, these are not very refined ways of controlling effort, as, for example, individual salmon fishers differ considerably in their ability. This effect is also apparent at the level of international legislation, illustrated by the fact that at the same time as old boats in the European fishing fleets are being decommissioned to meet the effort targets, the European Union is funding the building of new, large, efficient boats. Thus the effectiveness of effort controls is limited by efficiency changes in the harvesters.

Environmental fluctuation

Environmental fluctuations occur on a variety of scales, from major long-term climatic change which may have catastrophic effects on a global population, to more localized effects which may leave some populations at risk. Harvesting models are largely unable to predict such effects and rarely leave a '**safety zone**' to accommodate their impacts upon harvested populations. A classic example of the impact of a natural environmental fluctuation on a harvested population comes from the **Peruvian anchoveta**, *Engraulis ringens*. The usual pattern of oceanic currents in the Pacific results in an **upwelling of cold water** on the South American coast, giving rise to high productivity and large numbers of fish, including anchovetas. However, periodically an '**El Niño**' effect occurs, whereby this pattern is disrupted and the oceanic currents 'flip', switching off this upwelling reducing fish productivity markedly. In the late 1960s, the Peruvian anchoveta fishery was the world's largest, constituting over 15% of the global fish catch. This harvesting level was probably unsustainable even if the environment had remained constant. However, in 1972–73, a major El Niño event occurred, after which the catch was reduced to less than a sixth of the peak harvest. Another El Niño occurred in 1982, after a decade of further unsustainable exploitation of the fish stocks, which almost closed the fishery. A further El Niño occurred in 1997, causing a 70% catch reduction compared to 1996, at which point the Peruvian authorities banned anchovy fishing for seven months to limit further damage to this overexploited species.

Dynamic pool models

A serious limitation of the **surplus yield** models described above is that they do not consider the age or size structure of populations, which limits their predictive power, as the **mortality** rates and **reproductive output** of individuals is **age- and size-specific**. Dynamic pool models explicitly consider the recruitment, growth and mortality of different classes to improve model performance. One aspect of harvesting that can commonly be controlled is the size of individuals harvested, for example, via mesh size when net-fishing. Using a larger mesh allows more of the population to escape. By employing a **dynamic pool model** the impact of adopting various net sizes can be tested theoretically. This approach was taken in assessing the effect of varying both mesh size and fishing intensity on the Arctic cod population (*Fig. 3*).

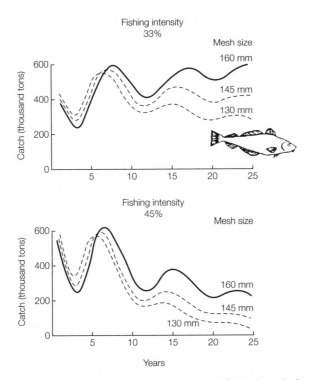

Fig. 3. Predictions of varying mesh size and fishing intensity for the Arctic cod population from a surplus yield model. Redrawn from Lubchenco, American Naturalist **112**, *23–39, 1978, University of Chicago Press.*

T2 FISHING AND WHALING

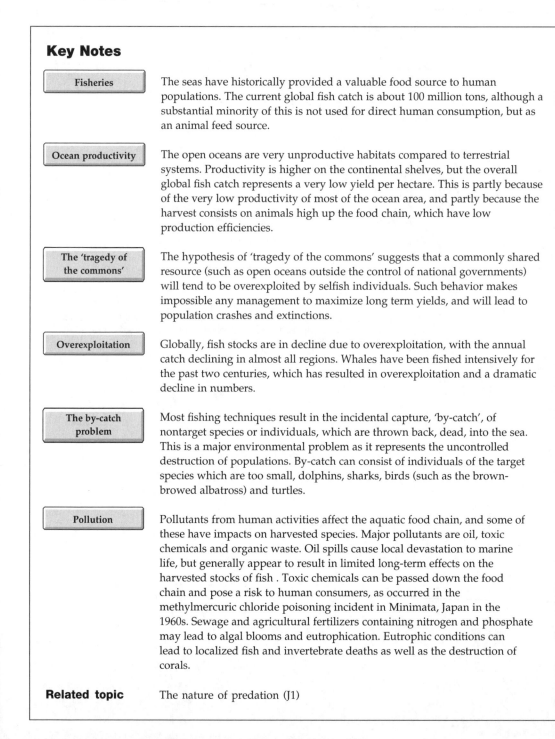

Key Notes

Fisheries	The seas have historically provided a valuable food source to human populations. The current global fish catch is about 100 million tons, although a substantial minority of this is not used for direct human consumption, but as an animal feed source.
Ocean productivity	The open oceans are very unproductive habitats compared to terrestrial systems. Productivity is higher on the continental shelves, but the overall global fish catch represents a very low yield per hectare. This is partly because of the very low productivity of most of the ocean area, and partly because the harvest consists on animals high up the food chain, which have low production efficiencies.
The 'tragedy of the commons'	The hypothesis of 'tragedy of the commons' suggests that a commonly shared resource (such as open oceans outside the control of national governments) will tend to be overexploited by selfish individuals. Such behavior makes impossible any management to maximize long term yields, and will lead to population crashes and extinctions.
Overexploitation	Globally, fish stocks are in decline due to overexploitation, with the annual catch declining in almost all regions. Whales have been fished intensively for the past two centuries, which has resulted in overexploitation and a dramatic decline in numbers.
The by-catch problem	Most fishing techniques result in the incidental capture, 'by-catch', of nontarget species or individuals, which are thrown back, dead, into the sea. This is a major environmental problem as it represents the uncontrolled destruction of populations. By-catch can consist of individuals of the target species which are too small, dolphins, sharks, birds (such as the brown-browed albatross) and turtles.
Pollution	Pollutants from human activities affect the aquatic food chain, and some of these have impacts on harvested species. Major pollutants are oil, toxic chemicals and organic waste. Oil spills cause local devastation to marine life, but generally appear to result in limited long-term effects on the harvested stocks of fish . Toxic chemicals can be passed down the food chain and pose a risk to human consumers, as occurred in the methylmercuric chloride poisoning incident in Minimata, Japan in the 1960s. Sewage and agricultural fertilizers containing nitrogen and phosphate may lead to algal blooms and eutrophication. Eutrophic conditions can lead to localized fish and invertebrate deaths as well as the destruction of corals.
Related topic	The nature of predation (J1)

Fisheries

For thousands of years the sea has provided human populations with a food source. The vastness of the oceans (comprising 78% of the Earth's surface) led to the belief, until this century, when the efficiency of fisheries increased, that fish stocks were inexhaustible. Even recently it has been proposed that exploitation of fish as a key protein source may relieve the pressure on agriculture to feed a growing global human population. The annual **global fish catch** is estimated to be about 90–100 million metric tons, although a substantial minority of this is not used for direct human consumption, but as an animal feed source and also as a soil fertilizer. The main components of this catch are summarized in *Table 1*.

There has also been a major fishery for the past two centuries for **whales**.

Table 1. Global annual catches of fishes and shellfish (1992, FAO data)

Fish group	% of global catch
Herrings, anchovies, sardines	25.0
Cod, haddock	13.0
Jacks, mullets	13.0
Tuna	5.4
Redfish, bass	7.2
Mackerel	4.1
Miscellaneous fish	16.7
Squid, octopus	3.4
Lobsters, crabs, shrimps	6.2
Mussels, oysters, clams	6.2

Ocean productivity

Although there are some productive marine fisheries, these are highly restricted in area. Highest productivity is concentrated on **continental shelves**, and the open oceans are **unproductive**. The global annual fish catch is equivalent to a return per hectare of ocean of only 0.004 GJ, compared to 0.02 GJ ha^{-1} for the poorest terrestrial agricultural systems. High-input agricultural systems (such as UK wheat or US maize) return around 100 GJ ha^{-1} yr^{-1}, more than ten thousand times the yield from the sea. This low productivity is the result of two factors: (i) the concentration of the majority of ocean productivity to a small part of the total ocean area, (ii) the **long food chains** with resultant **low production efficiency** typical of fish species which are predators (see Topic P2, *Fig. 1a*). Therefore, the oceans do not represent an untapped source of protein for a growing human population.

The 'tragedy of the commons'

Hardin (1968) presented an argument that suggests that overexploitation of a resource shared by many parties (a 'common') is inevitable. This argument is termed '**the tragedy of the commons**' and it states that each individual with access to a joint resource will try to maximize his own gain from the resource, and management of the resource to maximize long-term yield will not be possible. The costs of **selfish behavior** will be borne by the whole community, but the benefits will accrue to the selfish individual. Every individual will thus try to take as much resource as possible, and the resource will be **destroyed**. Open-ocean fisheries lie outside the control of national governments and are exploitable by all, so the overexploitation and near extinction of the large whales may be regarded as a demonstration of this model.

Overexploitation   There is abundant evidence of the **global decline** in the abundance of whale and fish stocks. Decline in whale stocks due to overfishing is shown in *Fig. 1*. The large size of whales has two unfortunate consequences: (i) the **economic rewards** for catches are high per individual whale (in the largest species, the blue whale, an individual may weigh up to 120 metric tons), and (ii) the **reproductive rates** of the populations are low. These features have resulted in overexploitation of all the major whale stocks. The development of the explosive harpoon and large, fast boats in the late nineteenth century was the start of the decline of large whale species. By the turn of the twentieth century, commercial whaling had seriously depleted the Northern Hemisphere whale stocks, and activities switched to the Southern Hemisphere, where further overexploitation occurred. Some whale populations are making recoveries due to a series of international agreements under the **International Whaling Commission (IWC)** which has constrained whaling in many species to indigenous peoples and for 'scientific' uses. Sharks and deep ocean species are also particularly sensitive to overexploitation due to their low reproductive rates.

All major marine fisheries show declines in catch, with the exception of the Indian ocean which has been relatively unexploited until recently (*Table 2*).

The decline of the cod (*Gadus morhua*) population of the North Atlantic is a further illustration of unsustainable harvesting practices (*Fig. 2*), and the annual

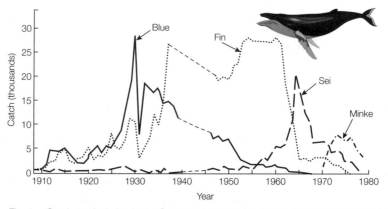

Fig. 1. Catches of whales in the Southern Hemisphere (after whaling had depleted all Northern Hemisphere stocks). The sequence of decline (blue, fin, sei, minke) reflects the declining size of whale species, the largest species being overexploited first.

Table 2. Change in catch for major fishing regions from peak year to 1992

Fishing region	% change since peak
Atlantic, North	-29.0
Atlantic, central	-28.0
Atlantic, South	-32.0
Mediterranean	-25.0
Pacific, North	-9.5
Pacific, central	-16.5
Pacific, South	-5.5
Indian	+5.5[a]

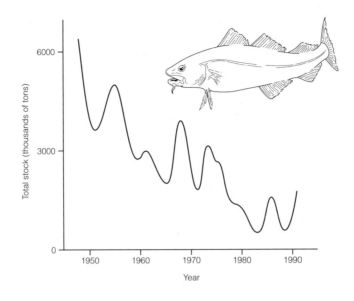

Fig. 2. The decline in cod stocks in the North Atlantic.

harvest is now far below that which the population could yield if allowed to recover and then was managed appropriately. There has been an even more dramatic decline of this species within the North Sea, indicating that the presence of a regulatory framework (the European Union Common Fisheries Policy) is insufficient to ensure sustainable harvesting. For many years political considerations have meant the allowable quotas have been set in excess of the levels recommended by fisheries' ecologists. This policy has unsurprisingly been disastrous and in January 2001 a 40 000 mile^2 (105 000 km^2) section of the North Sea was declared closed to cod fishing to attempt to allow the population to recover.

The decline of some fish populations has, as with whaling, shifted fishing pressure onto new species. In the North Atlantic, blue whiting (*Micromesistius poutassou*) was not commercially harvested prior to the 1980s, but overexploitation pushed this species into a decline within a decade.

The by-catch problem

Most fishing techniques result in the capture of species or individuals which are not the target of the harvest, called the **by-catch**. By-catch is usually discarded into the sea, dead. This is a very serious problem, as it represents the uncontrolled destruction of populations. By-catch can consist of individuals of the target species which are too small, dolphins, birds (such as the brown-browed albatross) and turtles.

The scale of the problem is enormous. The Food and Agriculture Organization (FAO) conservatively estimate that 27 million tons of by-catch are discarded annually. In a single year, 2 million tons of fish by-catch were thrown into the Gulf of Mexico, 16 million red king crabs were discarded in the Bering Sea (with only 16% of the total catch being kept) and in the North Sea 80 million cod were discarded. The most wasteful fisheries are those for shrimps and prawns, which account for only 4% of the annual global fish catch, but for 35%

of the global by-catch. Each landed ton of shrimp or prawns is caught at a cost of destroying over 5 tons of by-catch.

The by-catch of **below-regulation-size** individuals can be a high proportion of the total catch. In the North Sea, such discards comprised 40% of the haddock catch, 35% of the whiting catch and 45% of the cod catch. This process subverts the point of regulations for minimum sizes, which are designed to protect stocks.

Marine mammals also suffer from being incidentally caught in fishing operations. Yellow fin tuna nets formerly killed as many as 400 000 dolphins annually, although changes in technique, fuelled by the public awareness of the issue, have reduced this substantially. However, the method is still far from perfect and has moved the by-catch pressure onto turtles and sharks. Every 1000 nets set around dolphin herds catch 500 dolphins, 10 sea turtles and very few sharks, whilst the same number of nets set around logs and other floating objects (the modern technique) catch just two dolphins, but over 100 turtles and nearly 14 000 sharks. 'Dolphin-friendly' tuna is thus not particularly environment-friendly.

Seabirds are also victims of fishing. Every year, 40 000 albatross are drowned as a result of grabbing at squid used for bait on long lines being set for bluefin tuna. Diving seabirds such as auks, cormorants and eider ducks are also killed in nets, particularly monofilament nylon nets which are almost invisible in the water.

The scale of the destruction caused by by-catches can give rise to severe **community perturbations**. For example, over 1 million sharks have been taken as by-catch in the northwestern Atlantic swordfish fishery. The removal of sharks appears to have allowed the grey seal population to grow ten-fold, which in turn has increased the level of infections in cod of parasites which have seals as the primary host.

Pollution

Pollutants from a wide variety of sources have entered the marine environment, both from deliberate and accidental dumping. The major pollutants are oil, toxic chemicals and organic waste. **Oil spills** from grounded tankers cause headline news due to the local devastation to marine life, especially birds and marine mammals. The Exxon Valdez, which spilt 250 000 barrels of crude oil in the Gulf of Alaska in 1989, led to the deaths of around 400 000 seabirds. However, oil spills seem to result in limited long-term effects on the harvested stocks of fish except in the immediate vicinity of the incident.

Human sewage and other organic waste, in conjunction with agricultural fertilizers containing nitrogen and phosphate, may lead to **eutrophication** problems, in which **algal blooms** occur, which may result in **red tides** (caused by dinoflagellate algae). These are toxic and can cause localized fish and invertebrate mortality, as occurred in the upper Adriatic in 1989 and 1996. Although eutrophication is a serious ecological problem, destroying corals and reducing species diversity, the effects on fish harvests are generally localized. **Toxic chemicals** may pose a risk to fish and potentially to human consumers. The classic example of this comes from the **Minimata** incident in Japan in the 1960s, caused by the dumping of mercury in Minimata Bay since the 1930s. Through **biomagnification**, (see Topic U1) marine fishes accumulated high concentrations of **methylmercuric chloride**, which then entered the human food chain. Neurological disorders developed, especially in children, and eventually mercury release was halted.

U1 THE PEST PROBLEM AND CONTROL STRATEGIES

Key Notes

What is a pest?	Pests compete with humans for food or shelter, transmit pathogens, feed on humans or otherwise threaten human health, comfort or welfare. Weeds may be included in this definition. One of the most important characteristics of pests is the high degree to which they are normally regulated by their natural enemy. Pests are frequently species that have evaded their natural enemy, possibly due to their importation to new regions of the world.
The aim of pest control	Generally the aim of pest control is to reduce the pest population to a level at which no further reductions are profitable. This is known as the economic injury level (EIL) for the pest, or if social and amenity benefits are included, the aesthetic injury level (AIL). In the case of disease, total eradication can be justified on the basis that the saving of one single life far exceeds any economic costs. In practical pest control, the EIL is not as important as the control action threshold (CAT) – the pest density at which action should be taken in order to prevent an impending pest outbreak.
Types of pest control	Approaches to limit pest damage to food crops have been adopted for thousands of years, particularly using cultural control (e.g. altering the sowing date or avoiding repeatedly replanting the same crop in the same place). Biological control (the use of natural enemies – predators and parasites) has also had a long history. This century, chemical control became a key approach, although there are serious toxicity and other problems. The development of pest-resistant crop varieties is another valuable strategy.

Related topics	The nature of predation (J1)	Pesticides and problems (U2)
	Predator behavior and prey response (J2)	Biological control and integrated pest management (U3)

What is a pest?   A pest species is any species which are considered undesirable. The use of the term pest is therefore subjective. A more complete definition of **pest** is that pests compete with humans for food or shelter, transmit pathogens, feed on humans or otherwise threaten human health, comfort or welfare. Weeds may be included in this definition as they represent a plant which competes with other plants which have food, timber or amenity value. There are many species of pest representing a wide range of taxonomic groups. It is difficult to attribute certain general characteristics which are features of all pests. Pests have very varied life histories. Some pests have explosive bursts of population increases, rapidly reaching a level where vast damage is caused (e.g. rats, locusts).

However, other pest species cause huge damage and yet their population growth rate is relatively small (e.g. the codling moth, *Cydia pomonella* which lays only 40–50 eggs per year but is the most important pest of apples). One important characteristic of pests is the degree to which they are normally regulated by their **natural enemies** (predators and parasites). Pests are frequently species that have evaded their natural enemies, possibly due to their importation to new regions of the world, having left their natural enemy complex behind or because their natural enemies have been eliminated by man.

Weeds are often plant species which are adapted to the high-disturbance, low-competition environment that characterizes agricultural systems, having a **ruderal** life history strategy (see Topic M1).

The aim of pest control

Although in some instances the aim of pest control is to eradicate the pest species, generally the aim of pest control is to reduce the pest population to a level at which no further reductions are profitable. In other words, when the cost of further reducing the pest population exceeds the extra revenue obtained. This is known as the **economic injury level** (EIL) for the pest, or if social and amenity benefits are included, the **aesthetic injury level** (AIL). In the case of disease, total eradication can be justified on the basis that the saving of one single life far exceeds any economic costs. If the EIL of a pest is greater than zero, then total eradication is not profitable. If the population of the pest is generally below the level of EIL, then pest control would not be used as the costs would outweigh any benefits of control. However, if the EIL of a species is below the normal abundance of the species then it becomes a pest. Local eradication of a pest through biological control is rarely an aim as it will result in loss of the control agent.

The essence of the EIL concept is shown in *Fig. 1*. The value of a commodity varies with pest density as shown. At low pest densities the crop is affected to a negligible extent. However, above a threshold pest density the crop losses increase at an accelerated state until a pest density is reached beyond which the crop is valueless. This curve can now be matched against the cost of achieving any particular pest density. The cost of controlling the pest to lower and lower densities increases at an accelerated rate until at a density of zero (complete eradication) the cost is extremely high. The EIL is the density at which the difference between the two curves is greatest and represents the optimal strategy.

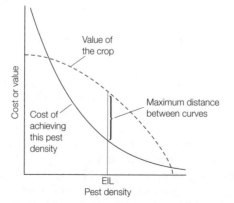

Fig. 1.   The definition of the economic injury level (EIL) is the pest density at which value exceeds cost by the greatest amount.

The relationship shown in *Fig. 1* represents an oversimplification; for example, the EIL changes over time, dependent on the life cycles of the pest and the commodity. Another shortcoming is the implied idea that control measures take effect immediately. Of course, if a population is to be kept below the EIL, measures must be taken at some density less than the EIL because it takes time for the control measure to be put into effect. Hence, in practical pest control the EIL is not as important as the economic threshold, or **control action threshold** (CAT) – the pest density at which action should be taken in order to prevent an impending pest outbreak. Again, there is never a single CAT as it varies with time and depends on the population levels of the pest's natural enemies. For example, the CAT values for the spotted alfalfa aphid (*Therioaphis trifolii*) on alfalfa are: 40 aphids per stem in spring; 20 aphids per stem in summer and autumn (however, if the population of ladybirds (a natural predator of the aphid) is more than one adult per 10 aphids then do not treat); 60 aphids per stem in overwintering populations.

These rules cannot be put into operation without a detailed understanding of the population ecology of the pest and a sufficiently accurate and frequent monitoring program to check on pest levels.

Types of pest control

Pest control in various forms has been used extensively since the development of agriculture some 10 000 years ago. Mixing crop species in a field, optimizing planting or harvesting times and avoiding repeated planting of the same crop at the same site are long-established **cultural** strategies for limiting pest damage. Encouraging local populations of the pest's predators and parasites has also been used for thousands of years. In Europe it was not until the seventeenth century that interest in biological control led to the wide use of a variety of natural pesticides. The period from 1750 to 1880 in Europe was a time of agricultural revolution, but it brought major pest-driven agricultural problems: the potato blight in Ireland, Scotland and Belgium (1840s); the outbreak of fungus leaf spot disease of coffee; and the invasion from America of the grape phylloxera (*Viteus vitifoliae*), which almost destroyed the French wine industry (1848–1878).

During this period there was a surge of interest in developing pest control techniques. The first book devoted to pest control was published in the early 1800s and the first pesticide spraying machine was produced in 1880. The first major success in biological control by importation of a natural enemy occurred in 1888 when the vedalia beetle (*Rodolia cardinalis*) was imported to California from Australia and New Zealand to control the cottony cushion scale (*Icerya purchasi*), a pest of citrus trees. In 1896 the first selective herbicide, iron sulfate, was introduced. Weed control during this time was dependent on crop rotation, the disturbance of soil and the removal of weed seeds by winnowing.

By the start of the twentieth century there were five main approaches to pest control, namely **biological control**, **chemical control**, **physical control**, **cultural control** and the use of **resistant varieties**. Pest control was revolutionized by the onset of the Second World War. Driven by the necessity to control vectors of human disease in the tropics, thousands of chemicals were screened for insecticidal properties. In the US, the compound dichlorodiphenyltrichloroethane (DDT) was produced, while in Germany the organophosphates were developed, and a third group of organic insecticides, the carbamates, were also discovered. These compounds initially targeted vectors of human disease, but during the rapid expansion of farming after the war these compounds were widely used.

At this time is was believed that high doses of these compounds could eradicate pests.

It was the publication of *Silent Spring* by Rachel Carson in 1962 that focused public attention on the problems associated with the use of simple organic compounds in pest control. By the 1970s, the widespread use of broad spectrum pesticides started to lose ground in favor of an integrated pest management approach.

U2 PESTICIDES AND PROBLEMS

Key Notes

Chemical insecticides and herbicides
Inorganic compounds were traditional herbicides, but due to problems with persistence and nonspecificity, compounds such as borates and arsenicals are rarely used. Chemical pesticides are generally used to treat a particular insect pest at a particular location. However, problems arise because these chemicals are toxic to other animals, and because many chemicals persist in the environment.

Chemical toxicity
Most chemical pesticides are toxic to a range of organisms, beyond the target pest species. Insecticides are particularly problematic, many affecting a wide range of vertebrates and invertebrates. The impact on nontarget species can cause ecological, economic and human health problems. Herbicides are often fairly unspecific in their action, and some may also be mammalian toxins.

Biomagnification
A further problem arises, particularly, in the use of chlorinated hydrocarbons, because of their susceptibility to biomagnification. Because these toxins cannot be metabolized or destroyed, they accumulate in the body of an individual. This results in an increasing concentration of insecticide in organisms at the higher trophic levels. These effects threaten natural predator populations and may pose a risk to the human food chain (especially if fish are affected).

Target pest resurgence and secondary pest outbreaks
Insecticides, unless highly specific, may decimate natural enemy populations, leading to a rapid increase in pest numbers after an initial decline – this is 'pest resurgence'. When natural enemies are destroyed, a number of potential pest species normally kept in check by their natural enemies may increase in number and become secondary pests.

Evolution of resistance
Evolved resistance to pesticides represents a serious threat to agricultural production. Pesticide resistance provides some of the best examples of evolution in action. Within a large population exposed to pesticide, a few genotypes may be unusually resistant and will possess a huge evolutionary advantage.

Related topics
The nature of predation (J1)
The nature of parasitism (K1)
The pest problem and control strategies (U1)

Biological control and integrated pest management (U3)

Chemical insecticides and herbicides

The use of inorganic compounds in pest control was common in the nineteenth and early twentieth centuries. They were usually metallic compounds of salts or copper, sulfur, arsenic or lead, effective only if ingested. Because of this, and the persistence of toxic metal residues, these compounds have largely been abandoned as agents of pest control. Naturally-occurring insecticidal plant

Table 1. The range of insecticides currently in widespread use

Insecticide	Example	Description
Pyrethroids	Permeathrin	These artificial compounds are replacing other organic insecticides due to their selectivity against pests.
Chlorinated hydrocarbons	DDT	Contact poisons which affect nerve impulse transmission. Use of these compounds was suspended in the 1970s but is still in use in poorer countries.
Organophosphates	Malathion	Derived from phosphoric acid, these nerve poisons are more toxic than chlorinated hydrocarbons but are less persistent in the environment.
Carbamates	Carbaryl	These compounds are derived from carbamic acid and are similar to organophosphates in their mode of action. However, most are toxic to bees and wasps.
Insect growth regulators	Methoprene	These chemicals mimic natural insect hormones and enzymes and hence interfere with insect growth and development. They are harmless to plants and animals.
Semiochemicals	Pheromones	These chemicals illicit a change in the behaviour of the pest rather than being toxins. They are based on naturally occurring substances. Pheromones act on members of the same species; allelochemicals act on members of another species. Sex attractants may be used to interfere with mating.

products, such as nicotine from tobacco and pyrethrum from chrysanthemums, have largely been superseded because of their instability upon exposure to light and air. Table 1 lists the commonly used insecticides.

Inorganic compounds were the traditional agents of herbicides, but due to problems with persistence and nonspecificity, compounds such as borates, arsenicals, ammonium sulfamate and sodium chlorate are rarely used unless semi-permanent sterility is required. Table 2 lists a range of herbicides used.

Problems arise with chemical pesticides due to their toxicity, and due to the ecological and evolutionary responses they give rise to.

Chemical toxicity Probably the most predictable problems stemming from the widespread toxicity of insecticides are the adverse effects on insects, fish, and mammals, including humans. For example, most broad-spectrum insecticides are highly toxic to honeybees and it has been estimated that much of the losses of bee colonies in California are attributed to pesticides. Losses of crop pollinators such as bees can have serious economic impacts. In addition, it has been estimated that 20 000–50 000 human deaths per year are caused through direct exposure to pesticides. Insecticides are commonly carelessly applied and spread beyond the target area through spray drift. This has led to incidents in which

Table 2. Examples of herbicides used

Herbicide	Example	Description
Organic arsenicals	DMSA	Used as spot treatments as they are non-selective. They upset growth by entering reactions in place of phosphate.
Phenoxy	2,4-D	These compounds are selective and work by stimulating plant growth to an unsustainable level causing plant death.
Substituted amides	Diphenamid	These compounds have diverse biological properties, active against a range of weeds.
Carbamates	Asulam	These compounds are herbicides as well as insecticides. they kill plants by preventing cell division and plant tissue growth.
Nitroanilines	Trifluralin	These are a group of preemergence herbicides in widespread use. They act by inhibiting root and shoot growth.
Substituted ureas	Monuron	These are a group of nonselective pre-emergence herbicides which inhibit photosynthesis.
Thiocarbamates	EPTC	This is a group of soil-incorporated pre-emergence herbicides, which selectively inhibit the growth of roots and shoots that emerge from wheat seeds.
Triazines	Metribuzin	These compounds represent the most important group of heterocyclic nitrogen herbicides. They strongly inhibit photosynthesis and can be used selectively or nonselectively.
Phenol derivatives	DNOC	This group, including nitrophenols, are contact chemicals with broad spectrum toxicity extending to plants, insects and mammals, acting by uncoupling oxidative phosphorylation.
Bipyridyliums	Diquat	These are powerful nonselective contact chemicals which destroy cell membranes.
Glyphosphate		This is a nonselective foliar-applied chemical active at any stage of plant growth.

cattle and sheep have been poisoned, birds and wild mammals have been killed, and illness caused in humans. Insecticides may also be toxic to plants and depress plant growth, which rather undermines the objective of improving crop production. Permethrin depresses photosynthesis by 80% in lettuce seedlings.

The potential animal toxicity of herbicides is generally not considered, as their specific plant biochemical target does not occur in animals. However, herbicides such as diquat and paraquat have high mammalian toxicity, to which there are no known antidotes. In the 1960s a controversy began over the effects to human health of 2,4,5-T and 2,4-D, which were used in combination (Agent Orange) between 1962 and 1970 to defoliate swamps and forests in South Vietnam. Low levels of 2,4,5-T caused birth defects in mammals and further studies suggested that 2,4-D was carcinogenic.

Biomagnification   A problem which arises, particularly in the case of chlorinated hydrocarbons is their susceptibility to **biomagnification**. This is an increasing concentration of insecticide in organisms at the higher trophic levels, as these toxins accumulate in the body tissues and cannot be denatured or metabolized. This results in the increasing concentration of the insecticide at each step in the food chain, until top predators suffer very high doses. A classic example of biomagnification is the use of 2,2-bis (f-chlorophenyl)-1, 1-dichloroethane (DDD) at concentration of 0.02 parts per million to control larvae of the midge, *Chaoborus astictopus*, around Clear Lake, California in 1949 (*Fig. 1*). After initial success, the spraying was repeated regularly until 1954, when the bodies of large numbers of a diving bird, the Western Grebe were found. The grebe's body fat contained 1600 ppm DDD and their diet of fish was loaded with similar levels of DDD. Another example of biomagnification includes the effects of DDT on the British spar-rowhawk (*Accipiter nisus*), which suffered a dramatic population crash during the 1960s, partly due to the DDT causing mothers to produce too thin eggshells, which easily broke before the hatch date. Biomagnification may also threaten the human food chain, especially where fish are affected.

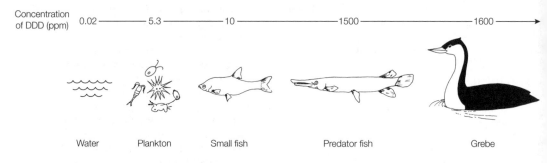

Fig. 1.   The biomagnification of DDD in Clear Lake, California. From Introduction to Integrated Pest Management, *Flint and Van den Bosch, 1981, with permission from Plenum Publishing Corp.*

Target pest resurgence and secondary pest outbreaks

Of particular importance are the effects of insecticides on the natural enemies of an insect pest. If the insecticide also affects the pest's natural enemy, then this may lead to a rapid increase in pest numbers some time after the initial drop in pest numbers caused by application of the pesticide. This effect occurs because a large number of pests and their natural enemies are killed. Under these conditions, any pests which survive will have a plentiful food supply resulting in a population explosion.

When natural enemies are destroyed, it is not only the target pest that might resurge but also a number of potential pest species which are not normally pests because they are kept in check by their natural enemies. Thus if a **primary pest** is treated with an insecticide that destroys a wide range of predators, other species may realize their potential and become **secondary pests**. An example of this concerns the insect pests of cotton in Central America. In 1950, when organic insecticides were applied, there were two primary pests, the boll weevil (*Anthonomous grandis*) and the Alabama leafworm (*Alabama argillacea*). The appli-cation of insecticides, applied five times a year, was initially very successful with large increases in yield. However, by 1955, secondary pests had emerged, the cotton bollworm (*Heliothis zea*), the cotton aphid (*Aphis gossypii*) and the false pink bollworm (*Sacadodes pyralis*). The application rate rose to ten per year and by the 1960s eight pests had emerged and 28 applications per year were being used.

Evolution of resistance

Evolved resistance probably represents the greatest problem of all. Pesticide resistance provides some of the best examples of evolution in action, providing clear demonstrations of the impact of intense natural selection on populations. *Figure 2* shows the extent of the problem and shows the growth in the number of insect species reported to be resistant to at least one insecticide. Within a large population subjected to pesticide, a few genotypes may be unusually resistant. Resistance will then spread very rapidly, especially if the majority of the population is exposed, and insecticide is reapplied regularly. The (partially) resistant individuals have a better chance of survival and breeding, and if the pesticide is applied repeatedly, each generation will contain a larger proportion of resistant individuals. Subsequent mutations may enhance resistance even further. This process has led to an exponential increase in the number of insect species resistant to insecticides, which now total over 500 species. Further, cross resistance can occur which provides species resistant to one pesticide with resistance to another, because the mode of action is common to both. This can extend further to multiple resistance in which the pest becomes resistant to a number of insecticides with different modes of action. For example, the house fly, *Musca domestica*, has developed resistance to almost every chemical used against it.

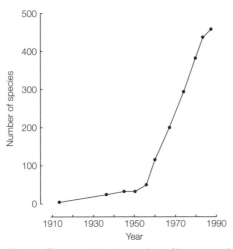

Fig. 2.   The growth in the number of insect species reported to be resistant to at least one insecticide.

Herbicide resistance was slower to evolve but it was early in the 1970s when multiple reports of triazine-resistant weeds appeared. Recently, multiple herbicide resistance has been reported. Canary grass (*Phalaris paradoxa*) has been found to be resistant to methabenzthiazuron, triazine and diclofopmethyl.

One solution to the problem of pesticide resistance is to develop strategies of resistance management. Approaches include: (i) reducing the frequency and population coverage with which a particular pesticide is used, thus weakening the strength of selection; (ii) using pesticides at a concentration high enough to kill heterozygous individuals carrying only one copy of a resistance gene. A common response to pest resistance has been the application of more pesticides, which leads to further resistance and the cycle continues. Despite pest resistance the production and sales of pesticides has continued to grow.

In spite of the increasing cost of pesticides the benefit-to-cost ratio produced has often remained in favor of pesticide use, with values ranging from $2.40 to $5.00 return for every $1 spent. However, these calculations omit the cost to the environment, which the farmer does not pay. In the UK, the cost of removing pesticides from the water supply, which is paid for by the public, is equivalent to £40 ($60) per kg of pesticide applied. Modern market demands may load the benefit-to-cost ratio in favor of chemical pesticides. For example, in the developed world, customers demand unblemished foodstuffs free of pests; this requires the use of pesticides. In third world countries the requirement to produce food at all costs means that pesticides banned in many parts of the world are still used extensively. Although pesticides have been invaluable in some instances (for example, more than one billion people have been freed from the risk of malaria), the requirement for alternatives to chemical pesticides is clearly evident. This will be discussed in the next topic.

U3 BIOLOGICAL CONTROL AND INTEGRATED PEST MANAGEMENT

Key Notes

Biological control

Biological control is the utilization of a pest's natural enemies in order to control that pest. There are four types of pest control. The introduction of a natural enemy from another geographical area is often termed classical biological control or importation. Inoculation is similar but requires the periodic release of a control agent where it cannot persist throughout the year. Augmentation involves the repeated release of an indigenous natural enemy in order to supplement an existing population. Inundation is the release of large numbers of a natural enemy, with the aim of killing those pests present at the time.

Microbial insecticides

Up until recently, insects have been the main agents of biological control against both pests and weeds, in all four types of biological control. Recently, increasing attention in the control of insect pests has focused on the use of insect pathogens, largely as microbial insecticides. The bacterium, *Bacillus thuringiensis*, is the only microbial pest control agent that has been commercialized worldwide. Viruses, in particular the baculovirus, have also been isolated which cause disease in insects and mites. There are around 100 genera of fungi which are pathogenic to insects and are potential control agents.

Genetic control and resistance

Autocidal control uses the pest itself to increase its own rate of mortality. This usually involves the release of sterile males, leading to a decrease in birth rate. Another type of genetic manipulation is to select plant varieties resistant to pests (and also herbicides). Resistant varieties have been selected and recently sophisticated genetic manipulations of plants have become possible. There are potential environmental and safety benefits from the development and use of such transgenic plants. However, there are public perception and legal problems associated with technique, and the problems of evolved resistance are as great as with chemical pesticides.

Integrated pest management

Integrated Pest Management (IPM) is a philosophy of pest management rather than a specific, defined strategy, combining physical, cultural, biological and chemical control and the use of resistant varieties. A considerable investment of time and effort is required to determine the optimum strategy, but there are long term environmental and economic benefits.

Related topics

The nature of predation (J1)
The dynamics of parasitism (K2)
The pest problem and control
  strategies (U1)

Pesticides and problems (U2)

Biological control The term 'biological control' has been applied to virtually all pest control measures except the nonselective application of chemical pesticides, but it is a term ideally restricted to the use of natural enemies in pest control. **Biological control** is the utilization of a pest's natural enemies in order to control that pest. There are four types of pest control:

- the introduction of a natural enemy from another geographical area, very often the area in which the pest originated from, to contain the pest below the EIL–this is often termed **classical biological control** or **importation**;
- **inoculation** is similar but requires the periodic release of a control agent where it cannot persist throughout the year, with the aim of providing control for only a few generations;
- **augmentation** involves the release of an indigenous natural enemy in order to supplement an existing population, and is therefore carried out repeatedly, usually to coincide with a period of rapid pest population growth;
- **inundation** is the release of large numbers of an natural enemy, with the aim of killing those pests present at the time, but with no expectation of providing long-term control. These are usually termed biological pesticides.

Up until now, insects have been the main agents of biological control against both pests and weeds, in all four types of biological control. The example of classical biological control marked the start of biological control. The cottony cushion scale insect, *Icerya purchasi*, a pest of Californian citrus orchard, was controlled by 1890 by a ladybird beetle, *Rodolia cardinalis*, and a dipteran parasitoid, *Cryptochaetum* sp, which were introduced from Australia. Inoculation as a means of biological control is widely practiced in the control of arthropod pests in glasshouses, where crops are removed, along with their pests and their natural enemies at the end of the growing season. The two species of natural enemy that are used most widely are *Phytoseiulus persimilis*, a mite which preys on spider mites, a pest of cucumbers and other vegetables, and *Encarsia formosa*, a chalcid parasitoid wasp of the whitefly and a pest of tomatoes and cucumbers. Around 500 million individuals of each species are released globally each year.

Microbial insecticides Recently, increasing attention in the control of insect pests has focused on the use of insect pathogens, largely as microbial insecticides. The bacterium *Bacillus thuringiensis* is the only microbial pest control agent that has been commercialized worldwide. When the microorganism sporulates, a spore and a large proteinaceous crystal are produced which release powerful toxins when ingested by insect larvae. Death occurs between 30 minutes and 3 days after ingestion. Importantly there is a range of varieties of *B. thuringiensis*, including those specific against Lepidoptera, Diptera and beetles. Their advantage is their selective toxicity; they are not toxic to man or the pest's natural enemies. Viruses have also been isolated which cause disease in insects and mites. Attention has largely focused on the baculovirus, a highly virulent virus whose host range is restricted to Lepidoptera and Hymenoptera. The drawback of these systems is that if several, unrelated pests are required to be controlled then a number of viruses must be released. This causes feasibility problems. There are around 100 genera of fungi which are pathogenic to insects. Commercially there are three main fungi used in pest control, *Beauveria bassianca*, produced against the Colorado beetle, *Metarhizium anisopliae*, produced against spittle bugs growing on sugar cane, and *Verticillium lecanii*, produced against aphids and whitefly in greenhouses. These fungi tend to have broader host ranges and are generally not reliant on ingestion. However, infection does rely on spore germination which only occurs at high humidities (>92% relative humidity) and this limits the range of habitats in which fungi can be effective. Nematodes can be included as agents of microbial pest control due to the association of the nematodes *Heterorhabditis* and *Steinernema* and the insect pathogen *Xenorhabdus*. In this association, the nematode firstly infects an insect host. Secondly, the bacteria escapes from the nematode, multiplies and kills the host. The nematode then feeds on the bacteria and decomposing flesh. The nematodes can survive for several months and represent the only biological product which is active in soils.

Fungal pathogens have also been used in the control of weeds. For example, the rust fungus, *Puccinia chondrillena*, was introduced in Australia to control the skeleton weed, *Chondrilla juncea*, where its density has since been reduced by 100-fold. Overall there have been documented biological control programs against over 100 weeds, with most involving herbivorous insects. The water hyacinth, *Eichhornia crassipes*, spread throughout South America after its introduction in 1894 and is presently controlled by the release of a weevil, *Neochetina eichhorniae*. Overall biological control has a number of successes, but also a number of failures. However, the cost of biological control is much lower than chemical pesticides with a return of investment of around $30 for every $1 invested. For successful biological control, detailed ecological and taxonomic understanding is required, but the rates of return are high.

Genetic control and resistance

There are a number of techniques which use genetic manipulation to kill pests. **Autocidal control** uses the pest itself to increase its own rate of mortality. This usually involves the release of sterile males, leading to a decrease in birth rate. These irradiated males mate with females producing infertile eggs. The method is expensive and to be successful, the females must mate infrequently and the males must be competitive with natural males. In addition, the insects must be amenable to mass rearing in the laboratory so they far outnumber the natural males, and finally, the target area should be isolated so that natural males from outside the area should not be able to undermine the program. The most notable success of this technique is the near extinction of the screw-worm fly, *Cochliomyia hominivorax*, which lays its eggs in fresh wounds of wildlife and livestock. If untreated the victim dies. The EIL is very low due to the high value of cattle. There is also no other form of treatment to control these pests.

Another type of genetic manipulation is to select plant varieties resistant to pests (and also herbicides). Resistant varieties have been selected for widespread use by farmers. For example, after the spotted alfalfa aphid, *Therioaphis trifolii*, devastated the California alfalfa crop in the 1950s a resistant variety was discovered and widely introduced. However, a more methodical approach is now employed for the selective breeding of resistant varieties. This can be a lengthy process, taking several years. Even at the end of the process, the pest itself may evolve a biotype which is virulent against the new variety. It is therefore preferable, but more difficult, to breed varieties in which resistance is determined by more than one gene as this makes it more difficult for the pest to make a rapid evolutionary response.

Recently, genetic manipulations of plants have become more sophisticated. For example, in 1987 the first success was reported of inserting a gene into a crop that conferred resistance against pests – the δ-toxin gene of *Bacillus thuringiensis* was inserted into tobacco plants, conferring resistance against Lepidoptera. This has since been followed by a number of other similar genetic manipulations in plants. Clearly, the potential benefits from the development and use of such transgenic plants is immense in terms of environmental and safety benefits and the reduced costs of application. However, there are public perception and legal problems associated with this technique. Also the continual application of a pesticide, even in a plant seems certain to lead to pest resistance. Further, the evolved resistance by pest species to varieties of the δ-toxin have already been recorded.

Integrated pest management

Integrated pest management (IPM) is a philosophy of pest management rather than a specific, defined strategy, combining physical, cultural, biological and chemical control and the use of resistant varieties. It is ecologically based, relying on mortality factors, including natural enemies and weather, aimed at controlling pests below the EIL and is based around monitoring of pest and natural enemy abundance. This involves considerable investment of time, effort and money in determining the optimum strategy, although to be successful the basis of this strategy must also be economic. To run an IPM program trained advisors must be available and time, money and effort must be given to this aspect before any economic return is seen. For this reason, actual examples of IPM programs are limited. However, as an example we can take the IPM program developed to control cotton pests in California. The cotton had been plagued by target pest resurgence and secondary pest outbreaks, along with the evolution of resistance, which all culminated in an increased frequency of pesticide application. The main pest was *Lygus hesperus*, which feeds on fruiting cotton buds, reducing yield. In addition, the cotton bollworm (*Heliothis zea*) was a secondary pest. The IPM program needed to reduce insecticide use to prevent secondary outbreaks. Subsequent studies showed that the lygus bug could only inflict serious damage during the budding season (June and July). Insecticide applications were only made during this time. Cultural control was made by introducing thin strips of alfalfa into the cotton field which attracted the lygus out of the cotton. The key to a successful IPM program is a good field monitoring system. In this case each field was sampled twice a week from the beginning of budding (mid-May) to the end of August. Plant development, pest and natural enemy data were all collected, leading to a successful pest control program. In this example, the yield decreased slightly compared to the conventional high-pesticide treatment, but the economic return was higher. A survey of IPM projects indicates that an increase in economic returns is common, and that decreased yields are not inevitable (*Table 1*).

Table 1. The effect of IPM programs compared to conventional high-pesticide regimes

Country and crop	Pesticide change(%)	Yield change(%)	Annual savings
Indonesia, rice	-38	+5	$ 75 million
Sri Lanka, rice	-26	+35	$ 1 million
Togo, cotton	-50	-2	$ 12 000
USA, 9 crops	Reduced (no figures)	+20	$ 578 million

V1 RARE SPECIES, HABITAT LOSS AND EXTINCTION

Key Notes

Rare species

The type of rarity depends on three attributes of the species in question (Rabinowitz, 1981): (i) size of geographical range (large *v*. small); (ii) habitat specificity (wide *v*. narrow); (iii) local population size (high *v*. low). Any species for which one or more of these attributes is low demonstrates some form of rarity. Narrow habitat requirements, low population density and endemism are all different facets of rarity. Many species are rare by virtue of their ecology while others are rare as a result of human activities. Species with poor dispersal ability or sedentary behavior are likely to become rare as a result of anthropogenic factors.

Genetic diversity in rare species

Decline in the number of individuals of a species will result in loss of genetic diversity and inbreeding depression. A minimum viable population (MVP) is required to maintain genetic variability and to ensure long-term species survival. An MVP of 250–500 individuals is accepted as the best estimate for inbreeding avoidance. However, frequently the protection of habitat is the conservation priority, overshadowing concern over genetic diversity.

Habitat loss and fragmentation

Numerous habitats, such as wetlands, grasslands and forests, have been either drastically reduced in area, modified or destroyed, usually for agriculture. Habitat degradation results in the decline or loss of species. Habitat fragmentation reduces the habitat area, increases the distance between remaining patches, and favors species found in edge habitats. Species are lost because of the overall loss of habitat area, and as a result of increasing insularization. Recently isolated patches may initially contain more species than the area can sustain over the long term and species will be lost from the patch – a process known as relaxation.

Extinction

Extinction is a natural process, although its pace and incidence have been increased through human activity. Historic extinctions are clustered in geological time (e.g. significant extinction occurred in the Permian and Pleistocene). The anthropogenic causes of extinction include hunting, habitat destruction and artificial species introductions. The IUCN categorizes species facing considerable risk of becoming extinct as 'endangered'. Small populations are more likely to become endangered as a direct or indirect result of normal population fluctuations. Demographic and genetic factors interact to cause extinction. The chance of a population becoming extinct can be predicted by Population Viability Analysis (PVA).

Related topics

Genetic variation (O1)
Island communities and
  colonization (Q2)

Conservation strategies (V2)

Rare species

Some species are naturally rare and are able to persist in nature in small populations. Rabinowitz in 1981 identified seven types of rarity depending on three characteristics:

(i)   size of geographical range (large $v.$ small);
(ii) habitat specificity (wide $v.$ narrow);
(iii) local population size (high $v.$ low).

The classification gives eight possible combinations and emphasizes that species do not have to have small local populations, or be highly habitat specific, to be rare. All combinations of these characteristics, except the large-wide-high combination illustrated by many common species, describe different types of rarity. For example, the New Guinea harpy eagle (*Harpyopsis novaeguineae*) has a large home range and a wide habitat range yet occurs at low densities. The osprey (*Pandion haliaetusi)* is widespread, but has narrow habitat requirements as it feeds exclusively on fish.

Tropical forests are the most diverse of terrestrial communities, holding around 50% of the Earth's species, but individual population densities tend to be very low. On Barro Colorado Island, in Central America, one-third of the tree species occur at a density of less than one individual per hectare. These rare species have very specific habitat preferences, some only occurring in forest gaps of a particular age and light regime. Tropical forests are also home to a large number of **endemic** species – native species found only in these locations. The tiny Rio Palenque reserve in Equador with an area of 0.8 km^2 has over 250 endemic plants. Islands also feature high levels of endemism; their isolation and the low levels of migration and genetic exchange with other populations allow unique biota to evolve.

Rarity is a multifaceted concept and species can show different degrees of rarity at global, regional and local levels. The red-backed shrike is a rare bird in Britain at the very edge of its ecological range with only a few breeding pairs occurring in East Anglia, yet it is common and widespread in Europe.

Although many species are rare by virtue of their ecology, many others are rare as a result of human activities. Some species are more susceptible to anthropogenic change and rarity than others. Species with poor dispersal ability or sedentary behavior are likely to be vulnerable to **habitat loss** and **habitat fragmentation**. Many butterflies, such as the marsh fritillary (*Euphydryas aurinia*), and insects associated with dead and decaying wood fit into this category. They are unable to move to another habitat patch if their patch is destroyed, for instance, by drainage or forest clearance.

Genetic diversity in rare species

A species which has declined to low numbers will suffer from a loss of genetic diversity, reducing the potential for adaptation to changing environmental conditions and new diseases. It may also experience inbreeding depression, an increase in the expression of deleterious alleles and reduced offspring fitness (see Topic O1). The cheetah (*Acinonyx jubatus*) is a solitary species with a low population density throughout its range and low overall population size. It appears to have been through a bottleneck and a period of inbreeding in the past which is apparent today in its low genetic variation. This homozygosity is thought to be responsible for the high level of sperm abnormality (70%), high juvenile mortality and increased susceptibility to diseases such as feline infectious peritonitis.

A **minimum viable population (MVP)** is required for the long-term survival of a species to ensure genetic variability is maintained. Many endangered populations are already below the MVP of 250–500 individuals accepted as the best estimate for inbreeding avoidance. Detailed and lengthy study is required for the accurate determination of MVP; a 50-year study of bighorn sheep in the southwestern USA was required to establish a MVP of 100. Often, habitat destruction is a real and more immediate threat and the protection of habitat is a common conservation priority, frequently overshadowing concern over genetic diversity.

Habitat loss and fragmentation

There are numerous examples of habitats that have been either drastically reduced in area, modified or destroyed. In the USA, for example, virtually 100% of natural grasslands have been lost since 1492. One of the most exploited terrestrial ecosystems has been wetlands. These once species-rich areas have been settled by humans or drained for agriculture. For example, 90% of wetlands in New Zealand have been lost since European settlement. Temperate native forests have been largely destroyed in Europe. Global forest loss can be monitored using satellite technology and has been estimated at 170 000 km^2 per year (for the period 1981 to 1990; see also Section S4).

A habitat may not be destroyed completely for species to disappear. A site or ecological interactions can be altered such that **habitat degradation** occurs which results in the decline or loss of species. The natural regeneration of the Brazil nut (*Bertholletia excelsa*) has been hindered because of the decline of the one bee species able to pollinate it and of the agouti, a large rodent which buries caches of the nuts. Similarly, in Hawaii a rare and endemic species of the silversword is doomed to extinction because its sole insect pollinator has become extinct. Thus, a single species can play a crucial role, having a kind of domino effect on the community that is disproportionate to its abundance.

Habitat fragmentation occurs where large tracts of habitat are broken up by agriculture or other development. The effect is to reduce the overall area of the habitat and the size of patches and to increase the distance between remaining patches. Some species are favored by a degree of fragmentation. Female meadow voles, *Microtus pennsyvanicus*, are more numerous in fragmented habitat because smaller, relatively isolated patches of the same habitat type make it easier to defend breeding territories. Fragmentation is more usually a detrimental process, which has been increasing inexorably throughout most of the world. Habitat quality will deteriorate in smaller fragments where there is a large boundary relative to the patch area and less undisturbed 'interior' habitat. Such patches tend to be dominated by 'edge' communities which favor the junction between two habitats. A study of heathland in the Poole Basin, UK has shown how fragmentation caused by road development results in significant changes in plant species composition, plant performance, and soil nutrient levels. As habitat quality declines the more susceptible animal species suffer first, particularly those vulnerable to predation when the edge-to-interior ratio increases. In Amazonian forests, raptors and large fruit- and nectar-feeding birds are the first to disappear as these are reliant on food sources found only in core habitat areas.

Some species are more able to survive in small, species-poor patches than others (see also Topic Q2) and this information can be used to predict which groups are most likely to succumb as habitats become fragmented. The Dorset heathland is an area of semi-natural shrubland in southern England which developed on freely drained, sandy, acidic soils following forest clearance. The vegetation is dominated by heather (*Calluna vulgaris*) and heaths (*Erica* spp.). Changes in land use have reduced its area to around 450 fragments totalling just 5% of the original habitat area. A model simulating the loss of species as fragmentation proceeds demonstrates that species with good powers of dispersal and little chance of becoming extinct locally are much less susceptible to fragmentation than poor colonizers prone to local extinction. However, above a certain point, the more resilient species are also lost rapidly. Species are lost because of the overall loss of habitat area, but also as a result of increasing **insularization**. As the distances between patches increase, colonization becomes more difficult, especially for the less mobile species.

Islands that have been isolated may have a residual number of species which is greater than the area can sustain over the long term. Over time the number of species will fall as extinction outweighs colonization – a process known as **relaxation** (see Topic Q2). The principle of relaxation can be used to predict extinction rates for large mammals on East African nature reserves. The average reserve of 4000 km^2 is estimated to lose between five and six of its 48 mammal species in the next 50 years.

Extinction

Extinction is a naturally occurring process which arises through a combination of population characteristics and random environmental factors. Some species have population attributes that render them prone to extinction. These are large body size in animals, small or restricted geographical range, habitat specificity, lack of genetic variability and inability to switch to alternative food sources.

Most past extinctions are clustered in geological time and even seem to show periodicity (*Fig. 1*). A major extinction occurred between the Cretaceous and the Tertiary; it is so distinct in the fossil record that it is used to delineate the boundary between these periods (known as the K-T boundary). *Figure 2* shows the changes with depth (time) in such a fossil section. One mass extinction occurred in the Permian (225 million years ago), when 90% of the world's shallow water marine invertebrates became extinct. More recently, the Pleistocene saw the extinction of many mammal species, including the woolly mammoth, mastodon and giant sloth. These losses were caused either by the advance and retreat of ice sheets, or overexploitation by Pleistocene hunters, or a combination of both.

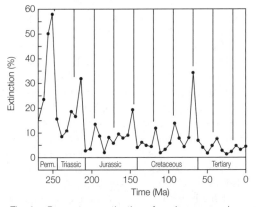

Fig. 1.   *Percentage extinction of marine mammal genera from the middle of the Permian to the present. The numerous peaks indicate clusters of extinctions. The vertical lines are spaced every 26 million years and appear to coincide with extinction peaks. Cyclical fluctuations in sunspot activity may be the cause of this apparent periodicity in extinctions. Reprinted from* Evolution, Skelton, 1993, *with permission of Addison Wesley Longman Ltd.*

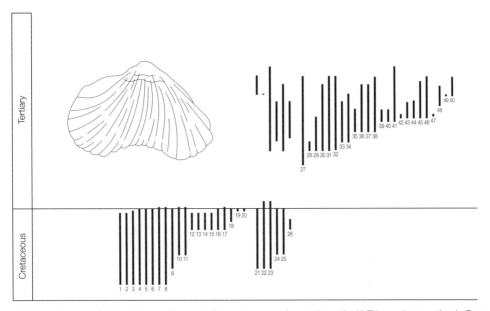

Fig. 2.   Ranges of 50 brachiopod (lampshell) species over about 15 m of a K-T boundary section in Denmark. There is a distinct break in the fossil record, indicating a mass extinction, and the species of fossil found above the boundary are largely different from those below. Reprinted from Evolution, Skelton, 1993, with permission of Addison Wesley Longman Ltd.

The incidence and pace of extinction have been greatly increased as a result of human influences. Humans have caused well over 75% of all extinctions since AD 1600. Direct human influences were notoriously responsible for the demise of the dodo (Raphus cucullatus), woolly mammoth (Mammuthus primigenius), Tasmania wolf (Thylacinus cynocephalus) and the giant moa (Diornid maximus), all of which were hunted to extinction. Several species have been brought to the brink of extinction by commercial exploitation. In Kenya and Uganda, between 85% and 89% of all elephants have been wiped out by poaching. Many whale species are endangered because of overfishing and nontarget species such as dolphins and seabirds are often also casualties in fishing nets.

A large number of species losses occur through habitat destruction; one estimate puts it at over 100 species per day. More subtly, habitat change through changed management, such as the removal of grazing, or through climatic change, can have a profound effect on particular species.

Artificial species introductions have been implicated in cases of extinction and can have a particularly devastating effect on native communities. The Fijian coconut moth (Levuana viridis), once a serious pest of coconut plantations, was driven to extinction by the deliberate introduction of the parasitic fly, Bressa remota. The fly then switched to other moths and has caused the extinction of several endemic species. The introduction of rats and other predators has been strongly implicated in the extinction of many of the native bird species of Hawaii and New Zealand – communities which have evolved in the absence of these common predators. The introduction of domestic cats, dogs, goats and pigs has also caused many disasters to island ecosystems.

Rarity per se does not mean that a species is threatened with extinction. The International Union for the Conservation of Nature and Natural Resources (IUCN) recognizes four categories of extinction risk (Table 1).

Table 1.   The IUCN classification of extinction risk

Risk category	Description
1	*Rare species* having small populations, usually within restricted geographical limits or localized habitats, or widely scattered individuals; they are at risk of becoming rarer but not of becoming extinct.
2	*Vulnerable species* are those which are under threat or actually decreasing in number.
3	*Endangered species* have very low population numbers and are in considerable danger of becoming extinct.
4	*Extinct species* are those that cannot be found in the areas they once inhabited, nor in other likely habitats.

A small population is more likely to become extinct than a large one, firstly, because normal population fluctuations resulting from variations in environmental conditions (a harsh winter, for example) are more likely to decimate the population, and secondly, because inbreeding depression will reduce the population's ability to produce viable offspring. This interaction between demographic and genetic factors acts as an **extinction vortex** pulling the population numbers lower.

It is not easy to predict the chance of a species becoming extinct as the data on population numbers are often not available. Even in well-studied cases, other factors such as environmental stochasticity, catastrophic events and the effects of inbreeding make population size alone an unreliable predictor of extinction. Population Viability Analysis (PVA) is one approach that uses computer modeling to predict a population's chance of extinction. The model incorporates information on the population dynamics of the species, genetic and behavioral factors and environmental effects, such as the state of the habitat, interactions with other species and human interference. This type of analysis predicts that there is a 25% chance that the eastern barred bandicoot will be extinct within the next 30 years.

This approach also helps in the management of endangered species by allowing the estimation of the minimum reserve area required to retain a viable population (see Topic V2).

V2 CONSERVATION STRATEGIES

Key Notes

Biodiversity	Biodiversity is a term encompassing the variety of organisms at all levels, from genetic variants belonging to the same species, to species diversity and including the variety of ecosystems. The conservation of biodiversity includes the preservation of genetic variation, the diversity of species and populations and also the life support properties of ecosystems, such as climatic and drainage effects.
The Earth Summit, Rio de Janeiro (1992)	The importance of biodiversity has been promoted by the International Convention on Biological Diversity, part of the Earth Summit held in Rio de Janeiro in 1992. The Convention was signed by 152 nations committing them by law to adopt ways and means for the conservation of biodiversity and to ensure equity of benefits from biological diversity. In addition, Agenda 21 is a blueprint to encourage sustainable development socially, economically and environmentally in the 21st century.
Strategies for conservation	Conservation strategies exist at a range of different levels of detail and scale, global, national, regional and local, to accommodate the markedly different political scales at which conservation objectives are directed. International strategies are essential for the conservation of globally threatened ecosystems and are led by the World Conservation Union. The Convention on the International Trade in Endangered Species (CITES) works to prevent illegal imports and exports. The Antarctic Treaty and the more recent Protocol on Environmental Protection to the Antarctic Treaty (1992) are examples of how political boundaries can be transgressed. At the national level, conservation objectives are set by governmental organizations and implemented through legislation, e.g. for the establishment of protected areas.
The design of nature reserves	In general, the larger the reserve, the greater the number of species in it. However, many small areas may contain more species in total than one reserve of the same area. The debate over the relative merits of a single large or several small reserves is known as the 'SLOSS' argument. The best compromise may be a network of small, linked reserves which allows dispersal and genetic interchange to take place between areas. Another way of maximizing the conservation value of a reserve is to surround the reserve with a buffer zone of the same habitat. The minimum area required to retain a viable population can be estimated by computer modeling.
Environmental assessment	Environmental assessment is the means by which the conservation value of a site may be assessed without detailed and time-consuming surveys of its entire biodiversity. Sites may be selected on the basis of a small number of conspicuous and sensitive species which are taken as indicative of the larger community. Environmental Impact Assessment (EIA) is the process of identifying, estimating and evaluating the environmental consequences of current or proposed development on an ecosystem.

Related topics

Island communities and colonization (Q2)	Rare species, habitat loss and extinction (V2)

Biodiversity

The emphasis today is on conserving **biodiversity**. Biodiversity is an abbreviation of 'biological diversity', and is a term encompassing the variety of organisms at all levels, from genetic variants belonging to the same species, to species diversity and including the variety of ecosystems. The conservation of biodiversity is not just about conserving certain popular animal species such as the panda or the north American spotted owl, or a few examples of wetlands and rainforests, but also refers to the conservation of *variety, interactions* between species and *processes* in ecosystems. This includes conserving genetic variation, the diversity of species and populations and also the life support properties of ecosystems, such as climatic and drainage effects.

The Earth Summit, Rio de Janeiro (1992)

The importance of biodiversity has been promoted by the international Convention on Biological Diversity, part of the Earth Summit held in Rio de Janeiro in 1992. The Convention was signed by 152 nations committing them by law to adopt ways and means of conserving biodiversity and to ensure equity of benefits from biological diversity. The USA was notable by its refusal to sign, but it was later ratified by a new administration. The Convention has spawned many international initiatives and projects while continuing meetings (e.g. Kyoto, 1997) develop its themes. It has laid the foundation for a unified and inclusive approach to global biodiversity involving the collaborative efforts of developed nations, which possess the financial means and scientific expertise required for effective conservation, and the developing world where most of our biodiversity resides.

Another product of the Earth Summit, **Agenda 21**, also addresses the conservation of biological diversity but is not legally binding. Agenda 21 is a blueprint to encourage sustainable development socially, economically and environmentally in the 21st century. It includes recommendations for strengthening the role of major groups, including women, in sustainable development and of indigenous people in their community. These groups are under-represented in the conservation arena yet have always played a very large part in practical conservation measures in developing countries and hold extensive knowledge about the value and use of indigenous wild species. The exploitation of indigenous knowledge by foreign interests is recognized by the Convention on Biological Diversity which refers to the sovereign rights of states over their natural resources and authority to determine access to genetic resources. There is an International Alliance of the Indigenous-Tribal Peoples of the Tropical Forests which has issued a Forest Peoples' charter setting out conservation policy based on the rights of these peoples to their forests.

Strategies for conservation

To achieve conservation of biological diversity it is usually necessary to establish protected areas, to reintroduce some species, to restore ecosystems and to manage or eradicate previously introduced plants and animals. Successful conservation strategies should be clear frameworks within which defined objectives can be both explained and realized. Conservation strategies must exist at a range of different levels of detail and scale to accommodate the markedly different political scales at which conservation objectives are directed. Global and national strategies meet the needs of national government, local strategies are required for local authorities. Such strategies form the basis of conservation legislation. **Nongovernmental organizations (NGOs)**, such as Greenpeace, establish strategies at a variety of scales according to their individual priorities and apply pressure on governments.

International strategies are essential for the conservation of globally threatened ecosystems and to allow transgression of national boundaries and priorities. The **IUCN (International Union for the Conservation of Nature)**, now called the **World Conservation Union**, provides a link between nongovernmental campaigning organizations, government agencies and sovereign states. It is an international and independent organization providing leadership and a common approach to conservation. The Convention on the International Trade in Endangered Species (**CITES**) has been successful in preventing the illegal import and export of many rare species and animal products, and has been credited with saving the elephant from extinction. However, many species, including the tiger, are becoming increasingly threatened because of a combination of consumer demand and habitat loss.

One region demonstrating the potential for international conservation agreements is Antarctica. The **Antarctic Treaty**, signed by those nations with territorial claims to the region, sets aside all sovereignty, bans all military activities and nuclear waste disposal and gives complete freedom for scientific investigation. More recently, Antarctic seals and marine wildlife have been given specific protection and mining has been banned. The **Protocol on Environmental Protection to the Antarctic Treaty** (1992) draws together the whole strategy into one document, including how environmental damage should be monitored and a requirement to report on the progress of species protection measures.

At the national level, the conservation strategy or strategies will reflect statutory responsibilities and provide the framework for the operation of governmental organizations. The establishment of local and national nature reserves and national parks, schemes to compensate farmers for reducing production and the management of habitats to maximize biodiversity are means by which a conservation strategy can be implemented.

The design of nature reserves

In general, the number of species in an area of habitat increases with the size of the area (see Topic Q2). Larger habitat patches will have a higher rate of colonization simply because they are bigger targets. They will also support larger populations which are less likely to become extinct, and tend to have a greater habitat and species diversity. Some large animals range over considerable areas and require a large area to maintain a minimum viable population (see Topic V1).

In spite of this, size is not everything in reserve design and it may be better to conserve many small reserves rather than one large one. In the Great Barrier Reef the number of species of fish found on 'patch-reefs' depends on the variety of habitats available, not on patch area. Although habitat diversity does tend to increase with area, the number of species in a habitat fragment can depend on other factors, including its history of disturbance, the number of species present at the time of isolation, and whether the conditions required by particular species prevail.

The debate over the relative merits of a single large or several small reserves has become known as the SLOSS argument. Some studies have shown that more species are found in small islands of habitat than in the equivalent continuous areas. This may be because of the inclusion of more habitat edge and a greater diversity of habitat as a result (see Topic Q2). *Table 1* outlines the major advantages of the two strategies.

A network of small reserves may have some of the advantages of one large one if the intervening habitat is not too hostile. These linked reserves (*Fig. 1*)

Table 1.   The relative advantages of several small and one large nature reserve

Advantages of several small reserves	Advantages of one large reserve
The communities in the reserves may be different because they are different sizes or because of chance.	Large population sizes protect against chance extinction and inbreeding.
More 'edge species' adapted to transitional habitats are favored.	Some habitats and species can only exist in large areas or habitat 'interiors', e.g. predators and other keystone species.
Periodic disturbance can help maintain diversity – this may be more frequent in small habitat patches.	Large reserves are less sensitive to catastrophic disturbance and pollution.
Many separate reserves provide some protection against disease and pathogens.	Disease can spread more easily through a single large reserve.
Small dispersed reserves are more likely to be encountered by colonists than one large reserve.	Large reserves will support species that are poor dispersers.

allow dispersal and genetic interchange to take place between areas and greatly reduce the chance of a species becoming extinct. Another way of maximizing the conservation value of a reserve is to have a **buffer zone**, which is a surrounding area of the same habitat. The buffer zone provides a source of immigrants and can protect the reserve from edge effects, thus maximizing the protection of species adapted to the interior of the habitat.

Population viability analysis (see Topic V1) can help in the estimation of the size of reserve necessary to retain a viable population. As part of the recovery plan for the northern spotted owl (*Strix caurina occidentalis*), computer models established that the minimum reserve area should contain 20 owl territories

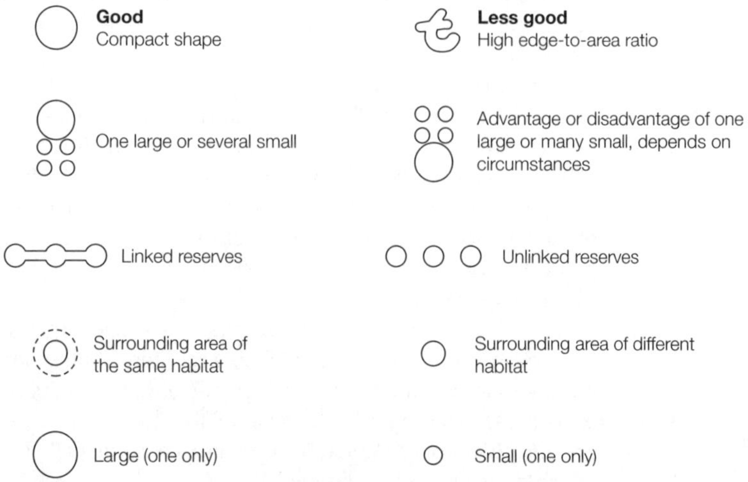

Fig. 1.   The importance of size and shape in setting up a nature reserve. Reprinted from Ecology, Chapman and Reiss, 1992, with permission from Cambridge University Press.

and that they should be close together in an area that is at least 30% suitable habitat. Such a model takes account of the fact that suitable habitat is discontinuous and interrupted by unsuitable (e.g. logged) areas and is a major advance on previous work.

Environmental assessment

Although the global priority promoted at the Earth Summit may be the conservation of biodiversity, biodiversity may be difficult to quantify in practice. It is not always possible to estimate the diversity of potential reserve sites as not all groups can be easily counted. Instead, sites may be selected on the basis of a small number of conspicuous species which are taken as indicative of the larger community. Birds are often selected and have the advantage of great popular appeal which may be valuable in winning the conservation of a site. Brown trout, *Salmo trutta*, are used as an indication of the quality of forest streams because they are sensitive to acidification.

The **Habitat Evaluations Procedure (HEP)** used by the US Fish and Wildlife Service identifies representative species whose occurrence indicates the presence of others with similar requirements. This information is taken together with habitat area and a **Habitat Suitability Index** to calculate an objective score for the site. A whole field of environmental assessment has been developed based around criteria for the evaluation and selection of sites for conservation.

Environmental Impact Assessment (EIA) is defined as the process of identifying, estimating, and evaluating the environmental consequences of current or proposed action. The term is interchangeable with Environmental Assessment (EA). The process permits the explicit analysis of the relationship between economic activity, e.g. the building of a new road, and the conservation of natural resources. It involves a number of common procedural steps:

1. Scoping – interpretation of the proposal and its sources of ecological disturbance; inventory of potential ecological receptors (organisms likely to be sensitive to the disturbance); determination of how stressor/receptor interactions should be studied.
2. Focusing – identification of the important species, habitats, or ecological processes; refinement of the scope of the study.
3. Impact Assessment – description and recording of the current condition of the ecosystem to be affected through baseline studies; impact prediction and assessment relative to the baseline.
4. Impact Mitigation – mitigation to redress significant adverse effects (e.g. reduction of the impact of the development to certain identified species).
5. Impact Evaluation – determination of the significance of predicted ecological impacts measured against specific criteria.
6. Monitoring and Feedback – monitoring to strengthen the knowledge base and to allow for corrective action to be taken in the light of unforeseen outcomes; feedback to assess proposal implementation.

If the impact of the proposed development has been assessed as very severe (e.g. likely to result in the loss of species of high conservation value), one option is for the project to be halted and another site found. In less severe cases, the impact on organisms can be reduced by regulating their exposure. To avoid wildlife mortality on roads, most major roads in the UK are fenced off. Deer fences are used in areas of high deer density, and badger and amphibian tunnels are used to maintain routes that have been bisected by new roads (*Fig. 2a*).

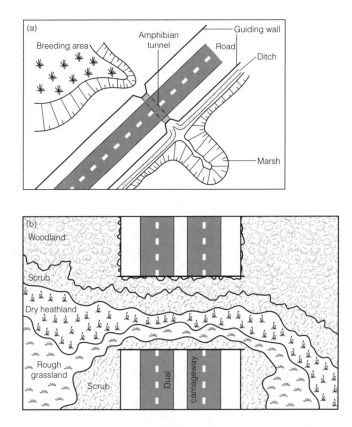

Fig. 2. (a) Connecting over-wintering and breeding areas for amphibians with a tunnel. (b) An ecoduct with several vegetation types.

Wildlife tunnels and bridges are similarly used to maintain migration routes across busy highways in Canada. Ecoducts are wildlife bridges vegetated as naturally as possible to maintain habitat continuity (*Fig. 2b*).

In cases where wildlife resources are lost or sites are damaged, compensation measures may be recommended. This could entail purchase and donation by the developer of an alternative wildlife habitat site, or, where feasible, the creation of new examples of the habitat that has been lost. The weakness of this approach is that the compensation site is often of lower ecological value than the habitat that has been lost or damaged.

V3 BIOLOGICAL RESOURCES AND GENE BANKS

Key Notes

Biological resources

The overexploitation of natural populations as biological resources leads to the decline of the resource and even population extinction and habitat destruction. Sustainable use ensures that habitats (such as forests) and populations (of game, for example) are conserved for future use. General plant biodiversity has an actual or potential commercial value because of the reliance of the pharmaceutical industry on botanical products. Biodiversity in crop plant species and their relatives is valuable because of the plant breeding opportunities such diversity offers.

Valuing biological resources

The economic value of a biological resource is the sum of the benefits it provides. The total economic value of a resource is the sum of its use value and nonuse value. Use values encompass the value of useful products such as timber. Nonuse values include the existence value and benefit to future generations. The benefits of biodiversity (e.g. of carbon storage by tropical forests) are often spread across national boundaries, but the costs of conserving biodiversity are usually borne by one nation. This means that in practice the value of biodiversity is underestimated and biodiversity is lost in favor of other land use which delivers direct and immediate benefits to the land owner.

Ecotourism

Ecotourists pay to experience the biodiversity of a country or national park. Ecotourism is a means of gaining economic benefit from biodiversity and can help to meet the cost of conservation. The disadvantages of ecotourism are 'spearheading', the 'leakage' of income away from the local area and overexpansion of tourism.

Gene banks

Where *in situ* conservation is not possible, the only option is for species or their DNA to be preserved by *ex situ* methods in museums, herbaria or zoos. Collections of living material in the form of zoo animals, botanical gardens and seeds, together with DNA collections have been termed gene banks. Plant species and crop varieties are readily stored in seed banks. The original wild strains of food crops could be very valuable in the future development of new crop varieties.

Zoos – captive breeding

Captive breeding programs aim to conserve threatened species in captivity with the ultimate aim of reintroduction into the wild. In order to make full use of the available gene pool to maximize genetic variation, individuals of a species held in zoos around the world are managed as a single population.

Ideally, captive numbers should be built up to the minimum viable population, the sex ratio should be maintained at 1:1 and breeding among distinct races and with domestic varieties should be avoided.

Successful release into the wild depends on factors such as habitat quality, area and protection from human interference. The reintroduction of captive-bred animals is fraught with difficulties, is expensive and tends to be used as a last resort after *in situ* methods have failed.

Related topics	Sex in ecology (N2)	Conservation strategies (V2)
	Rare species, habitat loss and	
	extinction (V1)	

Biological resources

Natural populations and ecosystems have been exploited by man for centuries as biological resources. Overexploitation leads to the decline of the resource and even population extinction and habitat destruction. Sustainable use, however, does not damage the resource and may ensure that habitats and populations are conserved for future use (see Topic V1). For example, grouse moors in Scotland are maintained by periodic burning and predators are controlled in order to maximize the population of red grouse for shooting. The income from shooting ensures that this important habitat and the many other species it supports are maintained. Shooting levels are controlled so as not to adversely affect the population in subsequent years.

Forests that are managed for timber are often sustainably harvested, i.e. trees that are felled are replaced with new saplings. Sustainable forestry, although it does not destroy the forest ecosystem and ensures that the timber resource is maintained, substantially alters it and almost always reduces biodiversity and conservation value. Managed forests are usually more even-aged and have a lower tree species diversity than natural forests. They also tend to have a poorly developed ground flora.

Plant biodiversity in general can be treated as a biological resource because of its actual or potential commercial value. Currently, between 25% and 50% of drugs are based on plant products. One of the best known medicinal plants, discovered in the 1940s and used in the treatment of leukaemia, is the rosy periwinkle. The vinca alkaloids, vincristine and vinblastine, produced by this plant can now be synthesized in the laboratory and the parent plant is no longer required to produce the drugs. Today there is renewed interest from multinational pharmaceutical companies in new botanical products, while countries with tropical forest ecosystems, where most of the world's floral biodiversity is found, are ready to charge royalties for access to this biological resource and for plant-based research (see Topic V1).

Biodiversity in crop plant species and their relatives is valuable because of the potential opportunities such genetic diversity offers to plant breeders. As well as maximizing yield, breeders may exploit this variation to create varieties with improved pest resistance, increased water use or nutrient efficiency, or other desirable features. However, crop species diversity has declined dramatically with the onset of modern agricultural techniques (*Table 1*). Lack of genetic diversity in a crop also makes that crop vulnerable to pathogen or pest damage

Table 1.   Selected examples of the decline in crop diversity

Country	Crop	Number of varieties lost
USA	Apples	6121 (86%)
	Peas	546 (95%)
	Maize	296 (96%)
	Broccoli	34 (100%)
Europe	Domestic animals	Since 1900, 50% of all varieties have become extinct.
	Wheat	345 (90%)
Philippines	Rice	Before 1970, 3500 varieties grown, now 5 dominate (>99% loss)

(see Topics O1 and N2). In the US, one variety of sweet potato accounts for 69% of production, whilst two varieties of pea account for 96% of total yield. A recent study concluded that US agriculture was 'impressively genetically uniform and [so] impressively vulnerable'.

Valuing biological resources It is possible to ascribe an economic value to biological resources by considering the totality of the benefits they provide. The difficulty lies in capturing the many components of value possessed by natural systems and in translating nonfinancial benefits into financial ones.

The **total economic value** of a resource is the sum of its **use value** and **nonuse value**. The components of use and nonuse values are summarized in *Table 2*. The direct use value of a tropical forest would encompass the value of timber and all forest products, including medicinal plants harvested by indigenous people. The option value includes the money pharmaceutical companies would pay to allow future research into the medicinal properties of these plants, and the bequest value is the benefit that future generations may experience because of new forest-based drugs.

Table 2.   Use and nonuse values for a biological resource, in this case a forest ecosystem

Use values	Nonuse values
Direct use value: the value of products actually harvested and used (e.g. timber)	Bequest value: the benefit that may accrue to future generations
Option value: the willingness to pay to safeguard the resource for future use	Existence or passive value: the value attached to the knowledge that the forest exists by people not benefiting directly from it
Indirect use value (e.g. the value of watershed protection afforded by a forest)	

Viewed in this way, it is apparent that the value of biodiversity is much greater than the value of useful products which can be bought and sold. There is also an existence value attached to distant ecosystems by people in other countries that may never experience, nor directly benefit from this biodiversity. Witness the sums paid by people in the west to organizations such as the World Wide Fund for Nature.

The **cost of conservation** is usually borne by one nation, yet there may be global benefits, for example, through climate regulation. If all beneficiaries are to contribute to the cost of conservation, national boundaries must be crossed. One recommendation at the Earth Summit (Rio, 1992) was that Brazil should levy a charge for carbon storage. Another approach is to undertake 'debt for nature' swaps: international conservation agencies pay back some of the debt of a developing country in exchange for conservation. Such global accounting goes against political convention and is still in its infancy.

Other difficulties lie in disengaging these values and in quantifying them. It is very difficult to tease apart option, bequest and existence values, for instance. A financial value is only that which someone is willing to pay and depends on who is asked – there is no market place for watershed protection. Furthermore, the total value of the ecosystem may exceed the sum of its individual functions and this approach may undervalue biodiversity. The alternative is to accept that economic value does not capture the **'intrinsic value'** of an ecosystem nor reflect the moral obligation many people feel towards the preservation of nature. In spite of its problems, the economic approach may help to persuade Governments of the need to fund conservation projects. It also helps explain why biodiversity is being lost; because the costs generated by its destruction are not incurred by those responsible for biodiversity loss.

Ecotourism

It is also possible to realize the **'attraction value'** of biodiversity through the income generated by 'ecotourists'. Ecotourists pay to experience the biodiversity of a country or national park and will pay more, and expect less by way of facilities, than conventional tourists. For example, the Monteverde cloud forest in Costa Rica receives 15 000 visitors per year who pay an entrance fee which contributes to the costs of running the reserve. At another park, Rara Avis, a combination of sustainable use – selective logging, farming pacas (small rodents) for food, rearing butterflies for export to collectors and growing medicinal plants – increases income and further reduces the dependency on outside funding.

Although ecotourism can undoubtedly help to meet the cost of conservation, it is not without its problems. **'Spearheading'**, the inadvertent opening-up and degradation of previously pristine areas, can occur (e.g. in Antarctica). The income generated does not always benefit the park or the local people but can 'leak' away to be absorbed by others. Finally, overexpansion is a real threat. In 1970, 12 000 visitors were permitted by the Ecuadorian government to visit the Galapagos Islands; in 1996 there were 80 000. The cause is almost certainly the $40 per person paid to a country short of other sources of dollar income.

Gene banks

Endangered biological resources should be preserved because of the benefits they may provide for future generations. Most conservation programmes aim to conserve species within their natural habitat (*in situ*). However, for some species this is not an option. The habitat has been destroyed or degraded, the risk from poaching is too high, or the number of individuals remaining is too few to sustain the species. These species, or failing that, their DNA which contains their genetic blueprint, must be conserved by *ex situ* methods in museums, herbaria or zoos if they are to be preserved at all. These collections of living material in the form of zoo animals, botanical collections and seeds, together with DNA collections have been termed **gene banks**. They provide the potential to breed the species in captivity and to release it when the conditions are more favorable, or to undertake research into their genetic properties and potential benefits to mankind, as crops, medicines and so on.

Seed banks

Kew Gardens in London has stored plant material since the 1850s and has developed a project to establish the **Millennium Seed Bank** which will store seeds from the semiarid tropics and subtropics and contribute to the cataloguing of global plant biodiversity. A US seed bank is focusing on the moist tropics. India also has a seed bank project to store the source of the original wild strains of food crops, such as rice, banana, beans and yam. These original varieties carry genetic information that could be very useful in the future development of new crop varieties. The ancestral genes could convey desirable characteristics such as resistance to new strains of disease or tolerance to novel growing conditions.

Seeds are cleaned and dehydrated prior to storage and then cooled to a temperature of –20°C. Cooling lowers the metabolic rate and allows the seeds to be stored for a longer time. The seeds do not stay viable indefinitely and must be periodically germinated in order to obtain fresh seeds. Species that do not produce seeds, have oily or hard seeds or seeds that do not tolerate heating and cooling are poor candidates for seed banks. These groups are rare but include commercially important species such as tea and coconut. Such plants should be given higher priority for *in situ* conservation. Methods for storing pollen and spores and for micropropagation of plant tissue are improving rapidly, adding to the range of techniques available for plant conservation.

Zoos – captive breeding

Ex situ conservation can prevent the immediate extinction of a species through captive breeding. As a last resort, all surviving individuals of a species can be brought into captivity. The Californian condor (*Gymnogyps californicanus)* and the black-footed ferret (*Mustela nigripes*) were saved from extinction in this way when their numbers fell to such a level (three individuals in the case of the condor) that they stood no realistic chance of survival *in situ*. Other examples of successful captive breeding include the golden lion tamarin (*Leontopithecus rosalia)* and the Arabian oryx (*Oryx leucoryx).*

Captive breeding programs aim to maintain maximal genetic variation in the threatened population. In order to make full use of the available gene pool, individuals of a species held in zoos around the world are managed as a single population. Information on births, deaths, mating and parentage is recorded in a studbook which is used to ensure that individuals whose genes are under-represented in the population are encouraged to breed. Ideally, captive numbers should be built up to the minimum viable population (see Topic V1) and the sex ratio should be maintained at 1:1. However, this can lead to very artificial conditions, for example, if the males of the species hold harems of females in the wild as is the case for Chapman's Zebra (*Equus burcelii chapmani*).

Unfortunately, by the time *ex situ* strategies are implemented the genetic variation in the natural population is often greatly reduced. As a result, most captive populations are based on a few founders and suffer from inbreeding depression (see Topic N2). Some captive populations have become 'contaminated' by the genes of domestic animals or through interbreeding among races that are distinct in the wild. The European bison carries genes from domestic cattle, while zoo populations of tigers (*Pantherus tigris*) and lions (*Pantherus leo*) are mixtures of subspecies. Modern breeding programmes recognize the importance of preserving the genetic integrity of the captive population.

The ultimate aim of captive breeding programmes is the reintroduction of animals into the wild. This will not succeed if the adverse factors originally responsible for the demise of the wild population (poaching for example) still prevail or if the habitat area is too small or of poor quality. Behavioral factors will also contribute to the success of reintroduction. For some species, such as primates, social interactions are very important to survival. The cohesiveness of social groups can be maintained by penning animals at the release site and only releasing when social hierarchies have stabilized. Release strategies need also to consider the timing of release to coincide with favorable environmental conditions.

The golden lion tamarin is a classic example demonstrating the pitfalls of reintroduction. In the early 1970s there were only a few hundred left in the wild in Brazil and their forest habitat was under threat. The management of the 80 animals held in captivity at that time was brought under the control of an international committee whose first aim was to retain genetic variation and increase captive numbers through breeding. In 1984, 14 captive-bred animals were released into the Poço des Antes Biological Reserve near Rio de Janeiro. Eleven died or were rescued suffering from disease, hunger, exposure or snakebite.

This failure highlighted the need to train the tamarins for release. Zoo-bred animals were taught how to forage effectively, to process whole fruits, live insects and leaves, to climb on flimsy branches and to drink water from bromeliads rather than from pools on the ground where they would be vulnerable to predators. They were also given a medical check to ensure that only the healthiest animals were released. Initially, after release, food was provided to allow acclimatization to the new environment. Released animals were also radio-tagged and monitored intensively to provide information on their behavior.

Between 1984 and 1991, a total of 91 golden lion tamarins were released into the wild; 33 survived and these went on to produce 38 surviving offspring. By 1991 the wild population had increased by 71 animals as a direct result of reintroduction. The cost per animal was $22 000. Although there have been some high profile cases, the successful release of captive-bred animals is very difficult and such programs are usually only undertaken as a last resort when *in situ* methods have failed.

W1 AIR, WATER AND SOIL POLLUTANTS

Key Notes

Air pollution	Air pollution is the transfer of harmful amounts of natural and synthetic materials into the atmosphere as a consequence of human activity. Pollutants can be added to the air directly (primary pollutants), or they can be created in the air (secondary pollutants) under the influence of solar radiation. The major air pollutants, which have documented environmental and health risks, include nitrogen oxides, sulfur dioxide, ozone, and particulates. Air pollution can also alter climates and the chemistry of soil, lakes and rivers.
Acid rain	Water bodies in northern temperate regions of Europe and North America have suffered from acidification due to 'acid rain'. Acid rain is a result of fossil fuel burning, which produces sulfur oxides (SO_x) and nitric oxide (NO) which may combine with atmospheric water to form sulfuric acid (H_2SO_4) and nitric acid (HNO_3), respectively. The term 'acid deposition' is more accurate as acid may also be deposited from the air in the form of snow, sleet and fog. Acid rain reduces the pH of soil and lakes, while acidification can also cause the death of trees and allow toxic metals (e.g. aluminium and mercury) to be leached from soils and sediments.
Water pollutants	Water pollution can be divided into one of four categories: (i) biological agents, (ii) dissolved chemicals, (iii) nondissolved chemicals, and (iv) heat. The eutrophication of aquatic ecosystems occurs due to an excess of inorganic nutrients. Organic matter in the water is broken down by microorganisms that deplete the oxygen levels, which may be quantified by the 'biochemical oxygen demand' (BOD). A particularly important class of organic water pollutants is the family of polychlorinated biphenyls (PCBs), a group of stable chlorinated compounds, that are highly toxic to vertebrates.
Soil pollution	A range of chemicals cause soil pollution problems, of which halogens (primarily solvents and pesticides) constitute the largest group. These chemicals are manufactured. The most complex group of compounds which are found polluting soils include polymers such as nylon, plastics and rubber. Bioremediation is a technique of utilizing microorganisms for the decontamination of polluted soils.
Related topics	The properties of water (D1) Greenhouse gases and global warming (W2) Soil formation, properties and classification (G3) Ozone depletion (W3)

Air pollution   **Air pollution** is the transfer of harmful amounts of natural and synthetic materials into the atmosphere as a direct or indirect consequence of human activity.

Air pollution is a complex problem because a pollutant can be any of a number of chemical substances existing in gaseous, liquid (aerosol) or solid form. Further, pollutants can be added to the air directly (**primary pollutants**), or they can be created in the air (**secondary pollutants**) from other pollutants under the influence of electromagnetic radiation from the sun. The major air pollutants are those produced in significant amounts and those having documented health and other environmental effects. The chemical composition and characteristics of some of the most important air pollutants are given in *Table 1*. While all these pollutants have effects, sources and control strategies in common, each is chemically unique.

Table 1. Molecular composition and characteristics of major air pollutants

Pollutant	Composition	Characteristics
Sulfur dioxide	SO_2	Colorless, heavy water-soluble gas with a pungent odor
Particulates	Variable	Solid particles or liquid droplets including fumes, smoke, dust and aerosols
Nitrogen dioxide	NO_2	Reddish brown gas, slightly water soluble
Hydrocarbons	variable	Many compounds of hydrogen and carbon
Carbon monoxide	CO	Colorless, odorless toxic gas, slightly water soluble
Ozone	O_3	Pale blue gas, water soluble, unstable, sweetish odor
Hydrogen sulfide	H_2S	Colorless gas with an offensive 'rotten egg' odor, slightly water soluble
Fluorides	Variable	Pungent, colorless, water-soluble gases
Nitric oxide	NO	Colorless gas, slightly water soluble
Lead	Pb	Metallic, can exist in a variety of chemical compounds with different characteristics
Mercury	Hg	Metallic, can exist in a variety of chemical compounds with different characteristics

Air pollutants have many different kinds of effects on humans and other parts of the natural world. Air pollutants can have a psychological effect; they can make the environment unpleasant, erode statues and damage property. Air pollution can also impair human health and the health of other organisms and can alter climates and the chemistry of soil, lakes and rivers. *Table 2* shows the proposed European Union limit values for some important air pollutants.

Acid rain

In the early 1970s it was noticed that lakes without any known source of acid (e.g. mine seepage) in Canada, the U.S. and Scandinavia were becoming increasingly acidic and that the fish populations of these lakes were being depleted. Acid from the sky was the only explanation, and indeed, monitoring demonstrated that the acidity of rainfall was well above the natural acidity of rain. This became known as **acid rain**. In fact, the term **acid deposition** is more accurate, as acid-forming materials may be deposited from the air in the form of snow, sleet, fog as well as in the form of rain. Acid rain should have been no surprise. For over a century we have been burning large quantities of oil and coal and smelting ore. Coal, and to a lesser extent oil contain sulfur. In the presence of oxygen and high combustion temperatures, sulfur compounds are oxidized to become sulfur oxides (SO_x). Sulfur dioxide is itself a poison, but it

Table 2. Proposed European Union limit values for some important air pollutants

	Averaging period	Limit value	Date by which limit value is to be met
Sulfur dioxide			
1. Hourly limit value for the protection of human health	1 hour	350 µg m^{-3} not to be exceeded more than 24 times per calendar year	1 January 2005
2. Daily limit value for the protection of human health	24 hours	125 µg m^{-3} not to be exceeded more than 3 times per calendar year	1 January 2005
3. Limit value for the protection of ecosystems, to apply away from the immediate vicinity of sources	calendar year and winter (1 October to 31 March)	20 µg m^{-3}	two years from coming into force of the directive
Nitrogen dioxide and nitric oxide			
1. Hourly limit value for the protection of human health	1 hour	200 µg m^{-3} NO$_2$ not to be exceeded more than 8 times per calendar year	1 January 2010
2. Annual limit value for the protection of human health	calendar year	40 µg m^{-3} NO$_2$	1 January 2010
3. Annual limit value for the protection of vegetation to apply away from the immediate vicinity of sources	calendar year	30 µg m^{-3} NO + NO$_2$	two years from coming into force of the directive
Lead			
Annual limit value for the protection of human health	calendar year	0.5 µg m^{-3}	1 January 2005
Particulate matter			
1. 24-hour limit value for the protection of human health	24 hours	50 µg m^{-3} PM$_{10}$ not to be exceeded more than 25 times per year	1 January 2005
2. Annual limit value for the protection of human health	calendar year	30 µg m^{-3} PM$_{10}$	1 January 2005

From O'Riordan (2000) Environmental Science for Environmental Management, 2nd Ed. ©Addison Wesley Longman 1994, 1999, reprinted by permission of Pearson Education Limited.

can also react with ozone, hydrogen peroxide and water vapor in the atmosphere to form sulfuric acid (H_2SO_4). Combustion at high temperatures in power plants and smelters also create oxides of nitrogen, mostly as atmospheric nitrogen combines with oxygen. Although nitric oxide (NO) is not very harmful as it does not readily dissolve, nitric oxide can combine with oxygen to form nitrogen dioxide:

$$2NO + O_2 \Rightarrow 2NO_2$$

Nitrogen dioxide is similar to sulfur dioxide; through various reactions with substances in the atmosphere, nitrogen dioxide is converted into nitric acid (HNO_3).

Acid deposition has recently begun to cause obvious damage to the animals and plants in susceptible waters and soils where rain falls most heavily. Intensification of the acid rain problem from the 1950s through the 1970s in the US was observed. In eastern Canada, acid rain has rendered hundreds of lakes fishfree. Crops and other plants have also been affected. Sweden is reported to have around 3000 dead lakes, lakes without fish or frogs and where algae are the only visible form of life. All living things have an optimal pH level and limits. Departure from near optimal pH means suboptimal reproduction, growth and survival. Acid rain changes the pH of soil and lakes while acidification can also cause toxic metals (e.g. aluminium and mercury) to be leached from soils and sediments. In addition, the detrimental effects of acid rain on forests was evident in Northern Europe in the 1980s.

Water pollutants   Because many kinds of chemicals dissolve in it, water is known as the universal solvent. This property of water makes its contamination inevitable in the technical sense. After water is purified by evaporation, condenses and begins to fall back to earth, it begins to pick up dissolved gases and particulates. Basically, **water pollution** can be defined as any human action that impairs the use of water as a resource. All water pollutants fall into one of four categories, biological agents, dissolved chemicals, nondissolved chemicals and heat.

Epidemics of waterborne diseases such as cholera and typhoid fever have occurred worldwide throughout history. In most cases, the problem of waterborne contamination can be stated as the contamination of drinking water with disease-causing organisms (**pathogens**) that have come from human wastewaters. However, microorganisms other than those found in human waste can also be a problem.

Eutrophication of aquatic ecosystems is caused by the oversupply of inorganic nutrients which fuel algal growth, resulting in algal blooms which cut out light to other plants, reduce oxygen levels and may be toxic to fish and other vertebrates. The key nutrients which may cause eutrophication are phosphates and nitrates. These substances can be added to aquatic ecosystems indirectly in the form of phosphorus and nitrate-containing organic matter, or they can be added as pollutants directly. Many detergents contain tripolyphosphates while as much as 25% of the nitrate and phosphate-containing fertilizers used in agriculture finds its way into water courses, contributing to eutrophication. Phosphate pollution is a particular problem as phosphorus is often the plant growth-limiting nutrient in aquatic communities. The addition of phosphate triggers an increase in plant growth. The subsequent decomposition of this organic matter leads to the problems of oxygen depletion.

The addition of organic matter provides energy and nutrients to the decomposers which then use up oxygen as they oxidize organic matter. **Biochemical oxygen demand** (BOD) is a quantitative expression of this oxygen-depleting impact. It is an expression of how much oxygen is needed for microorganisms to oxidize that organic matter. Organic matter will also undergo chemical oxidation and there is therefore also a **chemical oxygen demand**. In the extreme, large amounts of organic matter may result in the total depletion of oxygen. This would make it impossible for any species requiring oxygen to live. Fish and zooplankton will die. The only species to survive will be those bacteria that can live in the absence of oxygen, **anaerobes**. For most aquatic systems, dissolved oxygen should never be lower than 3 ppm at any time and should be above 5 ppm for most of the day. Trout require at least 5 ppm oxygen while certain scavenger fish like carp can survive in water containing only 1 ppm oxygen.

Many other chemicals that get added to water are poisons. Among the inorganic toxic chemicals found in water supplies are arsenic (which comes from insecticides), cadmium (from electroplating), cyanide and mercury. These metals interfere with many essential enzymes in man and other organisms. Metal toxicity problems are often compounded by the biological magnification of biometallic compounds in aquatic ecosystems. A particularly important class of organic water pollutants, one that has generated considerable discussions is the family of **polychlorinated biphenyls** (PCBs), a group of stable chlorinated compounds which are used in a variety of industrial processes. Although these compounds do not affect the BOD of water they are extremely toxic. The maximum PCB level accepted in fish to be eaten by humans in only 2 ppm. Among the things that affect water physically are undissolved solids. These compounds decrease water quality by clogging waterways and making water cloudy. Such solids can also cause physical problems for gill breathers (fish). Also, by adsorption, suspended solids can concentrate metals and toxins.

Water has a high heat capacity, and for this reason many industrial processes are located on rivers, where water can be used to remove waste heat. Heat affects life in water in a variety of ways. Firstly, every organism has a temperature tolerance range; at some point in the life cycle of every organism there is a most-temperature-sensitive stage (e.g. hatching eggs). For example, trout eggs take 165 days to hatch in cool water (10°C), but will hatch within 1 month if kept at 12°C. They will not hatch at all if temperatures reach 15°C. Changing water temperatures can therefore result in changed species composition. Heat also disrupts and changes the chemistry of the abiotic environment. Heat increases the solubility of certain chemicals while generally decreasing the solubility of gases. Therefore, at higher temperatures, oxygen solubility decreases. Also at higher temperatures, metabolic activity increases. This would increase decomposition processes and speed up oxygen utilization.

Soil pollution

A range of chemicals can be found polluting soils. One-carbon (C1) materials include carbon monoxide, cyanides and halogenated methanes. Numerous microbial species present in the soil use these compounds. Aliphatic hydrocarbons from crude oils and fuels are major environmental pollutants in soils. Short-chain, saturated, unbranched compounds are most readily degraded: methylation, as well as long chains and branching results in lower breakdown rates. Compounds with acrylic and aromatic rings are constituents of petroleum products, pesticides and wood preservatives as well as being found naturally in wood as lignin and in waxes. Halogens constitute the largest group of environmental priority soil pollutants. They have been manufactured primarily as solvents and pesticides. They include pentachlorophenol, a wood preservative and polychlorinated biphenyls. Pesticides in this class include ethylene dibromide, heptachlor, alochlor and lindane.

Nitrogen-containing pollutants include the azo and aniline dyes, aniline herbicides (propanil), and explosives (TNT). These, as well as the sulfur-containing compounds in surfactants and detergents, are generally considered to be biodegradable. The most complex group of compounds which are found polluting soils include polymers such as nylon, plastics and rubber. These compounds contribute much to present pollution. Soil microorganisms have long been the Earth's garbage control agents. The use of microorganisms for the removal of pollutants is termed **bioremediation.**

W2 GREENHOUSE GASES AND GLOBAL WARMING

Key Notes

Carbon dioxide concentrations in the atmosphere	Carbon dioxide (CO_2) is the main vehicle of carbon flux between atmosphere, oceans and biota. Fossil fuels (coal, oil and natural gas) present in the lithosphere lay dormant until recent centuries. The concentration of CO_2 in the atmosphere has increased from about 280 parts per million (ppm) in 1750 to about 350 ppm in 1990 and is still rising, with the main reason for the increase being the combustion of fossil fuels. Estimates of future CO_2 emissions, and the concentration to be expected in the atmosphere vary, but it appears likely that the concentration will rise to a mean of about 550 ppm by the year 2050.
The greenhouse effect	The 'greenhouse effect' is a theory which proposes that pollution by common anthropogenic (i.e. created by humans) pollutants such as CO_2 and methane may lead to an increased global temperature. Over the last century, CO_2 concentrations have risen and global air temperatures have increased by 0.4–0.7°C, which supports this theory. The atmosphere formed around the Earth insulates the planet from the full effects of heat loss by trapping heat in the atmosphere using greenhouse gases, which include water vapor as well as CO_2 and other anthropogenic pollutants. Doubling of the atmospheric CO_2 concentration from its present level is predicted to lead to a further warming of around 3.5°C.
Related topics	Solar radiation and climate (C1) Ozone depletion (W3) Air, water and soil pollutants (W1)

Carbon dioxide concentrations in the atmosphere

Photosynthesis and respiration are the two opposing processes driving the global carbon cycle. It is predominantly a gaseous cycle, with CO_2 as the main vehicle of flux between atmosphere, oceans and biota. Historically, the lithosphere played only a minor role; **fossil fuels** (coal, oil and natural gas) lay dormant until recent centuries. The concentration of CO_2 in the atmosphere has increased from about 280 (ppm) in 1750 to about 350 ppm in 1990 and is still rising. Since 1958 the atmospheric CO_2 concentration has risen from 315 ppm to 350 ppm in 1990 (*Fig. 1*). The principal cause of the increase in atmospheric CO_2 in recent years has been the combustion of fossil fuels, which in 1990 released about 5.5×10^9 tons of carbon into the atmosphere.

The exploitation of rainforest has also caused a significant release of CO_2. The burning that follows most forest clearance quickly converts some of the vegetation to CO_2, while decomposition of the remaining vegetation releases CO_2 over a longer period. If forests have been cleared to provide for permanent agriculture, the carbon content of the soil is reduced through decomposition of

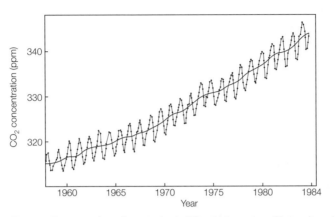

Fig. 1. Concentration of atmospheric CO_2 at Mauna Loa Observatory, Hawaii, showing seasonal cycle and the long term increase in atmospheric CO_2 concentration. Reprinted from Ecology, Second Edition, King, 1989, with permission from Thomas Nelson & Sons Ltd.

soil organic matter and/or by erosion. In total it is estimated that about 1.0×10^9 tons year^{-1} of carbon are released through changes in tropical land use. The total amount of carbon released each year to the atmosphere as a result of man's activities around 6.5×10^9 tons compared with 100×10^9 tons released into the atmosphere naturally by respiration of the world's biota. The observed increase in atmospheric CO_2 is 2.9×10^9 tons year^{-1}. Much of the remainder of the additional CO_2 is removed by dissolution in the oceans. It is possible that terrestrial communities act as a net sink for atmospheric CO_2, and terrestrial ecosystems may be able to assimilate the extra carbon as biomass; it is already known that increasing the atmospheric concentration of CO_2 in controlled environments increases the rate of photosynthesis and growth of many plant species. Estimates of future CO_2 emissions and the concentration to be expected in the atmosphere vary, but it appears likely that the concentration will rise to a mean of about 550 ppm by the year 2050.

The greenhouse effect

Ten out of the 16 years between 1980 and 1995 were among the hottest on record; accurate records have been kept since 1881. Global climate change resulting from human activity is one of the most contentious topics in ecology. The major source of generated power comes from the burning of fossil fuels. Fossil fuels are organic in nature and when they are burned, CO_2 is released into the atmosphere. The fear is that pollution by common **anthropogenic** (created by humans) pollutants, such as CO_2 and methane, may lead to increased global temperature. Accurate measurements of CO_2 concentrations in air have been recorded for the last 116 years, and during this time period increases in CO_2 levels have been recorded along with an 0.4–0.7°C rise in global temperature. Further, historical measurements of global ice volumes and gaseous compositions have shown a correlation between global CO_2 concentrations and temperature (Fig. 2). This close correlation supports the argument that industrial emissions, in particular gases released by the burning of fossil fuels, will lead to further climate change over the next century.

On the basis of the distance of the earth from the sun, global temperatures should be 33°C lower than they are, with an average temperature of –18°C. However, from the time of planetary fusion, the earth has been radiating heat.

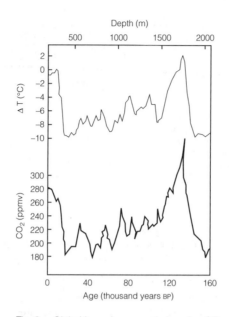

Fig. 2.   *Global ice volumes and changing CO_2 concentrations over the past 160 000 years.*

The atmosphere formed around the earth insulates the planet from the full effects of heat loss. This atmosphere allows much of the incoming radiation to strike the earth. A large proportion of this light energy is transferred into heat energy, and the heat energy held by the earth during the day is irradiated back at night. Much of the energy transformed to heat is trapped by atmospheric gases, particularly water vapor, CO_2, O_2, methane and nitrous oxides. The warming of the Earth attributable to atmospheric gases is termed the **greenhouse effect**. The natural thermal insulation of the greenhouse gases raises the average global temperature from –18°C to 15°C. The greenhouse effect is thought to be increased as a result of perturbations by human activity of the five gases, and have been added to by a group of synthetic chemicals, notably chlorofluorocarbons, hydrofluorocarbons, perfluorocarbons and sulfurhexafluoride. *Table 1* lists the evolution of the main greenhouse gases over the past 200

Table 1.   *The key greenhouse gases and their change over time*

	CO_2	CH_4	N_2O	CFC-11	HCFC-22 (a CFC substitute)	CF_4 (a perfluoro-carbon)
Pre-industrial concentration	280 ppm	700 ppb	275 ppb	zero	zero	zero
Concentration in 1994	358 ppm	1720 ppb	312 ppb	268 ppt	110 ppt	72 ppt
Annual concentration change	+1.5 ppm	10 ppb	+0.8 ppb	zero	+5 ppt	+1.2 ppt
Annual rate of change	+0.4%	+0.6%	+0.25%	zero	+5%	+2%
Atmospheric lifetime (years)	50–200	12	120	50	12	50 000

ppm = parts per million (10^6), ppb = parts per billion (10^9), ppt = parts per trillion (10^{12}). *From O'Riordan (2000)*
Environmental Science for Environmental Management, 2nd Ed. ©Addison Wesley Longman 1994, 1999, reprinted by permission of Pearson Education Limited.

Table 2. Global warming potential of selected greenhouse gases

Gas	Chemical formula	Lifetime (years)	Gobal warming potential (years)		
			20	100	500
Carbon dioxide	CO_2	140	1	1	1
Methane	CH_4	12	56	21	6.5
Nitrous oxide	N_2O	120	280	310	170
HFC group	C-H-F	1.5–50	5000	3000	500
Sulphur hexafluoride	SF_6	3200	16 300	24 900	35 900
Perfluorocarbons[1]	C-F	3–10 000	6000	8000	14 000

From O'Riordan (2000) Environmental Science for Environmental Management, 2nd Ed. ©Addison Wesley Longman 1994, 1999, reprinted by permission of Pearson Education Limited.

years and *Table 2* summarizes the relative global warming potential of the main greenhouse gases. Doubling of the atmospheric CO_2 concentration from its present level is predicted to lead to a further warming of around 3.5°C. The effects of this would be profound, as polar ice would melt and the oceans would expand with heat, raising sea levels and resulting in large-scale changes to the global climate. In addition, global warming may cause organisms to migrate in order to seek optimal temperatures (organism tracking), or to adapt to the changed conditions or face extinction. This movement of organisms may include tropical diseases, which may as a result of the climate change move from tropical to temperate climates (e.g. malaria).

W3 OZONE

Key Notes

What is ozone	Ozone is a highly reactive oxygen molecule containing three oxygen atoms, O_3. There is a naturally occurring high altitude (stratospheric) layer of ozone which is important to life as it absorbs potentially damaging ultraviolet radiation. Low-altitude (tropospheric) ozone is toxic, and is produced by photochemical smog resulting from fossil fuel emissions.
The importance of the ozone shield	Prior to the development of the stratospheric ozone layer which forms the 'ozone shield', the evolution of terrestrial life was inhibited. DNA efficiently absorbs UV light, which seriously disrupts DNA replication, causing reproductive failure and death. Relatively small increases in UV radiation can cause mutations during the replication process that may result in the production of cancerous cells. A major concern is the potential damage caused by UV light to plants, which could reduce primary productivity, and therefore affect whole ecosystems.
Stratospheric ozone depletion	Chlorofluorocarbons (CFCs) can degrade ozone in the stratosphere. Attention was focused on this problem when the British Antarctic Survey demonstrated a strong depletion of the Antarctic ozone shield. It is estimated that a single chlorine atom can breakdown 100 000 ozone molecules. If the trend of ozone loss continues it is predicted that the ozone shield will be depleted by a further 10% by 2050; this may lead to an additional 300 million cases of skin cancer.
Related topics	Solar radiation and climate (C1) Air, water and soil pollutants (W1)

What is ozone

Ozone (O_3) is a molecule that contains three atoms of oxygen, with the atoms held together in their configuration by electrical attraction. This electrovalent bond is much weaker than the covalent bond that holds oxygen (O_2) so the molecule is always susceptible to having one of its oxygen atoms stripped by an 'oxygen-seeking' molecule. The reactivity of the chemical is in large part determined by the strength of the bonds between atoms. The weak bonds in the ozone molecule make it an even more highly reactive gas than oxygen.

What allowed life to leave the oceans and become terrestrial was not the presence of oxygen, but the presence of **stratospheric ozone**. Many organisms can live without oxygen, so the biological significance of developing an oxygen-rich atmosphere lay not in the oxygen itself, but in a secondary chemical reaction that could take place only in the presence of oxygen. High in the atmosphere, oxygen molecules were transformed by ultraviolet light from their normal configuration of O_2 to ozone. The major subdivisions of the atmosphere and the location of the ozone layer are shown in *Fig. 1*. The oxygen molecules high in the stratosphere are broken down through dissociation, resulting in the absorption of UV light in the wavelengths 180–240 nm:

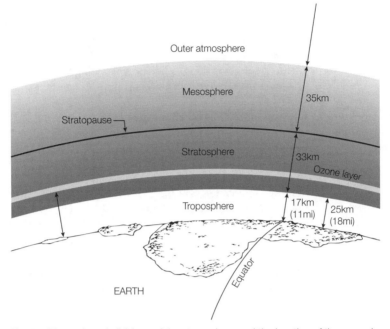

Fig. 1.   The major subdivisions of the atmosphere and the location of the ozone layer. From Environmental Science, 3rd edn, Kupchella and Hyland, 1993. Reprinted with permission from Pearson Education Inc., Upper Saddle River, NJ.

$$O_2 + \text{ultraviolet light} \Rightarrow O + O$$

The free oxygen then joins an O_2 molecule to form ozone:

$$O + O_2 \Rightarrow O_3$$

As the oxygen concentration in the atmosphere remains constant at 21%, it is likely that natural degradation of ozone must occur to maintain the ozone:oxygen equilibrium. Small amounts of ozone are broken down in the troposphere, but much more is broken down to oxygen as a result of another photochemical reaction involving UV radiation in the spectrum 200–320 nm:

$$O_3 + \text{ultraviolet light} \Rightarrow O_2 + O$$

The singlet oxygen can then pull an oxygen free from an ozone molecule to form two oxygen molecules:

$$O + O_3 \Rightarrow O_2 + O_2$$

These reactions of ozone absorb incoming UV radiation. Ultraviolet light arriving in the atmosphere may be divided into three categories according to wavelength. **UV-A** has the longest wavelength, and **UV-C** the shortest. Ozone absorbs more than 99% of the UV-C wavelengths, about half of the lower energy **UV-B** and little of the relatively harmless UV-A. If the equilibrium of the reactions that absorb UV radiation is perturbed, a change in the influx of ultraviolet radiation to the surface of the earth will occur.

Ozone also occurs at low altitudes, in the troposphere. Naturally, **tropospheric ozone** levels are low. However, ozone is produced by **photochemical smog** which is a result of the action of sunlight on fossil fuel pollutants, the nitrogen oxides (NO_x). Because ozone is highly reactive, it causes cell damage to living organisms and is thus a general toxin. Large cities with low rainfall and protected from the wind, such as Mexico City and Los Angeles, suffer acutely from photochemical smog, and in these places ozone pollution may be a health risk.

Thus, high-level stratospheric ozone is highly beneficial to terrestrial life, but low-level tropospheric ozone is an undesirable toxin.

The importance of the ozone shield

Life was able to evolve in water before there was an **ozone shield** because water reflects ultraviolet light. However, no terrestrial life could occur because DNA absorbs UV light, especially between 280 nm and 320 nm wavelengths. UV radiation disrupts DNA replication to the point of reproductive failure or death. Consequently, an ozone shield was a prerequisite for terrestrial life. Relatively small increases in UV radiation can cause mutations during the replication process that may result in the production of cancerous cells. Because the cells most exposed to UV radiation are the skin cells, exposure to UV radiation is associated with skin cancer and also eye cataracts. As well as humans, other mammals, birds and lizards can also be affected. However, perhaps the major concern is the potential damage caused by UV light to photosynthetic systems, as this may reduce primary productivity, and therefore affect entire ecosystems.

Stratospheric ozone depletion

Chlorofluorocarbons (CFCs) have been widely used as supercoolants (e.g. Freon), as well as in aerosols, and insecticides. Research has shown that CFCs can rise into the stratosphere where they may degrade ozone. Attention was focused on this problem when the British Antarctic Survey demonstrated a strong depletion of the Antarctic ozone shield. The South Pole receives the least radiant energy of any place on earth and in the winter, the water in the stratosphere above Antarctica exists as ice crystals. CFCs and ozone molecules are held on the surface of the ice crystals. With the arrival of Spring (September and October) in Antarctica, the ice clouds return to water vapor and the CFCs and ozone are released, leading to the sudden degradation of ozone by CFCs. Although CFC starts as a stable configuration, in the presence of ultraviolet light the bonding of the molecule is altered leaving one of the chlorine atoms only loosely attached. As the modified CFC encounters an ozone molecule, a chlorine atom breaks off and strips an oxygen atom from ozone:

$$Cl^- + O_3 \Rightarrow ClO + O_2$$

The chlorine monoxide that results is still highly active and as it encounters another ozone molecule, it will strip one of the oxygen molecules:

$$ClO + O_3 \Rightarrow ClO_2 + O_2$$

The chlorine dioxide is then broken down by ultraviolet light to leave a free chlorine atom and an oxygen molecule:

$$ClO_2 + \text{ultraviolet light} \Rightarrow Cl^- + O_2$$

It is estimated that in this way, a single chlorine atom can break down 100 000 ozone molecules. In October 1993, the Antarctic ozone shield was thinned by about 70% of its normal value. The concentration of ozone at the South Pole fell to 96 Dobson units (DU), while ozone concentrations in the midlatitudes were around 400 DU. However, the Antarctic hole was relatively short-lived, and had disappeared by the summer. Changes in the polar wind pattern caused the hole to break up. In addition, a smaller thinning of the ozone layer was discovered at the Arctic during April 1994. A 10–20% ozone depletion was seen with levels down to 240 DU in 1993. Globally it appears that about 4–5% of the Earth's ozone has been degraded since the early 1980s. If the trend of ozone loss continues it is predicted that the ozone shield will be depleted by a further 10% by 2050. It has been estimated that this will lead to an additional 300 million cases of skin cancer.

X1 SOIL EROSION AND AGRICULTURE

Key Notes

Overview	Agricultural practices have a considerable impact on many ecosystems globally, and give rise to a variety of ecological problems. Four key areas in which agriculture impacts upon ecology are (i) the consequences of pest control (Section U), (ii) reduction in biodiversity (Topic V3), (iii) soil erosion, and (iv) the effects of nutrient, water and energy use. These last two subjects are covered in this section.
The causes of soil erosion	Soil erosion and the loss of soil fertility as a result of poor farming practices are serious problems worldwide. The removal of vegetation cover from soils, the use of large fields without boundaries to slow water movement and inappropriate ploughing techniques all fuel soil erosion.
The costs of soil erosion	Most modern agricultural systems are based on short-term economic gain and as soil erosion usually occurs very gradually it may not appear serious over the short term. Erosion reduces productivity of the land by depleting it of organic matter and nutrients, reducing its water-holding capacity and limiting rooting depth. As much as 4 billion tons of top soil, valued at £10 billion is lost annually from wind and water erosion. As much as 30% of the world's farmland has been lost to soil erosion in the past 50 years, whilst food demand increases due to human population growth. Soil erosion not only reduces the biological productivity of soils, but also may cause flooding.
Soil conservation practices	A number of methods can be employed to reduce soil erosion. By ploughing a field at right angles (contour ploughing) to the slope, furrows follow the contours of the land rather than the slope. Also, planting crops on areas of bare field (cover crops) helps to prevent soil erosion. No-till farming consists of planting a narrow slit trench without ploughing the soil. These systems, along with crop rotation, can all be used to reduce soil loss and maintain fertility. However, progress has been slow in establishing these practices.

Related topics	Soil formation, properties and classification (G3)	Pesticides and problems (U2)
	Primary and secondary production (P2)	Biological resources and gene banks (V3)
	Food chains (P3)	Nutrient, water and energy use (X2)

Overview

Few ecosystems globally escape the impact of human activity, and one pervasive influence is agricultural activity, which necessarily involves gross changes to the local ecosystem, and may have more wide-reaching effects. The five key areas in which agriculture impacts upon ecology are (i) the consequences of pest control (Section U), (ii) reduction in biodiversity (Topic V3), (iii) soil erosion, and (iv) nutrient, water and energy budgets. These latter two subjects are covered in this section.

The causes of soil erosion

Good soil produces good vegetation which provides food and habitat for animals. Ninety percent of all human food is produced on land, from soils varying widely in quality, nature and extent. As an essential component of terrestrial ecosystems, soil sustains the primary producers (living vegetation) and the decomposers (microorganisms, herbivores, carnivores). It also provides the major sinks for heat energy, nutrients, water and gases. A major concern of good land management is keeping soil in place and maintaining its fertility. Once lost soil is irreplaceable – 1 cm of soil may take 500 years or more to form, yet can be lost within a year. **Soil erosion** and the loss of soil fertility as a result of poor farming practices are serious problems worldwide. *Table 1* highlights the major processes of soil degradation. Indigenous systems are insufficient alone to prevent agricultural land from continuing to lose productive soil, water and nutrient resources. This is partly because not all farmland is protected by conservation measures, but also because not all erosion arises from farmland. Both roads and urban areas concentrate water flows and nonagricultural areas are also subject to erosion.

Farmers may not be conserving soil and water due to lack of local knowledge or skills, or they may be unwilling to invest in conservation measures if the economic costs are greater than the benefits. In the UK, a major cause of soil erosion has been the shift in recent years towards the cultivation of winter cereals. The high price of wheat has encouraged winter cultivation in fragile environments, which has led to a large increase in soil erosion. Soil erosion can be in the order of 30–100 tons per hectare in fields where hedges have been removed. Erosion is greatest where there is little vegetation cover, such as during winter, and when slopes are long, such as in big fields. To farmers, erosion reduces the biological productivity of soils and the capacity to sustain productivity in the future. Soil erosion may also cause widespread flooding as soils become less able to retain water.

Table 1. Process of soil degradation

- *Water erosion.* Splash, sheet and gully erosion, as well as mass movements such as landslides.

- *Wind erosion.* The removal and deposition of soil by wind.

- *Excess of salts.* Processes of the accumulation of salt in the soil solution (salinization) and of the increase of exchangeable sodium on the cation exchange of soil colloids (sodication or alkalinization).

- *Chemical degradation.* A variety of processes related to leaching of bases and essential nutrients and the build-up of toxic elements; pH-related problems such as aluminum toxicity and P-fixation are also included.

- *Physial degradation.* An adverse change in properties such as porosity, permeability, bulk density and structural stability; often related to a decrease in filtration capacity and plant-water deficiency.

- *Biological degradation.* Increase in rate of mineralization of humus without replenishment of organic matter.

**The costs of
soil erosion**

Most agricultural systems are based on short-term economic gain and as soil erosion occurs so gradually that it is almost imperceptible it may not, in the short-term, appear serious. Fifteen tons of soil lost from a hectare of land in a single storm will diminish soil depth by 1 mm. However, almost 1000 tons of soil are ejected from the Mississippi River into the Gulf of Mexico every minute. According to the US Department of Agriculture, soil loss from erosion threatens the productivity of one-third of croplands in the US. *Figure 1* compares the rates of soil erosion from two valleys in New Hampshire, USA: an undisturbed soil and forest-cleared soil. Soil erosion is a major problem globally, in the prairie and steppe soils of North America and Eurasia, and in arid and fragile soils across South America, Africa and Asia. This erosion affects more than just soil depth. Erosion reduces productivity of the land by depleting it of organic matter and nutrients, reducing its water-holding capacity, and limiting rooting depth. As much as 4 billion tons of top soil are lost annually from wind and water erosion. This translates into 8.2 million tons of nitrogen, 0.6 million tons of phosphorus and 2 million tons of potassium lost annually from US cropland. The value of this nutrient loss is put at around $15 billion annually. As much as 30% of the world's farmland has been lost to soil erosion in the past 50 years. At the same time, global food demand inexorably increases due to the growth of the human population.

Eroded soil is not simply a loss of a valuable resource, it can cause air, water and land pollution. The great soil erosion event of the south and southwest USA in the 1930s lead to the area being called the '**Dust Bowl**' from the level of air pollution.

Certain agricultural practices also **diminish soil fertility**. Soil that is continually cropped or grazed will eventually lose its fertility because the normal cycle that returns mineral-rich dead plant matter to the soil is disrupted. Agricultural practices also degrade soil in other ways. **Soil compaction** that results from the use of heavy machinery can destroy the structure of soil, restrict water infiltration and lead to increased erosion.

The problem of soil erosion is greatest in the Third World where extensive areas of land are cleared for food production. Soil erosion from forest removal and cultivation on slopes are major problems both in Latin America and Africa. Many problems in Third World countries make good land management diffi-

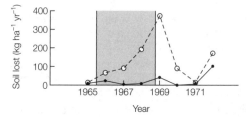

Fig. 1. A comparison of soil erosion from two valleys in Hubbard Brook Forest, New Hampshire, U.S.A. ●, *undisturbed control forest;* ○, *forest clear-felled in 1965. Redrawn from Bormann and Likens,* Pattern and Processes in a Forest Ecosystem, *1979, Springer-Verlag New York.*

cult. Poor workers must farm poor soils and crop intensively. Population growth puts stress on the need for increased agricultural production as well as the clearing of land for cropping and firewood supplies. However, despite these problems, some efforts to improve erosion control have been successful. In Ethiopia, walls of rock and earth are built across hillsides to catch eroding soil. This forms natural terraces to help reduce further erosion. In Australia, widespread use of fencerows and hedgerows as physical barriers helps prevent wind erosion of agricultural fields.

Soil conservation practices

It has been known for some time that a number of methods can be employed to reduce soil erosion. **Soil conservation** is defined as any set of measures which control or prevent soil erosion, or maintain soil conservation. **Water conservation**, especially in drier zones, is closely related to soil conservation, so that measures to conserve water act through increasing plant-available water and maintaining soil fertility. Good land management practice has been shown to reduce soil erosion and maintain crop productivity (*Fig. 2*). By ploughing a field at right angles to the slope, furrows follow the contours of the land rather than the slope. By using this **contour ploughing** method, water erosion is reduced. Also, by planting crops on areas of bare field helps to prevent soil erosion. If leguminous plants are used as **cover crops**, nitrogen will be fixed and the nitrogen content of the soil increased. **No-till farming** is a system that consists of planting a narrow slit trench without ploughing the soil. By reducing soil disturbance, soil erosion is reduced. These systems, along with crop rotation can all be used to reduce soil loss and fertility. *Table 2* provides more details on the modern approaches to soil and water conservation. However, although these techniques are widely known, progress has been slow in establishing these practices. In addition, no-till farming often uses herbicides, invoking other problems.

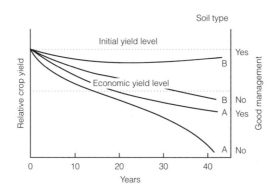

Fig. 2. Soil productivity (crop yield) in relation to different soils (A and B), level of management and soil degradation over 40 years. Reprinted from O'Riordan et al ©Addison Wesley Longman 1994, 1999, reprinted by permission of Pearson Education Limited.

Table 2. Modern techniques of soil and water conservation

Technique	Description	Resource implications
Tillage practices		
Strip tillage	Conditioning soil along narrow strips in or adjacent to seed rows, leaving rest of soil undisturbed	Special equipment; possible crusting/ infiltration problems in untilled parts
Basin listing, or tied ridging	Formation of contour bunds (earth banks) with constructed 'ties' or banks between bunds. Listing done by machinery and tied ridging usually by hand	Need for dedicated machinery and/or considerable extra labor. Danger of waterlogging
Conservation tillage	Technique of light harrowing and retention of crop residues at surface	Special machinery. Crop residues not available for other uses, e.g. forage, fuel
Minimum/zero tillage	Use of herbicides, then direct drilling into residues. Very little disturbance to the soil	Expensive chemicals and machinery. Danger of pollution and soil compaction. Saving on conventional tillage costs
Land formation techniques		
Contour bunds	Earth banks up to 2 m wide across slope to form barrier to runoff and break slope into shorter segments. Varieties: narrow or wide-based; contour or graded	Additional labor and/or equipment. Varieties have different costs, e.g. narrow-based loses about 14% of plantable area
Terraces	Earth embankments and major reformation of surface. Three main types: diversion, retention and terrace	Large labor and equipment requirement. Continual maintenance
Terracettes	Small constructions usually to harvest water. Known by many local names, e.g. fish-scale terraces in China for rows of perennial crops; eyebrow terraces for tree planting in semi-arid zone	Labor and continuing maintenance requirement
Stabilization structures		
Gabions	Stone and rock-filled bolsters to protect vulnerable surfaces, e.g. bridges, culverts	High cost, transport of stone, continual maintenance
Gully control dams	Usually constructed of brushwood across a gully	Materials, labor and maintenance. Often fail in heavy storms

From O'Riordan (2000) Environmental Science for Environmental Management, *2nd Ed. ©Addison Wesley Longman 1994, 1999, reprinted by permission of Pearson Education Limited.*

X2 NUTRIENT, WATER AND ENERGY USE

Key Notes

Increasing agricultural productivity	The direct application of inorganic nutrients to crops or soils is a simple route to increase crop yield. Average yield per hectare has increased from 1.1 tons in 1950 to 2.3 tons in 1986. To maintain 1987 consumption levels in the year 2000 will require at least a 25% increase in average grain yields. However, there are diminishing returns of fertilizer application; one ton of fertilizer added to US soil 20 years ago increased the world grain harvest by 15 to 20 tons; today the same amount of fertilizer would result in only a 6- to 10-ton increase.
Effects of fertilizer use on the phosphorus cycle	More than 13×10^6 tons of phosphorus are dispersed annually over agricultural land as fertilizer and a further 2×10^6 tons as an additive to domestic detergents. In many lakes worldwide, the input of large quantities of phosphorus (and nitrogen) from agricultural run-off produce ideal conditions for high phytoplankton activity. In such cases of eutrophication (enrichment), the lake water becomes turbid because of dense populations of phytoplankton (often blue-green species) and large water plants are outcompeted. In addition, decomposition of the large phytoplankton biomass may lead to low oxygen concentrations which kill fish and invertebrates.
The effect of agriculture on the nitrogen cycle	Deforestation, and land clearance in general leads to substantial increases in nitrogen flux in streamflow and N_2O losses to the atmosphere. The agricultural practice of planting legume crops, with their root nodules containing nitrogen-fixing bacteria, contributes further to nitrogen fixation. The production of nitrogenous fertilizers (more than 50×10^6 tons year^{-1}) is of particular significance because an appreciable proportion of fertilizer added to land finds its way into streams and lakes, leading to eutrophication.
Strategies to limit nutrient loss	The practices of both agriculture and forestry necessarily involve the removal of biomass and with it, the removal of minerals. This loss of nitrogen causes two problems. Firstly, the farmer needs to supplement the crops with manure to ensure high productivity. Secondly, the nitrogen leached from land creates problems elsewhere by contributing to the eutrophication of lakes and finding its way into drinking water. The management of livestock on permanent grasslands (a managed animal–plant community) leads to the inefficient cycling of nitrogen.
Desertification	One way to increase food production is by irrigation of soil creating new areas for agriculture. However, irrigation often leads to a decline in water availability. Another problem is irrigation of dry areas where evaporation is high results in soils accumulating salt (salinization). The process of improper management of arid or semiarid land to the extent that it is no longer suitable

for range or cropland is called desertification. Actual global losses to desertification are estimated to be around six million hectares annually. The process of desertification can also occur naturally.

Meat versus grain for human consumption

The second law of thermodynamics tells us that far more people can be supported on grain than on the meat of grain-eating animals. Approximately 40% of the grain consumed in the world is consumed by livestock. Efficiencies are such that 5 kg of vegetable protein suitable for human consumption, but fed to livestock instead, results in only 1 kg of livestock protein for human consumption. In addition, vast quantities of fish protein are used to feed animal stock, which is a very inefficient use of resource.

A new role for traditional agriculture

Throughout the world the principles of sustainable agriculture are being studied. Traditional agriculture practices are characterized by few external inputs (e.g. chemicals, irrigation), the effective accumulation and cycling of nutrients, diversity in cropping and protection of the soil. Legume-based crop rotations are common in traditional agriculture to build up nitrogen reserves in the soil. Shifting cultivation, a practice by which an area is cleared and farmed and then, as yields decrease, a new area is cleared, is another traditional method of agriculture commonly practiced. Without question, the enhancement of these traditional methods holds the key to widespread adoption of sustainable and higher-yielding techniques.

Related topics

Sources and cycles (G1)   Food chains (P3)
Plants and consumers (G2)   Soil erosion and agriculture (X1)
Components and processes (P1)
Primary and secondary production (P2)

Increasing agricultural productivity

Whereas one hectare of land supported 2.6 people in 1970, a hectare will have to support four people in the year 2000. One way to increase yield is to use **fertilizer**. *Figure 1* shows how fertilizer input per capita increased, while grain area per capita decreased between 1950 and 1986. During this time average

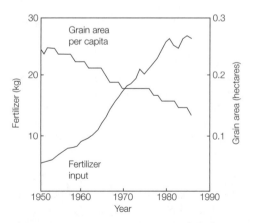

Fig. 1.   Illustration of how fertilizer input per capita increased as grain area per capita decreased between 1950 and 1986. From Environmental Science, 3rd edn, Kupchella and Hyland, 1993. Reprinted with permission from Pearson Education Inc., Upper Saddle River,

yield per hectare increased from 1.1 tons in 1950 to 2.3 tons in 1986. This was as a result of increased irrigation and improved crop varieties, as well as increased fertilizer use. To maintain 1987 consumption levels in the year 2000, with a projected population of six billion, will require at least a 25% increase in average grain yields. Also, production is maintained by the use of petro-chemicals to run tractors and other machinery. However, eventually the law of diminishing returns will prevail. The annual increase in the use of fertilizer has dropped worldwide from 6% to 3%. A ton of fertilizer added to US soil 20 years ago increased the world grain harvest by 15 to 20 tons; today the same amount of fertilizer would result in only a 6- to 10-ton increase. In addition, the continuing use of fertilizers causes its own environmental problems.

Effects of fertilizer use on the phosphorus cycle

More than 13×10^6 tons of phosphorus are dispersed annually over agricultural land as fertilizer and a further 2×10^6 tons as an additive to domestic detergents. Much of the former reaches the aquatic system as agricultural run-off, whereas the latter arrives as domestic sewage. Phosphorus run-off is currently a major problem in the Chesapeake Bay area of the eastern USA. In addition, deforestation and many forms of land cultivation increase erosion in catchment areas and contribute to artificially high amounts of phosphorus in run-off water. In all, human activities account for about two thirds of the annual river outflow of phosphorus to the oceans.

An increase in the phosphorus input to the oceans on this scale is insignificant, but as the more concentrated water passes through rivers, estuaries and particularly lakes, its influence can be profound because phosphorus limits aquatic plant growth. In many lakes worldwide, the input of large quantities of phosphorus (and nitrogen) from agricultural run-off and sewage produces ideal conditions for high phytoplankton activity. In such cases of **eutrophication** (enrichment), the lake water becomes turbid because of dense populations of phytoplankton (often blue-green species) and large water plants are outcompeted and disappear along with their associated invertebrate populations. In addition, decomposition of the large phytoplankton biomass may lead to low oxygen concentrations which kill fish and invertebrates. The outcome is a productive community, but with low biotic diversity. The solution to this problem is to reduce nutrient input by diverting sewage, changing agricultural practices or by chemically stripping phosphorus from treated sewage before it is discharged. Where phosphate loading has been reduced in deep lakes, a reversal of the above trends may occur. However, in shallow lakes phosphorus stored in the sediment may continue to be released and physical removal of sediment may be required.

The effect of agriculture on the nitrogen cycle

Deforestation, and land clearance in general lead to substantial increases in **nitrogen flux** in streamflow and N_2O losses to the atmosphere. The agricultural practice of planting legume crops, with their root nodules containing nitrogen-fixing bacteria, contributes further to nitrogen fixation; in fact the amount of fixed nitrogen produced by human activities is of the same order of magnitude as that produced by natural nitrogen fixation. The production of nitrogenous fertilizers (more than 50×10^6 tons year^{-1}) is of particular significance because an appreciable proportion of fertilizer added to land finds its way into streams and lakes. As already discussed, the artificially raised concentration of nitrogen contributes to the process of cultural eutrophication.

Strategies to limit nutrient loss

The practices of both agriculture and forestry necessarily involve the removal of biomass, and with it the removal of minerals. These resources are then added elsewhere. The act of **cultivation** and **harvesting** therefore reduces the biomass of plants on an area of land and the amount of mineral resources held in living tissues. Minerals which are then released by decomposition are free to leach away in drainage waters. Nitrogen is especially likely to be lost in this way, as nitrate ions move freely in soil water. This loss of nitrogen causes two problems. Firstly, the farmer needs to supplement the crops with manure to ensure high productivity and secondly, the nitrogen leached from land creates problems elsewhere by contributing to the eutrophication of lakes and finding its way into drinking water. The nitrogen economy of farming systems varies widely and the rate of nitrate leaching depends on the climate. The wetter a climate the more expensive it is to apply nitrogen fertilizer because more is lost to leaching instead of being available to the crop. The nitrogen economy of agricultural land can be controlled in a variety of ways:

- If actively-absorbing plant root systems are present throughout the year, at least some of the nitrate ions released by decomposition may be recycled through the vegetation instead of leaching away in drainage waters. Therefore, if annual cereal crops can be sown in autumn and make some growth in winter, nitrogen cycling is likely to be more efficient than if the land is left bare in winter.
- Any treatment that allows organic matter to accumulate will reduce the processes leading to the liberation of nitrate ions. For example, incorporating the stubble left after a cereal crop has been harvested reduces nitrification dramatically. This is because the incorporated organic matter has a high ratio of carbon to nitrogen (100:1) and the ratio in the bodies of decomposers is much narrower (10:1). Hence, as material is consumed by decomposers, freely available mineral nitrogen is quickly absorbed into their tissues instead of being released to leaching.
- If the land is to be irrigated, water addition must be controlled so that it only makes up deficiencies in the soil's water-holding capacity – excess will drain away carrying nitrates with it.
- If nitrogen fertilizer is to be applied, it should be applied when required by the crop and can be applied as a spray to the foliage.
- Any soil conditions that hinder decomposition, such as poor drainage or low pH, will hold the nitrogen in an organic form which is less likely to be leached. This regime could not be applied in planted soils but could be used in set aside land.

The management of livestock on permanent grasslands (a managed animal–plant community) leads to the inefficient cycling of nitrogen. Extensive loss of nitrogen occurs in intensive livestock farming through denitrification, leaching, volatilization and surface run-off. In addition, intensive systems lead to problems of waste disposal. These problems arise because materials harvested over large areas are concentrated in small areas. The excreta of cattle, pigs and humans have a nitrogen content of around 2.4% and the droppings from battery hens have a nitrogen content as high as 6%. Dung may be dried and used as fuel, but mostly it contains too much water. Other problems are that if sludge is to be returned to the land, the product is bulky and expensive to transport and may create problems as nitrates are released and enter water supplies. Oceans may be used for waste disposal and human and other effluent can be disposed of at sea.

Many of the features of the nitrogen economy of agricultural systems are also found in managed forests. However, for a large part of its growth cycle, a forest provides complete canopy cover and an extensive root system. The critical period in terms of nitrate loss is following felling and the period when a new crop is becoming established.

Water use problems

The water utilization demands of agricultural production can have serious ecological consequences. The most dramatic and tragic example is the devastation of the **Aral Sea** in Kazakhstan and Uzbekistan in the former USSR. This enormous inland sea formerly covered 64 500 km², but has been reduced by 45% in area and by 66% in volume by the excessive extraction of water required to **irrigate** the cotton crop in an area which was formerly desert and semidesert. The remaining water is highly contaminated by agricultural effluent, both fertilizers and pesticides, which has decimated the fish stocks. Water shortages are globally widespread, even where rainfall volume is moderate or even high, as water storage may be limited. In the US, 25% of irrigated cropland is watered through the unsustainable overuse of groundwater. Consequently, water tables are declining by 15–120 cm per year. Even in the relatively wet UK, excessive water extraction in some areas (e.g. East Anglia) threatens wetland and freshwater habitats.

Desertification

A key problem caused by irrigation of dry areas where evaporation rates are high is the accumulation of salts in the soil surface (**salinization**), causing yield reductions and eventually sterility. Salinization can be avoided if soil salt levels are low or moderate and if investment is made in drainage systems, to remove salt-rich run-off. However, because of poor irrigation practices in arid areas, more than 125 000 hectares of cropped land are lost annually to salinization. Notable problems have occurred in the Indus and Euphrates river valleys, around the Aswan Dam in Egypt and in the western USA. Salinization can also have further impact on ecosystems. The Kesterton Wildlife Refuge in the San Joaquin Valley in California was fed with salt-rich agricultural run-off. After only three years, this internationally important wildlife site was so contaminated by salts (especially selenium salts, which are particularly toxic to many animals) that it was designated a toxic waste site. The Colorado River carries very high levels of salt due to the irrigation of desert soils. To meet Mexican objections to the water quality (which was killing all plants it was applied to), a desalination plant was built in Yuma, Arizona. The operation of this plant is not economic, but the costs are borne by the US taxpayers, not the farmers.

The process of improper management of arid or semiarid land to the extent that it is no longer agriculturally productive is **desertification**. High stock densities may over-consume the vegetation and expose fragile soils. It is a serious and growing problem that results from climate and human activities. In addition, desertification may occur naturally. Approximately 30% of the Earth's land suffers from some degree of desertification. Actual global losses to desertification are estimated to be around six million hectares annually.

Meat versus grain for human consumption

The second law of thermodynamics tells us that far more people can be supported on grain than on the meat of grain-eating animals. However, the economics of current demand worldwide will probably hinder a decline in **meat production**. The pressure for food is coming not only from growing populations in the Third World but also from rising demand for meat in industrialized nations. The consumption of grain in affluent countries is proportionally four or five times that of poorer countries owing to the use of grain for poultry and livestock to produce eggs, milk and meat. Approximately 40% of the grain consumed in the world is consumed by livestock. In the US 70% of grain is consumed by livestock; in China this figure is 20% and in India only 2%. Efficiencies are such that 5 kg of vegetable protein suitable for human consumption but fed to livestock instead results in only 1 kg of livestock protein for human consumption. The loss of carbohydrates is even greater. A forage-fed-only system for livestock would release around 135 million tons of grain for human consumption, an amount capable of feeding 400 million people. Another highly wasteful and inefficient use of resources is the use of vast quantities of high quality fish protein used globally in animal feeds. For example, the huge harvest of Peruvian anchoveta (see Topic T2), which in the 1960s comprised 15% of the world's fish catch, is solely used in animal foodstuffs rather than for human consumption.

A new role for traditional agriculture

There is now a growing awareness of the value of examining traditional agricultural practices as another way of maintaining yields on existing land. In many parts of the world, farming has taken place on the same land for many, many years without depleting the land resource. Throughout the world the principles of **sustainable** (or **organic**) agriculture are being studied. Traditional agriculture practices are characterized by few external inputs (e.g. chemicals, irrigation), the effective accumulation and cycling of nutrients, diversity in cropping and protection of the soil. They are generally based on sound ecological principles and operate in concert with natural ecological cycles and systems. With few inputs, the cost of sustainable agriculture is low. In the Third World, the advantages of this method include low cost and familiarity. A good example of traditional agriculture is the practice of agroforestry, used in the Sahel region of Africa. Acacia trees are grown in the crop fields, interspersed between crops. Farmers have realized that the yields of crops around the acacia trees were the highest. Ecologically what is happening is that the acacia trees fix nitrogen, draw nutrients up from the deep soil to the surface and improve soil texture.

Legume-based crop rotations are common in traditional agriculture to build up nitrogen reserves in the soil. The traditional farming method in many Third World countries has been **shifting cultivation**, a practice by which an area is cleared and farmed for a while and then, as yields decrease, a new area is cleared and the old area allowed to regenerate. Without question the enhancement of these traditional methods holds the key to widespread adoption of sustainable and higher-yielding techniques.

Crop rotation is a method of pest control that has been successfully used for a long time, primarily against insects, nematodes and fungi in soil that attack roots and that survive in the soil, or on crop residues, from one season to the next. It involves growing alternating crops, or a sequence of several crops, each of which is susceptible to different soil pests. This technique has been shown to be important in reducing the susceptibility of high-yielding crop monocultures currently in widespread use.

FURTHER READING

General Reading

Begon, M., Harper, J.L. and Townsend, C.R. (1996) *Ecology – Individuals Populations and Communities*, 3rd Edn. Blackwell Science, Oxford.

Cockburn, A. (1991) *An Introduction to Evolutionary Ecology*. Blackwell Science, Oxford.

Cox, C.B. and Moore, P.D. (1993) *Biogeography – An Ecological and Evolutionary Approach*, 5th Edn. Blackwell Science, Oxford.

Krebs, C.J. (1994) *Ecology*, 4th Edn. Harper Collins, New York.

Krebs, J.R. and Davies, N.B. (1993) *An Introduction to Behavioural Ecology*, 3rd Edn. Blackwell Science, Oxford.

Nebel, B.J. and Wright, R.T. (1996) *Environmental Science*, 5th Edn. Simon & Schuster, New Jersey.

Smith, R.L. (1996) *Ecology and Field Biology*, 5th Edn. Harper Collins, New York.

More specific references

Section A
Allen, T.F.H. and Starr, T.B. (1982) *Hierarchy: Perspectives for Ecological Complexity*. University of Chicago Press, Chicago.

Gould, S.J. and Lewontin, R.C. (1979) Spandrels of San Marco and the Panglossian paradigm – a critique of the adaptationist program. *Proc. R. Soc. London Ser. B* **205**, 581–591.

Jordan, C.F. (1981) Do ecosystems exist? *Am. Nat.* **106**, 237–253.

Lewontin, R.C. (1970) The units of selection. *Ann. Rev. Ecol. Syst.* **1**, 1–8.

Margalef, R. (1963) On certain unifying principles in ecology. *Am. Nat.* **47**, 357–374.

Odum, E.P. (1964) The new ecology. *Bioscience* **14**, 14–16.

Section B
Dodson, S.I. (1970) Complementary feeding niches sustained by size-selective predation. *Limnol. and Oceanog.* **15**, 131–137.

Kettlewell, H.B.D. (1961) The phenomenon of industrial melanism in Lepidoptera. *Ann. Rev. Entomol.* **6**, 245–262.

Spooner, G.M. (1947) The distribution of *Gammarus* species in estuaries: part I. *J. Mar. Biol. Assoc. UK* **27**, 1–52.

Whittaker, R.H., Leven, S.A. and Root, R.B. (1973) Niche, habitat and ecotype. *Am. Nat.* **107**, 321–338.

Section C
Colinvaux, P. (1993) *Ecology 2*. John Wiley, New York.

Cox, C.P. and Moore, P.D. (1980) *Biogeography: An Ecological and Evolutionary Approach*. Blackwell Science, Oxford.

Geiger, R. (1955) *The Climate Near the Ground*. Harvard University Press, Cambridge, MA.

Lamb, H.H. (1972) *Climate: Present, Past and Future, Vol 1*. Methuen, London.

MacArthur, R.H. and Connell, J.H. (1966) *The Biology of Populations.* John Wiley, New York.

Rosenberg, N.J. (1974) *Microclimate: The Biological Environment.* John Wiley, New York.

Woodwell, G. and Whittaker, R. (1968) Primary production in terrestrial ecosystems. *Am. Zool.* **8**(1), 19–30.

Section D

Cook, R.B. (1984) Man and the biogeochemical cycles: interacting with the elements. *Environment* **26**(7), 10–15.

Leith, H. (1975) Primary productivity in ecosystems: comparative patterns of global analysis. In *Unifying Concepts in Ecology* (W.H. van Dobben and R.H. Lowe-McDonnell eds). Junk, The Hague.

Strahler, A.N. (1969) *Physical Geography,* 3rd Edn. John Wiley, New York.

Stumm, W. and Morgan, J.J. (1981) *Aquatic Chemistry: An Introduction Emphasizing Chemical Equilibria in Natural Waters,* 2nd Edn. John Wiley, New York.

Section E

Barnes, H. (1957) The northern limits of *Balanus balanoides* (L.). *Oikos* **8**, 1–15.

Bartholomew, G.A. (1982) Body temperature and energy metabolism. In *Animal Physiology* (M.S. Gordon ed). MacMillan, New York.

Cox, C.B., Healey, I.N. and Moore, P.D. (1976) *Biogeography,* 2nd Edn. Blackwell Scientific Publications, Oxford.

Gilbert, L.E. (1984) Biology of butterfly communities. In *The Biology of Butterflies* (R. Vane-Wright and P. Ackey eds). Academic Press, New York.

Hainsworth, F.R. (1981) *Animal Physiology,* Addison-Wesley, Reading, MA.

Heirich, B. (1977) Why have some animals evolved to regulate a high body temperature? *Am. Nat.* **111**, 623–640.

Larcher, W. (1980) *Physiological Plant Ecology,* 2nd Edn. Springer-Verlag, Berlin.

Lewis, D.H. (1974) Microorganisms and plants. The evolution of parasitism and mutualism. *Symp. Soc. Gen. Microbiol.* **24**, 308–315.

Section F

Ehleringer, J. and Mooney, H.A. (1984) Photosynthesis and productivity of desert and Mediterranean climate plants. In *Encyclopedia of Plant Physiology* (O.L. Lange, P.S. Nobel, C.B. Osmond and H. Ziegler eds), 12D, pp. 205–231. Springer-Verlag, Berlin.

Fitter, A.H. and Hay, R.K.M. (1987) *Environmental Physiology of Plants,* 2nd Edn. Academic Press, London.

Holmes, M.G. (1983) Perception of shade. In *Photoperception by Plants* (P.F. Wareing and H. Smith eds). The Royal Society, London.

Larcher, W. (1980) *Physiological Plant Ecology,* 2nd Edn. Springer-Verlag, Berlin.

Mooney, H.A. and Gulmon, S.L. (1979) Environmental and evolutionary constraints on the photosynthetic characteristics of higher plants. In *Topics in Plant Population Biology* (O.T. Solbrig, S. Jain, G.B. Johnson and P.H. Raven eds), pp. 316–337. Columbia University Press, New York.

Orsham, G. (1963) Seasonal dimorphism of desert and mediterranean chamaephytes and its significance as a factor in their water economy. In *The Water Relations of Plants* (A.J. Rutter and F.W. Whitehead eds), pp. 207–222. Blackwell Scientific Publications, Oxford.

Pearcy, R.W., Bjorkmann, O., Caldwell, M.M., Keeley, J.E., Monson, R.K. and Strain, B.R. (1987) Carbon gain by plants in natural environments. *Bioscience* **37**, 21–29.

Szarek, S.R., Johnson, H.B. and Ting, I.P. (1973) Drought adaption in Opiuntia basilasis. Significance of recycling carbon through Crassulacean acid metabolism. *Plant Physiol.* **52**, 539–541.

Whately, J.W. and Whately, F.R. (1980) *Light and Plant Life.* Edward Arnold, London.

Section G

Ascaso, C., Glavan, J. and Rodriguez-Oascual, C. (1982) The weathering of calcareous rocks by lichens. *Pedobiologica* **24**, 219–229.

Broecker, W.S. and Peng, T. (1982) *Tracers in the Sea,* Lamont-Doherty Geological Observatory, Columbia University, New York.

Buckman, H.O. and Brady, N.C. (1969) *The Nature and Properties of Soil,* 7th Edn. Collier-Macmillan, Toronto.

Elwood, J.W., Newbold, J.D., O'Neill, R.V. and van Winkled, W. (1983) Resource spiralling: an operational paradigm for analysing lotic ecosystems. In *Dynamics of Lotic Ecosystems* (T.D. Fontaine and S.M. Bartell eds). Ann Arbor Science Publishers, Ann Arbor, MI.

Evans, F.C. (1956) Ecosystem as the basic unit of ecology. *Science,* **123**, 1127–1128.

Lindberg, S.E., Lovett, G.M., Richter, D.D. and Johnson, D.W. (1986) Atmospheric deposition and canopy interactions of major ions in a forest. *Science,* **231**, 141–145.

Moss, B. (1989) *Ecology of Fresh Waters.* Blackwell Scientific Publications, Oxford.

Muir, A. (1961) The podzol and podzolic soils. *Adv. Agron.* **13**, 1–56.

Ramade, F. (1981) *Ecology of Natural Resources.* John Wiley & Sons, Chichester.

Rosswell, T.H. (1983) *The Major Biogeochemical Cycles and their Interactions,* pp. 46–50. John Wiley & Sons, Chichester.

Sexstone, A.J., Parkin, T.B. and Tiedje, J.M. (1985) Temporal response of soil denitrification rates to rainfall and irrigation. *Soil Sci. Soc. Am. J.* **49**, 99–103.

Ward, J.V. (1988) Riverine-wetland interactions. In *Freshwater Wetlands and Wildlife* (R.R. Sharitz and J.W. Gibbons eds). US Department of Energy, Office of Science and Technology Information, Oak Ridge, TN.

Waring, R.H., Rogers, J.J. and Swank, W.T. (1981) Water relations and hydrologic cycles. In *Dynamic Properties of Forest Ecosystems* (D.E. Reichle ed), pp. 205–264. Cambridge University Press, Cambridge.

Section H

Battensweiler, W. (1993) Why the larch bud moth cycle collapsed in the subalpine larch-cembran pine forests in the year 1990 for the first time since 1850. *Oecologia* **94**, 62–6.

Bellows Jr., T.S. (1981) The descriptive properties of some models for density dependence. *J. Anim. Ecol.* **50**, 139–156.

Bryant, J.P., Wieland, G.D., Clausen, T. and Kuropat, P.J. (1985) Interactions of snowshoe hares and feltleaf willow (*Salix alaskensis*) in Alaska. *Ecology* **66**, 1564–1573.

Elliot, J.M. (1984) Numerical changes and population regulation in young, migratory trout *Salmo truta* in a Lake District stream 1966–83. *J. Anim. Ecol.* **53**, 327–350.

Harper, J.L., Rosen, R.B. and White, J. (1986) The growth and form of modular organisms – preface. *Philos. Trans. R. Soc. London Ser. B* **313**, 3–5.

Grant, P.R. and Grant, B.R. (1992) Demography and genetically effective sizes of two populations of Darwin's finches. *Ecology* **73**, 766–784.

Gleick, J. (1987) *Chaos: Making a New Science.* Viking, New York.

Keith, L.B. (1983) The role of food in hare population cycles. *Oikos* **40**, 385–395.

Nobel, J.C., Bell, A.D. and Harper, J.L. (1979) The population biology of plants with clonal growth. 1. The morphology and structural demography of *Carex arenaria*. *J. Ecol.* **67**, 983–1008.

Richards, O.W. and Waloft, N. (1954) Studies on the biology and population dynamics of British grasshoppers. *Anti-locust Bulletin* **17**, 1–182.

Sinclair, A.R.E., Gisline, J.M., Krebs, C.J., Smith, J.N.M. and Boutin, S. (1988) Population biology of snowshoe hares. III. Nutrition, plant secondary compounds and food limitation. *J. Anim. Ecol.* **57**, 787–806.

Section I

Connell, J.H. (1961) the influence of interspecific competition and other factors on the distribution of the barnacle, *Chthamalus stellatus*. *Ecology* **42**, 710–723.

Gill, F.B. and Wolf, L.L. (1975) The economics of feeding territoriality in the goldenwinged sunbird. *Ecology* **56**, 333–345.

Horseley, S.B. (1977) Allelopathic inhibition of black cherry by ferns, grass, goldenrod and aster. *Can. J. For. Res.* **7**, 205–216.

Lenington, S. (1980) Female choice and polygyny in female red-winged blackbirds. *Behaviour* **28**, 347–361.

Lonsdale, W.M. and Watkinson, A.R. (1982) Light and self-thinning. *New Phytol.* **90**, 431–435.

Mackauer, N. (1986) Growth and developmental interaction in some aphids and their hymenopteran parasites. *J. Insect Physiol.* **32**, 275–280.

Muller, C.H. (1970) Phytotoxins as plant habitat variables. *Recent Adv. Phytochem.* **3**, 105–121.

Shorrocks, B., Rosewell, J., Edwards, K. and Atkinson, W. (1984) Competition may not be a major organizing force in many communities of insects. *Nature* **310**, 3120.

Tilman, D., Mattson, M. and Langer, S. (1981) Competition and nutrient kinetics along a temperature gradient: an experimental test of a mechanistic approach to niche theory. *Limnol. Oceanog.* **26**, 1026–1033.

Waser, P.M. (1985) Does competition drive dispersal? *Ecology* **66**, 1170–1175.

Weins, J.A. (1977) On competition and variable environments. *Am. Sci.* **65**, 590–597.

Weller, D.E. (1987) A revaluation of the –3/2 power ruke of plant self-thinning. *Ecol. Monogr.* **57**, 23–43.

Section J

Fryxell, J.M. and Doucett, C.M. (1993) Diet choice and functional response of beavers. *Ecology* **74**, 1297–1306.

Goss-Custard, J. (1977) The energetics of prey selection by redshank *Tringa totanus*. *J. Anim. Ecol.* **46**, 1–35.

Heard, D.C. (1982) The effect of wolf predation and snow cover on musk ox group size. *Am. Nat.* **139**, 190–204.

Inoye, R.S., Byrers, G.S. and Brown, J.H. (1980) Effects of predation and competition on survivorship, fecundity and community structure of desert animals. *Ecology* **61**, 1344–1351.

Janzen, D.H. (1971) Seed predation by animals. *Ann. Rev. Ecol. Syst.* **2**, 465–492.

Krebs, J.R., Kacelnik, A. and Taylor, P. (1978) Optimal sampling by foraging birds: an experiment with great tits (*Parus major*). *Nature* **275**, 27–31.

Levin, D.A. (1976) The chemical defences of plants to pathogens and herbivores. *Ann. Rev. Ecol. Syst.* **7**, 121–159.

Lindstrom, E.R., Andren, H., Angelstam, P. and Cederlund, G. (1994) Disease reveals the predator. *Ecology* **75**, 1042–1049.

McNaughton, S.J., Tarrants, J.L., McNaughton, M.M. and Davis, R.T.H. (1985) Silica as a defense against herbivory and a growth promotor in African grasses. *Ecology* **66**, 528–535.

Suhonen, J. (1993) Predation risk influences the use of foraging sites by tits. *Ecology* **74**, 1174–1203.

Section K

Dobson, A.P. and Hudson, P.J. (1986) Parasites, disease and the structure of ecological communities. *Trends Ecol. Evol.* **1**, 11–15.

Hudson, P.J. (1986) The effect of a parasitic nematode on the breeding production of red grouse. *J. Anim. Ecol.* **55**, 85–92.

Jaenike, J. (1992) Mycophagous *Drosophila* and their nematode parasites. *Am. Nat.* **139**, 893–906.

Jarvinen, A. (1985) Predation causing extended low densities in microtine cycles; implications from predation on hole-nesting passerines. *Oikos* **45**, 157–158.

Minchella, D.J., Leathers, B.K., Brown, K.M. and McNair, J.N. (1985) Host and parasite conteradaptation: an example from a freshwater snail. *Am. Nat.* **126**, 843–854.

Møller, A.P. (1991) Parasite load reduces song output in a passerine bird. *Anim. Behav.* **41**, 723–730.

O'Brien, S.J. and Evermann, J.F. (1988) Interactive influence of infectious disease and genetic diversity in natural populations. *Trends Ecol. Evol.* **3**, 254–259.

Petrie, M. and Møller, A.P. (1991) Laying eggs in others' nests: intraspecific brood parasitism in birds. *Trends Ecol. Evol.* **6**, 315–320.

Section L

Feinsinger, P. (1983) Coevolution and pollination. In *Coevolution* (D. Futuyma and M. Slatkin eds). Sinauer, Sunderland, MA.

Howe, H.F. (1980) Monkey dispersal and waste of neotropical fruit. *Ecology* **61**, 944–959.

Howe, H.F. and Smallwood, J. (1982) Ecology of seed dispersal. *Annu. Rev. Ecol. Syst.* **13**, 201–238.

Janzen, D.H. (1979) How to be a fig. *Ann. Rev. Ecol. Syst.* **10**, 13–51.

Martin, M.M. (1970) The biochemical basis of the fungus–attine ant symbiosis. *Science* **169**, 16–20.

Section M

Fowler, C.W. (1981) Density dependence as related to life history strategy. *Ecology* **62**, 602–610.

Harvey, P.H. and Zammuto, R.M. (1985) Patterns of mortality and age at first reproduction in natural populations of mammals. *Nature* **315**, 319–320.

Ims, R.A. (1990) On the adaptive value of reproductive synchrony as a predator swamping strategy. *Am. Nat.* **136**, 485–498.

Janzen, D.H. (1976) Why bamboos wait so long to flower. *Ann. Rev. Ecol. Syst.* **7**, 347–391.

Leverlich, W.J. and Levon, D.A. (1979) Age specific survivorship and reproduction in *Phlox drummondi*. *Am. Nat.* **113**, 881–903.

Lloyd, D.G. (1987) Selection of offspring size at independence and other size v. number strategies. *Am. Nat.* **129**, 800–817.

McNaughton, S.J. (1975) r and K selection in *Typha*. *Am. Nat.* **109**, 215–261.

Pemberton, J.M., Albon, S.D., Guinness, F.E. and Clutton-Brock, T.H. (1991) Countervailing selection in different fitness components in female red deer. *Evolution* **4**, 93–103.

Section N   Bourke, A.F.G. and Franks, N.R. (1995) Kin selection, haplodiploidy and the evolution of eusociality in ants. In *Social Evolution in Ants*, pp. 69–106. Princeton University Press, Princeton, NJ.

Clarke, A.B. (1978) Sex ratio and local resource competition in a prosiman primate. *Science* **201**, 163–165.

Davies, N.B., Hatchwell, B.J., Robson, T. and Burke, T. (1992) Paternity and parental effort in dunnocks *Prunella modularis*: how good are male chick-feeding rules? *Anim. Behav.* **43**, 729–745.

Catchpole, C.K. (1987) Birdsong, sexual selection and female choice. *Trends Ecol. Evol.* **2**, 94–97.

Foster, W.A. and Treherne, J.E. (1981) Evidence for the dilution effect in a selfish herd from fish predation on a marine insect. *Nature* **293**, 466–467.

Hoogland, J.C. (1982) Prairie dogs avoid extreme inbreeding. *Science* **215**, 1639–1641.

Kenward, R.E. (1978) Hawks and doves: factors affecting success and selection of goshawk attacks on wood pigeons. *J. Anim. Ecol.* **47**, 449–460.

Maynard Smith, J. (1991) Theories of sexual selection. *Trends Ecol. Evol.* **6**, 146–151.

Sage, R.D. and Wolf, J.O. (1986) Pleistocene glaciations, fluctuating ranges and low genetic variability in a large mammal (*Ovis dalli*). *Evolution* **40**, 1092–1095.

Section O   Bush, G.L. (1975) Modes of animal speciation. *Ann. Rev. Ecol. Syst.* **6**, 339–364.

Cain, A.J. and Sheppard, P.M. (1954) Natural selection in *Cepaea*. *Genetics* **39**, 89–116.

Coyne, J.A., Orr, H.A. and Futuyma, D.J. (1989) Do we need a new species concept? *Syst. Zool.* **37**, 190–200.

Mayr, E. (1952) Processes of speciation in animals. In *Mechanisms of Speciation* (C. Barigozzi ed), pp. 1–19. Alan Liss, New York.

Paterson, H.E.H. (1985) The recognition concept of species. In *Species and Speciation* (E.J. Vrba ed), pp. 21–29. *Transvaal Museum Monograph no. 4*, Transvaal Museum, Pretoria, South Africa.

Section P   Heal, O.W. and MacLean, S.F. (1975) Comparative productivity in ecosystems – secondary productivity. In *Unifying Concepts in Ecology* (W.H. van Dobben and R.H. Lowe-McConnell eds), pp. 89–108. Junk, The Hague.

Odum, E.P. (1983) *Basic Ecology*. W.B. Saunders, Philadelphia.

Rodin, L.E. *et al.* (1975) Primary productivity of the main world ecosystems. In *Proceedings of the First International Congress of Ecology*, pp. 176–181. Centre for Agricultural Publications, Wageningen.

Whittaker, R.H. (1975) *Communties and Ecosystems*, 2nd Edn. Macmillan, London.

Woodwell, G. (1970) The energy cycle of the Earth. In *The Biosphere*. Freeman, New York.

Woodwell, G.M. and Whittaker, R.H. (1968) Primary production in terrestrial ecosystems. *Am. Zool.* **8** (1), 19–30.

Section Q   Bazzaz, F.A. (1975) Plant species diversity in old-field successional ecosystems in southern Illinois. *Ecology* **56**, 485–488.

Bowers, M.A. and Brown, J.H. (1982) Body size and coexistence in desert rodents: chance or community structure? *Ecology* **63**, 391–400.

Diamond, J.M. (1975) Assembly of species communities. In *Ecology and Evolution of Communities* (M.L. Cody and J.M. Diamond eds), pp. 342–444. Harvard University Press, Cambridge, MA.

Frank, D.A. and McNaughton, S.J. (1991) Stability increases with diversity in plant communities: empirical evidence from the 1988 Yellowstone drought. *Oikos* **62**, 360–362.

McNaughton, S.J. (1977) Diversity and stability of ecological communities: a comment on the role of empiricism in ecology. *Am. Nat.* **111**, 515–525.

O'Neil, R.V. (1976) Ecosystem persistence and heterotrophic regulation. *Ecology* **57**, 1244–1253.

Pyke, G.H. (1982) Local geographic distributions of bumblebees near Crested Butte, Colorado: competition and community structure. *Ecology* **63**, 555–573.

Townsend, C.R., Hildrew, A.G. and Francis, J.E. (1983) Community structure in some southern English streams: the influence of physicochemical factors. *Freshwater Biology* **13**, 521–544.

Schoenly, K., Beaver, R.A. and Heumier, T.A. (1991) On the trophic relations of insects: a food-web approach. *Am. Nat.* **137**, 597–638.

Simberloff, D.S. and Wilson, E.O. (1970) Experimental zoogeography of islands: a two-year record of colonization. *Ecology* **51**, 934–937.

Soulé, M.E., Wilcox, B.A. and Holtby, C. (1979) Benign neglect: a model for faunal collapse in the game reserves of East Africa. *Biol. Conserv.* **15**, 259–272.

Thomas, C.D. and Harrison, S. (1992) Spatial dynamics of a patchily distributed butterfly species. *J. Appl. Ecol.* **61**, 437–446.

Tilman, D. and Downing, J.A. (1994) Diversity and Stability in grasslands. *Nature* **637**, 363–365.

Williams, G.R. (1981) Aspects of avian island biogeography in New Zealand. *J. Biogeog.* **8**, 439–56.

Section R

Chesson, P.L. (1986) Environmental variation and the co-existence of species. In *Community Ecology* (J. Diamond and T.J. Case eds), pp. 22–239. Harper Row, New York.

Noble, J.C. and Slatyer, R.O. (1979) The effects of disturbance on plant succession. *Proc. Ecol. Soc. Aus.* **10**, 135–145.

Paine, R.T. and Levin, S.A. (1981) Intertidal landscapes: disturbance and the dynamics of pattern. *Ecol. Monogr.* **51**, 145–178.

Sale, P.F. (1979) Recruitment, loss and extinction in a guild of territorial reef fishes. *Oecologica* **42**, 159–177.

Sousa, M.E. (1979) Experimental investigation of disturbance and ecological succession in a rocky intertidal algal community. *Ecol. Monogr.* **49**, 227–254.

Tilman, D. (1988) *Plant Strategies and the Dynamics and Structure of Plant Communities*. Princeton University Press, Princeton.

Section S

Breymeyer, A. and Van Dyne, G. (eds) (1980) *Grasslands Systems Analysis and Man*. Cambridge University Press, Cambridge.

Carpenter, S.A., Kitchell, J.F. and Hodgson, J. (1985) Cascading trophic interactions and lake productivity. *BioScience* **35**, 634–639.

Denslow, J.G. (1987) Tropical rainforest gaps and tree species diversity. *Ann. Rev. of Ecol. System.* **18**, 431–451.

Dugan, P. (ed) (1993) *Wetlands in Danger: a World Conservation Atlas.* Oxford University Press, Oxford.

Evenardi, M., Non-Meir, I. and Goodall, D. (eds) (1986) *Hot Deserts and Arid Shrublands of the World. Ecosystems of the world 12A and 12B.* Elsevier Scientific, Amsterdam.

Fenchel, T. (1987) Marine plankton food chains. *Ann. Rev. Ecol. Syst.* **19**, 19–38.

Grassle, J.F. (1985) Species diversity in deep sea communities. *Trends Ecol. Evol.* **4**, 12–15.

Jefferies, R.L. and Davy, A.J. (eds) (1979) *Ecological processes in coastal environments.* Blackwell Science, Oxford.

Meyer, J.L. (1990) A blackwater perspective on river ecosystems. *BioScience* **40**, 643–651.

Polunin, O. and Walters, M. (1985) *A Guide to the Vegetation of Britain and Europe.* Oxford University Press, Oxford.

Reichle, D.E. (ed) (1981) *Dynamic Properties of Forest Ecosystems.* Cambridge University Press, Cambridge.

Sinclair, A.R.E. and Norton-Griffiths, M.L. (eds) (1979) *Serengeti: dynamics of an Ecosystem.* Chicago University Press, Chicago, IL.

Whittaker, R.H. and Woodwell, G.M. (1969) Structure and function, production and diversity of the oak-pine forest at Bookhaven, New York. *Ecology* **57**, 155–174.

Section T

Breen, P.A. and Kendrick, T.H. (1997) A fisheries management success story: the Gisborne, New Zealand, fishery for red rock lobsters (*Jasus edwardsii*). *Mar. Freshwater Res.* **48**(8), 1103–1110.

Cook, R.M., Sinclair, A. *et al.* (1997) Potential collapse of North Sea cod stocks. *Nature* **385**(6616), 521–522.

Costanza, R., d'Arge, R. *et al.* (1997) The value of the world's ecosystem services and natural capital. *Nature* **387**(6630), 253–260.

Crowder, L.B. and Murawski, S.A. (1998) Fisheries bycatch: implications for management. *Fisheries* **23**(6), 8–17.

Jelmert, A. and Oppenberntsen, D.O. (1996) Whaling and deep-sea biodiversity. *Conserv. Biol.* **10**(2), 653–654.

Kerr, S.R. and Ryder, R.A. (1997) The Laurentian Great Lakes experience: a prognosis for the fisheries of Atlantic Canada. *Can. J. Fish. Aquat. Sci.* **54**(5), 1190–1197.

Masood, E. (1997) Fisheries science: all at sea when it comes to politics? *Nature* **386**(6621), 105–106.

Roberts, C.M. (1997) Ecological advice for the global fisheries crisis. *Trends Ecol. Evol.* **12**(1), 35–38.

Stone, C.D. (1997) The crisis in global fisheries: can trade laws provide a cure? *Environ. Conserv.* **24**(2), 97–98.

Uyoung, O.R., Freeman, M.M.R. *et al.* (1994) Subsistence, sustainability, and sea mammals – reconstructing the international whaling regime. *Ocean Coast. Man.* **23**(1), 117–127.

Section U

Carson, R. (1962) *Silent Spring.* Houghton Mifflin, Boston.

Copplestone, J.F. (1977) A global view of pesticide safety. In *Pesticide Management and Insecticide Resistance* (D.L. Watson and A.W.A. Brown eds), pp. 147–155. Academic Press, New York.

Flint, M.L. and van den Bosch, R. (1981) *Introduction to Integrated Pest Management*. Plenum Press, New York.

Headley, T.J. (1975) The economics of pest management. In *Introduction to Insect Pest Management* (R.L. Metcalf and W.L. Luckmann eds), pp. 75–89. John Wiley & Sons, New York.

Hill, D.S. (1987) *Agricultural Insect Pests of Temperate Regions and their Control*. Cambridge University Press, Cambridge.

Lockhart, J.A.R., Holmes, J.C. and MacKay, D.B. (1982) The evolution of weed control in Britrish agriculture. In *Weed Control Handbook: Principles*, 7th Edn (H.A. Roberts ed), pp. 37–63. Blackwell Scientific Publications, Oxford.

Metcalf, R.L. (1980) Changing role of insecticides in crop protection. *Annu. Rev. Entomol.* **25**, 219–256.

Miller Jr., G.T. (1988) *Environmental Science*, 2nd Edn. Wadsworth, Belmont.

Pimental, D., Krummel, J. *et al.* (1978) Benefits and costs of pesticide use in US food production. *BioScience* **28**, 777–784.

Putwain, P.D. (1989) Resistance of plants to herbicides. In *Weed Control Handbook: principles*, 8th Edn. Blackwell Scientific Publications, Oxford.

Waage, J.K. and Greathead, D.J. (1988) Biological control: challenges and opportunities. *Phil. Trans. R. Soc. London, Ser. B* **318**, 111–128.

Ware, G.W. (1983) *Pesticides Theory and Application*. Freeman, New York.

Section V

Berger, J. (1990) Persistence of different sized populations: an empirical assessment of rapid extinctions in Bighorn sheep. *Conserv. Biol.* **4**, 91–98.

Fowler, C. and Mooney, P. (1990) *The Threatened Gene: food, Politics and the Loss of Genetic Diversity*. The Lutterworth Press, Cambridge.

Lande, R. and Barrowclough, G.F. (1987) Effective population size, genetic variation, and their use in population management. In *Viable Populations for Conservation* (M.E. Soule ed), pp. 87–123. Cambridge University Press, Cambridge.

Masden, T., Stille, B. and Shine, R. (1996) Inbreeding depression in an isolated population of adders, *Viperus berus*. *Biol. Conserv.* **75**, 113–118.

O'Brien, S.J., Roelke, M.E. and Marker, L. (1985) Genetic basis for species vulnerability in the Cheetah. *Science* **227**, 1428–1434.

Pimm, S.L., Jones, H.L. and Diamond, J. (1988) On the risk of extinction. *Am. Nat.* **132**, 757–785.

Quinn, J.F. and Harrison, S.P. (1988) Effects of habitat fragmentation and isolation on species richness: evidence from biogeographic patterns. *Oecologia* **75**, 132–140.

Soulé, M.E. (ed) (1986) *Conservation Biology: the Science of Scarcity and Diversity*. Sinauer, Sutherland, MA.

Thompson, N.B. (1988) The status of loggerhead *Lepidochelys kempi* and green, *Chelonia mydas* sea turtles in US waters. *Mar. Fish. Rev.* **50**, 16–23.

Tudge, C. (1988) Breeding by numbers *New Sci.* **119**(1628), 68–71.

Wheeler, B.D. (1988) Species richness, species rarity and conservation evaluation of rich-fen vegetation in lowland England and Wales. *J. Appl. Ecol.* **25**, 331–353.

Wilcove, D.S., McLellan, C.H. and Dobson, A.P. (1986) Habitat fragmentation in the temperate zone. In *Conservation Biology: The Science of Scarcity and Diversity* (M.E. Soulé ed), pp. 237–256. Sinauer, Sunderland, MA.

Section W   Ausubul, J.H. (1991) A second look at the impacts of climate change. *Am. Sci.*
                79, 210–221.

                Bacastow, R.B. and Keeling, C.D. (1981) Atmospheric carbon dioxide concen-
                tration and the observed airborne fraction. In *Carbon Cycle Modelling*
                (B. Bolin ed), pp. 103–112. John Wiley & Sons, New York.

                Blair, G. (1984) Chasing acid rain. *Science* **84**, 64.

                Cohn, J.P. (1987) Chlorofluorocarbons and the ozone layer. *BioScience* **37**,
                647–650.

                Cowling, E.B. (1982) International aspects of acid deposition. In *Acid Rain: a
                Water Resources Issue for the 80s* (R. Hermann and A.I. Johnson eds).
                Proceedings of the American Water Resources Association, International
                Symposium of Hydrometerology, Bethesda, MD.

                French, H.F. (1990) *Clearing the Air: a Global Agenda*. Worldwatch Institute,
                Washington DC.

                Galloway, J.N., Dianwu, Z. and Likens, G.E. (1987) Acid rain: China, United
                States, and a remote area. *Science* **236**, 1559–1562.

                Mooney, H.A., Vitousek, P.M. and Matson, P.A. (1987) Exchange of materials
                between terrestrial ecosystems and the atmosphere. *Science* **238**, 926–932.

                Strain, B.R. (1987) Direct effects of increasing atmospheric CO_2 on plants and
                ecosystems. *Trends Ecol. Evol.* **2**, 18–21.

Section X   Duynisveld, W.H.M., Strebel, O. and Bottcher, J. (1988) Are nitrate leaching
                from arable and nitrate pollution of groundwater avoidable? In *Ecological
                Implications of Contemporary Agriculture* (H. Eijsackers and A. Quispel eds).
                Ecol. Bull. **39**, 116–125.

                National Research Council (1989) *Alternative Agriculture*. National Academy
                Press, Washington DC.

                Pimentel, D., Allen, J. *et al.* (1987) World agriculture and soil erosion. *BioScience*
                37, 277–283.

                Russell, E.W. (1973) *Soil Conditions and Plant Growth*, 10th Edn. Longman,
                London.

                Sprent, J.I. (1987) *The Ecology of the Nitrogen Cycle*. Cambridge University Press,
                Cambridge.

                Tanner, T. (ed) (1989) *Aldo Leopold: the Man and His Legacy*. Soil and Water
                Conservation Society, Ankeny, Iowa.

INDEX